Introduction to Metamorphic Textures and Microstructures

Introduction to Metamorphic Textures and Microstructures

Second edition

A.J. Barker

Lecturer in Geology
University of Southampton, UK

LONDON AND NEW YORK

First published in 1990 by Chapman & Hall

Published 2013 by Routledge
2 Park Square, Milton Park, Abingdon, Oxfordshire OX14 4RN
711 Third Avenue, New York, NY 10017

Routledge is an imprint of the Taylor and Francis Group, an informa business

First issued in hardback 2015

© 1998 A.J. Barker

The right of A.J. Barker to be identified as author of this work has been asserted by him in accordance with the Copyright, Designs and Patents Act 1988.

All rights reserved. No part of this publication may be reproduced or transmitted in any form or by any means, electronic or mechanical, including photocopying, recording or any information storage and retrieval system, without permission in writing from the publisher.

A catalogue record for this book is available from the British Library.

ISBN 978-0-7487-3985-1 (pbk)
ISBN 978-1-138-13829-2 (hbk)

Preface

Technological advances over recent years, and a wealth of new research providing refreshing new interpretations on many metamorphic microstructures, encouraged me to embark on this considerably expanded and fully updated second edition.

My aim in producing this book on metamorphic textures and microstructures has been to provide a detailed introduction to the thin-section description and interpretation of metamorphic rocks. Although primarily written for the advanced undergraduate student, it should provide a useful first source of reference for any geologist dealing with metamorphic rocks. It is intended that the text should be both well-illustrated and comprehensive, but at the same time concise and affordable.

The book is comprised of three parts. Part A provides an introduction to metamorphism and metamorphic rocks, and compared to the first edition has been enlarged to include more detail on the basic interrelationships between equilibrium assemblages, mineral chemistry and microstructures in the interpretation of metamorphic reactions. Part B introduces the fundamental textures and microstructures of metamorphic rocks, with emphasis on the conditions and processes responsible for their formation. This section includes a chapter on fabric development (including the use of SEM), followed by chapters covering topics such as crystal nucleation and growth, inclusions, intergrowths and retrogression. Part C examines interrelationships between deformation and metamorphism. It includes an extensive chapter on strain-related microstructures, and others on the controversial but important topics of porphyroblast—foliation relationships, and shear-sense indicators. The penultimate chapter is on the topic of fluids and veining, and in the final, completely redesigned, chapter, the characteristic reaction textures and microstructures associated with particular environments of metamorphism and specific $P-T-t$ trajectories are discussed. This is done with reference to many of the classic areas of metamorphism from around the world. Appendices giving mineral (and other) abbreviations (Appendix I), a glossary of terms (Appendix II), and a list of key mineral assemblages for the major compositional groups of rocks at each metamorphic facies (Appendix III) form an integral part of the book.

A comprehensive list of references is given at the end of each chapter. Wherever possible, the intention has been to refer to specific references relevant to individual topics covered in the text, as well as referring to key review papers on all the major subjects. It is important to recognise that this book can provide only a glimpse into the complexities of textural and microstructural development of metamorphic rocks, and so for a more comprehensive insight into a particular topic the cited references are strongly recommended. My knowledge of

Preface

metamorphic rocks has grown considerably during the writing of this second edition, and it is my hope that your knowledge and enthusiasm will be enhanced by reading it. Without doubt, the interpretation of metamorphic rocks and their microstructures is a complex topic, and there are still many unanswered questions. I hope that this text provides a useful introduction to this fascinating area of study, and unlocks a few of the secrets of the rocks that you are studying.

Andy Barker
Southampton
March 1997

Acknowledgements

Acknowledgement is given once again to all those postgraduate and undergraduate students who over the years have contributed to what is now a most intriguing and diverse array of thin sections. The study of this worldwide collection of metamorphic rocks has without doubt broadened my knowledge of their textural and microstructural features. It provided the background that induced me to write the original edition, and in the seven years since initial publication, my enthusiasm has been maintained through the continued interest shown by students in providing me with new and interesting thin sections to look at.

Once again, I would like to acknowledge the classic works of Harker, Spry and Vernon, whose books greatly stimulated my own interest in metamorphic textures and microstructures. Additionally, I would like to acknowledge the Metamorphic Studies Group, whose interesting programme of conferences and field meetings has, over the years, afforded me excellent discussion with a truly international list of contributors and participants. Without doubt, this has maintained my enthusiasm for the subject and greatly enhanced my understanding of metamorphic rocks and metamorphic processes.

Special thanks are extended to Barry Marsh for developing and printing the photographs, and to Chris Forster, who converted many of the line drawings of the first edition into computerised images. All new diagrams were computerised and/or hand drawn by the author. I should also like to thank those people who supplied thin sections, and those authors who supplied original photographs, or permitted me to reproduce diagrams from their publications. Also, thanks to Barbara Cressey for advice and assistance with SEM work.

I would also like to thank the many people who gave me feedback on the first edition, or have spared some of their time to review chapters of this new edition. The comments and suggestions have all been useful in shaping and improving on early drafts of the text. The people I would particularly like to thank are Kate Brodie, Giles Droop, Scott Johnson, Cees Passchier, Stephen Roberts, Doug Robinson, Joan Soldevila, Peter Treloar, Rudolph Trouw and Ron Vernon. Last but by no means least, I offer many thanks to Linda, for her continued patience and encouragement, especially during the final stages of writing and production.

AJB

Contents

Preface		v
Acknowledgements		vii
Part A Introduction to metamorphism and metamorphic rocks		1

1 Environments and processes of metamorphism — 3
- 1.1 Environments of metamorphism — 3
  - 1.1.1 Regionally extensive metamorphism — 4
  - 1.1.2 Localised metamorphism — 6
- 1.2 The limits of metamorphism — 7
- 1.3 An introduction to chemical processes of metamorphism — 8
  - 1.3.1 Equilibrium assemblages and the phase rule — 8
  - 1.3.2 The energy of the system — 10
  - 1.3.3 Reaction types — 11
  - 1.3.4 Reaction rates — 14
  - 1.3.5 Diffusion — 15
  - 1.3.6 Fluid phase — 16
- 1.4 Physical processes acting during metamorphism — 16
  - 1.4.1 Volume changes during reaction — 16
  - 1.4.2 Deformation processes on the macro- and microscale — 16
  - 1.4.3 Crystal defects and surface energy — 17
- 1.5 Deformation–metamorphism interrelationships — 18
- References — 18

2 Facies concept and petrogenetic grids — 21
- 2.1 Metamorphic facies, grade and zones — 21
- 2.2 Petrogenetic grids — 23
- References — 25

3 Compositional groups of metamorphic rocks — 27
- 3.1 Pelites — 27
  - 3.1.1 Medium-pressure 'Barrovian' metamorphism — 28
  - 3.1.2 Low-pressure assemblages — 31
  - 3.1.3 High-pressure assemblages — 32

Contents

	3.2	Metacarbonates and calc-silicate rocks	32
	3.3	Quartzofeldspathic metasediments	33
	3.4	Metabasites	33
	3.5	Metamorphosed ultramafic rocks	36
	3.6	Meta-granitoids	37
	References		37

Part B Introduction to metamorphic textures and microstructures — 39
  Definition of texture and microstructure — 39
  Equilibrium and equilibrium assemblages — 39
  Reference — 40

4 Layering, banding and fabric development — 41
  4.1 Compositional layering — 41
  4.2 Introduction to stress, strain and fabric development — 42
  4.3 Classification of planar fabrics in metamorphic rocks — 43
  4.4 Processes involved in cleavage and schistosity development — 45
  4.5 Processes involved in formation of layering in gneisses and migmatites — 51
    4.5.1 The nature and origin of gneissose banding — 51
    4.5.2 The nature and origin of layered migmatites — 53
  References — 55

5 Crystal nucleation and growth — 57
  5.1 Nucleation — 57
  5.2 Growth of crystals — 60
  5.3 Size of crystals — 61
  5.4 Absolute growth times — 65
  5.5 Shape and form of crystals — 65
  5.6 Twinning — 75
    5.6.1 Introduction — 75
    5.6.2 Primary twins — 76
    5.6.3 Secondary twins — 77
  5.7 Zoning — 79
  References — 83

6 Mineral inclusions, intergrowths and coronas — 85
  6.1 Growth of porphyroblasts to enclose residual foreign phases — 85
  6.2 Exsolution textures — 89
  6.3 Inclusions representing incomplete replacement — 93
  6.4 Symplectites — 93
  6.5 Coronas (of high-grade rocks) — 98
  References — 99

7 Replacement and overgrowth — 101
  7.1 Retrograde metamorphism — 101

		7.1.1	Environments of retrograde metamorphism	101
		7.1.2	Textural features of retrogression	103
		7.1.3	Specific types of retrogression and replacement	105
	7.2	Overgrowth textures during prograde metamorphism		110
	References			112

Part C Interrelationships between deformation and metamorphism — 115

8 Deformed rocks and strain-related microstructures — 117
	8.1	Deformation mechanisms		117
	8.2	Inter- and intracrystalline deformation processes and microstructures		117
		8.2.1	Defects	117
		8.2.2	Dislocations	119
		8.2.3	Creep mechanisms	123
		8.2.4	Grain boundaries	125
		8.2.5	Recovery	128
		8.2.6	Recrystallisation	129
		8.2.7	Crystallographic-preferred orientations	131
	8.3	Fault and shear zone rocks and their microstructures		132
		8.3.1	Deformation of quartzitic and quartzofeldspathic rocks	136
		8.3.2	Deformation of mafic rocks	137
		8.3.3	Deformation of carbonate rocks	139
		8.3.4	Distinguishing between schists and mylonites	141
	8.4	The influence of deformation on metamorphic processes		142
	8.5	The influence of metamorphism on deformation processes		144
	References			145

9 Porphyroblast–foliation relationships — 149
	9.1	Thin-section 'cut effects'		149
	9.2	Porphyroblast growth in relation to foliation development		151
		9.2.1	Recognition and interpretation of pre-foliation (pre-tectonic) crystals	151
		9.2.2	Recognition of syntectonic crystals	154
		9.2.3	Recognition and interpretation of post-tectonic crystals	160
		9.2.4	Complex porphyroblast inclusion trails and multiple growth stages	160
	References			161

10 Shear-sense indicators — 163
	10.1	Introduction	163
	10.2	Vein asymmetry and sense of fold overturning	163
	10.3	S–C fabrics, shear bands and mica-fish	165
	10.4	Differentiated crenulation cleavages	168
	10.5	Spiralled inclusion trails	169
	10.6	Mantled porphyroclasts and 'rolling structures'	169
	10.7	Strain shadows	171

Contents

	10.8	Grain-shape fabrics and crystallographic preferred orientations	176
	References		177

11 **Veins and fluid inclusions** — 179
- 11.1 Controls on fluid migration and veining — 179
- 11.2 Initial description and interpretation of veins — 179
- 11.3 The 'crack–seal' mechanism of vein formation — 183
- 11.4 Interpretation of fibrous veins — 185
  - 11.4.1 Syntaxial fibre veins — 185
  - 11.4.2 Antitaxial fibre veins — 185
  - 11.4.3 Composite fibre veins — 186
  - 11.4.4 'Stretched' (or 'ataxial') crystal fibre veins — 187
- 11.5 Veins and melt segregations at high metamorphic grades — 187
- 11.6 Fluid inclusions — 187
- References — 194

12 **Deciphering polydeformed and polymetamorphosed rocks** — 197
- 12.1 Polymetamorphism — 197
- 12.2 Local and regional complications — 198
- 12.3 P–T–t paths — 199
  - 12.3.1 Introduction — 199
  - 12.3.2 Orogenic metamorphism — 203
  - 12.3.3 Orogenic metamorphism with a subsequent thermal overprint — 209
  - 12.3.4 Granulite facies P–T–t paths — 212
  - 12.3.5 Blueschist facies P–T–t paths — 215
  - 12.3.6 Eclogite facies P–T–t paths — 218
- 12.4 Final comments — 220
- References — 221

Appendix I: Abbreviations — 227
- Mineral abbreviations — 227
- Additional abbreviations — 228

Appendix II: Glossary — 229

Appendix III: Key mineral assemblages — 247
- Zeolite facies — 247
- Sub-greenschist facies — 247
- Greenschist facies — 248
- Epidote–amphibolite facies — 248
- Amphibolite facies — 248

Granulite facies	249
Eclogite facies	249
Blueschist facies	250
Albite–epidote hornfels facies	250
Hornblende hornfels facies	250
Pyroxene hornfels facies	251
Sanidinite facies	251

References for Appendices I–III and the plates   253

Index   255

Plates   between 130–131

Part A

Introduction to metamorphism and metamorphic rocks

Chapter one

Environments and processes of metamorphism

The process of metamorphism is one of **change**, and within the mineral assemblage and texture of a metamorphic rock is a memory of that change. The transformations are brought about by geological processes from the global plate tectonic level to the more localised scale. In view of this, recovering the memory of the change locked in metamorphic rocks helps to constrain geological processes well back into the Earth's history.

Locked within the mineralogies of metamorphic rocks is much information about changing P–T conditions. Locked within the textures of metamorphic rocks is further information on metamorphic process, and principally interaction with deformation, that ultimately records plate tectonic movements. As metamorphic rocks recrystallise in the solid state, they can, in favourable circumstances, record a memory of events operating over many millions of years. This book concentrates primarily on recovering the memory of metamorphic processes locked into the textures of these rocks.

In formal terms, metamorphism can be defined as "The mineralogical, chemical and structural adjustment of solid rocks to physical and chemical conditions which have generally been imposed at depth below the surface zones of weathering and cementation, and which differ from the conditions under which the rocks in question originated' (Bates & Jackson, 1980).

1.1 Environments of metamorphism

There are various environments in which metamorphism occurs (Fig. 1.1), the most major of which are linked to processes operating at constructive and destructive plate margins, and are closely interrelated to igneous activity. Such metamorphism is thus of regional extent, although, following the approach of Miyashiro (1973, 1994) and Bucher & Frey (1994), 'regional metamorphism' in the traditional sense linked to mountain-building processes is termed orogenic metamorphism. This is to distinguish it from several other types of metamorphism that, being linked to plate tectonic processes, are of regional extent. There are other types of metamorphism that are of localised extent, linked to geological processes of a non-global character.

The main styles of metamorphism can be classified as follows (numbers are cross-referenced to Fig. 1.1):

Regionally extensive metamorphism (in relative order of importance/abundance)

(1) orogenic metamorphism (traditionally referred to as regional metamorphism);
(2) ocean-floor metamorphism;

Environments and processes of metamorphism

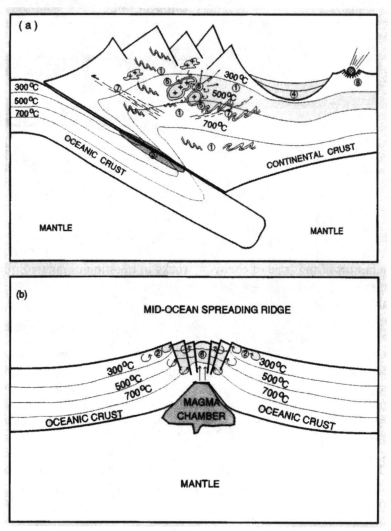

FIG. 1.1 Environments of metamorphism: (a) a schematic illustration of oceanic crust subducting beneath continental crust at a convergent plate margin; (b) a schematic illustration of a mid-oceanic spreading ridge (divergent plate margin). Different metamorphic environments are numbered as follows: (1) orogenic metamorphism; (2) ocean-floor metamorphism; (3) subduction zone metamorphism; (4) burial metamorphism; (5) contact metamorphism; (6) hydrothermal metamorphism; (7) shear-zone metamorphism; (8) shock metamorphism.

(3) subduction-zone metamorphism;
(4) burial metamorphism.

Localised metamorphism

(5) contact metamorphism;
(6) hydrothermal metamorphism;
(7) shear-zone metamorphism;
(8) shock metamorphism.

1.1.1 Regionally extensive metamorphism

Orogenic metamorphism

Major belts of orogenic metamorphism (traditionally referred to as regional, or dynamothermal, metamorphism) occur on all continents, and extend over distances of hundreds to thousands of kilometres. Areas of such orogenic metamorphism include major tracts of the Caledonian–Appalachian orogenic belt, more recent and present-day orogenic mountainous zones such as the Alps, Rockies, Andes and Himalayas, as well as vast areas within Precambrian cratonic blocks, such as those of Africa, Australia, India, Brazil, Canada and Scandinavia. These places record the sites of former areas of continental thickening (i.e. mountain belts), and such orogenic metamorphic belts account for the vast majority of metamorphic rocks seen at the Earth's surface.

Orogenic metamorphism occurs during active deformation, over a broad range of pressure (P) and temperature (T) conditions and variable geothermal gradients. At low P–T in the middle to upper crust, extensive slate belts form, whereas in mid-crustal regions metamorphic belts are dominated by schist, marble, amphibolite and quartzite. These metamorphic rocks have generally experienced polyphase deformation and metamorphism in response to lateral and vertical motions associated with convergent plate tectonic movements. In consequence, these metamorphic rocks retain a memory of these forces in the form of strong planar, linear or combined planar–linear fabric (e.g. cleavage and schistosity; Chapter 4) and are usually extensively folded. At deeper levels in the crust in the presence of hydrous fluids, high-grade gneisses form, and partial melting may occur to form rocks of mixed metamorphic and igneous appearance, termed migmatites. In the absence of aqueous fluid, rocks dominated by anhydrous mineral assemblages, such as granulites, develop, and at deepest crustal levels eclogites can form.

Although there are areas of orogenic metamorphism that have been interpreted in terms of high heat flow during extension-related crustal thinning (e.g. Weber, 1984; Sandiford & Powell, 1986), most orogenic metamorphism is associated with collisional orogenesis and crustal thickening, which generates regionally elevated P–T conditions. The duration of such orogenic metamorphism is estimated to be of the order of 10–50 Ma (e.g. England & Thompson, 1984; Bucher & Frey, 1994), although the metamorphic history experienced by any given rock in such an environment is often recorded as a punctuated series of short-lived events of less than 10 Ma duration during a protracted history (e.g. Barker, 1994).

Ocean-floor metamorphism

At mid-oceanic ridges (Fig. 1.1(b)), ocean-floor metamorphism is an important process. The combination of high heat flow and sea water percolating into fractured oceanic crust causes metamorphism of the primary basalt assemblages. This occurs in the upper parts of the oceanic crust, typically in the sheeted dyke complex and above, but not in the gabbros and ultramafic rocks. Once metamorphosed, ocean-floor spreading transports such rocks away from the spreading ridge to be replaced by new mafic material generated at the ridge, which is in turn metamorphosed. This means that although the metamorphism occurs as a localised style of hydrothermal metamorphism, the continual spreading away from the ridge and provision of fresh material to be metamorphosed means that most of the ocean-floor crust has been metamorphosed. Rocks that have experienced ocean-floor metamorphism show little evidence of any foliation (except in localised fault/shear zones), but commonly exhibit extensive veining.

Subduction-zone metamorphism

At convergent margins, the subduction of cold oceanic crust and overlying sediments against an adjacent plate results in an environment of high shear strain and low geothermal gradient, so that rocks record a high-P/low-T imprint. To preserve such high-P/low-T mineral assemblages requires rapid uplift, during which process the rocks are often tectonically dismembered. This leaves fragmented areas of high-P/low-T metamorphism possessing a strong tectonic fabric ('blueschists'), within a complexly faulted zone. The circum-Pacific margins (e.g. California and Japan) preserve some of the best examples of subduction-zone (blueschist facies) metamorphism, but remnants of blueschist facies metamorphism have been identified from a growing number of locations worldwide.

Burial metamorphism

Burial metamorphism (Coombs, 1961) is the term used to describe incipient metamorphism developed in thick basinal sequences in the absence of major deformation (Fig. 1.1(a)). These rocks characteristically lack any form of foliation, show incomplete mineralogical transformation and preserve many of their original textural features. The type area for burial metamorphism is the South Island of New Zealand, but other areas displaying this style of metamorphism have also been recognised.

1.1.2 Localised metamorphism

Contact (or thermal) metamorphism

The most common example of localised metamorphism, that occurring in the immediate vicinity of an igneous intrusion, is considered as a separate case and is referred to as contact metamorphism. Unlike regional low-P/high-T examples of orogenic metamorphism, the contact metamorphic environment is usually one involving very limited synmetamorphic deformation, such that contact-metamorphosed rocks commonly show little in the way of a foliation, unless the rock already possessed one, or experienced subsequent deformation. Concentrically arranged zones of contact metamorphism define a thermal (contact) aureole around the intrusion (Fig. 1.1(a)). These zones, with their characteristic mineral assemblages and textures, are the result of thermally induced transformations in the country rock. For a comprehensive insight into the subject of contact metamorphism, the edited volume of Kerrick (1991) provides an excellent starting point.

Hydrothermal metamorphism

Ocean-floor metamorphism is a type of hydrothermal activity giving rise to regionally extensive metamorphism. There are many other examples of hydrothermal metamorphism (i.e. the interaction of hot, largely aqueous, fluids with country rock) of localised rather than regional extent. A hydrothermal fluid can originate from various sources, including those of igneous or metamorphic origins. It is usually transported via fractures or shear zones, and at some distance either near to or far from the source, the fluid interacts with the host rock to cause hydrothermal alteration or metamorphism. Many cases of such metamorphism are intimately associated with mineralising fluids and ore deposits.

Shear-zone metamorphism

Faults and shear zones represent localised environments of stress-induced dynamic metamorphism, where textural and mineralogical transformations take place in an environment of high or very high shear strain and variable strain rates, and over a wide range of P–T conditions according to the depth within the crust. As described in Chapter 8 (Sections 8.4 & 8.5), the deformation processes may enhance metamorphic transformations, while metamorphic processes may assist the deformation of the rock concerned. Although at high crustal levels brittle fracturing dominates, at deeper levels where ductile deformation occurs it is crystal–plastic processes that are the most significant. Such processes involve grain-size reduction, and produce rocks with a strong planar and linear fabric, termed mylonites.

Shear zones are commonly environments in which mineral assemblages formed under high

P–T conditions start to alter to lower-temperature assemblages in response to increased shear stress in the presence of a fluid. The presence of a fluid ($H_2O \pm CO_2$) is usually essential, since the lower-temperature alteration assemblages typically comprise hydrous minerals such as micas, serpentine and chlorite, or carbonate minerals such as calcite, dolomite and ankerite.

Shock (or impact) metamorphism

This rare and unusual type of metamorphism ('dynamic metamorphism') is caused by meteorite impact, and consequently is an extremely short-lived event in a localised area. For a few microseconds, the area of impact experiences extreme P–T conditions, varying up to 1000 kbar/5000°C. At the highest P–T conditions the impacted rocks are vaporised, but at slightly less extreme conditions quartz and feldspar can melt to produce a highly vesicular glass containing coesite and stishovite, the high-P and extreme high-P polymorphs of SiO_2 (Fig. 7.1). Notable cases of impact metamorphism are recorded at Ries Crater, Germany, and Meteor Crater, Arizona, USA. It is outside the scope of this book to cover this type of metamorphism in any detail, but for further insight reference should be made to the publications by Englehardt & Stöffler (1968), Grieve (1987), Bischoff & Stöffler (1992) and White (1993).

1.2 The limits of metamorphism

The precise limits of metamorphism within the Earth's crust and upper mantle are not sharply defined. At the low-temperature end of the scale there is a rather blurred transition from diagenesis to metamorphism, over the range 100–200°C. In the past, petrologists paid little attention to the very low grade metamorphic rocks because they are generally fine-grained and difficult to examine using standard optical microscopy. In contrast, coarser-grained higher-grade metamorphic rocks, with their varied and interesting assemblages, provide a much more attractive proposition. However, technological advances in recent decades, providing much greater resolution than the petrographic microscope (e.g. X-ray diffraction, scanning electron microscopy (SEM) and transmission electron microscopy (TEM)) have shown that regular mineralogical transformations do occur at low grade and that fruitful studies of low-grade metamorphic rocks can also be made. Understanding of the mineralogical changes and processes operating in these rocks has advanced considerably in recent years. For further details on low-temperature metamorphism, the various chapters in the book edited by Frey (1987) are recommended.

At the high-temperature end of the scale there is overlap between the realm of metamorphic rocks and that of igneous rocks. Again, there is no precise boundary. At depth in the Earth's crust, metamorphic rocks of suitable composition can commence melting at temperatures as low as about 630°C, to produce granitoid lenses and layers interleaved with high-grade metamorphic rocks. In other cases, most notably when an aqueous fluid is absent, metamorphic rocks will continue to recrystallise in the solid state to temperatures of the order of 1000°C. In the upper mantle, solid-state *metamorphic* processes in ultramafic rocks can take place at still higher temperatures. Since granitoid melt can form at temperatures as low as 630°C, it is clear that the realms of igneous and metamorphic rocks show considerable overlap in the range c.630°–c.1100°C. The essential distinction is that metamorphic processes are dominantly solid-state. Even in migmatites, where there is considerable partial melting, as long as the bulk of material remains solid, the rock is considered to be metamorphic.

Since hydrothermal processes operate in geothermal regions around the world to produce metamorphic minerals, the low-pressure limit of crustal metamorphism can be considered as being about 1 bar. The upper pressure limit of crustal metamorphism is less easily defined, but on the basis of experimental and theoretical considerations it is estimated that certain coesite-bearing rocks indicate stability at pressures of at least 30 kbar (3.0 GPa). This indicates that crustal rocks can be buried to mantle depths in excess of 100 km and subsequently be exhumed. The process allowing such metamorphism to develop is presently a stimulating area of the subject.

The precise type and style of metamorphism is controlled by a number of variables such as temperature, pressure, whole-rock chemistry, fluid chemistry, fluid flux, strain rate and so on. Pressure is a fairly simple function of depth in the crust, and for approximately every 3 km depth of burial, the lithostatic pressure increases by 1 kbar (= 0.1 GPa). The temperature experienced by a given rock within the crust throughout its history is a more complex function of the geothermal gradient and geotherm for the given region. The conductive properties of different rocks in a given segment of crust play an important role, and the precise nature of the geotherm can be affected and disrupted by many factors, including thrusting, uplift, erosion and localised intrusive activity.

1.3 An introduction to chemical processes of metamorphism

The aim of this book is primarily the recognition and interpretation of textures and microstructural features of metamorphic rocks, rather than providing an in-depth petrology text. For a comprehensive treatment of metamorphism and metamorphic petrology, the reader is referred to the texts of Yardley (1989), Philpotts (1990), Spear (1993), Bucher & Frey (1994), Kretz (1994) and Miyashiro (1994). However, in order to understand the textural and microstructural features of metamorphic rocks, it is of course necessary to introduce some elementary principles and concepts relating to metamorphism. **Abbreviations for minerals** used throughout this book are listed in Appendix I, and are primarily based on Kretz (1983) and Bucher & Frey (1994).

1.3.1 Equilibrium assemblages and the phase rule

A rock can be considered as a chemical **system** comprised of a number of chemical **components**, and for most rocks these can be considered in terms of oxide components. Pure quartzite is a simple system comprised of just one component (SiO_2), whereas a pelitic schist is a much more complex system comprised of many components (SiO_2, Al_2O_3, K_2O, Na_2O, FeO, MgO, MnO, TiO_2, CaO and H_2O). The different components are contained within various minerals of the system, which are termed **phases**. As well as the solid phases (minerals), there is often a fluid phase present in the intergranular region during metamorphism. For most metamorphic rocks this fluid is dominantly a mix of H_2O and CO_2, with small amounts of CH_4 and N_2 (high-density gas). Based on the components of the rock (system), and assuming enough time for reactions to proceed, the system will develop a stable mineral (phase) assemblage, according to the prevailing P–T–X(fluid) conditions. This is the **equilibrium** assemblage. The equilibrium assemblage will change as P–T conditions change. For example, during the heating of a mudrock by contact metamorphism, the clay minerals of the original assemblage break down to form micas such as muscovite, which at higher temperatures is itself involved in reactions to form andalusite; and at the highest temperatures andalusite will be succeeded by sillimanite. These phase changes can all occur without any chemical modification of the system (i.e. the system is isochemical). Therefore, according to the prevailing P–T

Chemical processes of metamorphism

conditions a given rock will have a specific equilibrium assemblage. During prograde (increasing P/T) metamorphism, assemblages close to equilibrium are usually maintained, but during retrograde (decreasing P/T) metamorphism, equilibrium is usually not reached. This is because the decline of P–T conditions may be at a faster rate than that at which reactions can proceed. It is this feature, in fact, that allows metamorphic rocks to survive uplift to the Earth's surface, where they are metastable.

From field and thin-section observation, and with the knowledge of the chemistry of a given rock, it is clear to all metamorphic geologists that the more chemically complex a system is, the more phases are present in the assemblage. For example, a chemically complex rock such as a pelitic schist may have a phase assemblage comprising Qtz + Pl + Ms + Bt + Grt + Ilm, whereas a pure quartzite from the same locality and equilibrated to the same P–T conditions will have an assemblage consisting solely of quartz. The **Gibbs Phase Rule** allows us to evaluate the maximum number of phases that can exist in a particular system at equilibrium. It enables assessment of whether an observed assemblage considered to be at equilibrium satisfies the rule. The simplified phase rule can be expressed as:

$$F = C - P + 2,$$

where F is the number of degrees of freedom (or variance) of the system, C is the number of chemical components and P the number of phases. The '2' in the equation relates to the number of intensive or independent variables. For metamorphic rocks the two intensive variables are pressure and temperature. Based on the above equation, the maximum number of phases in an equilibrium assemblage is $C + 2$ (i.e. for equilibrium F must be ≥ 0).

P–T phase diagrams show the stability fields of different mineral assemblages for a certain chemical system, and in the case of Fig. 1.2 it is for one consisting only of Al_2O_3, SiO_2 and H_2O

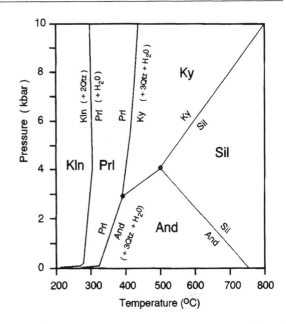

FIG. 1.2 The Al_2O_3–SiO_2–H_2O (ASH) system (based on data of Berman, 1988), showing stability fields for kaolinite, pyrophyllite and Al_2SiO_5 phases (modified after Bucher & Frey, 1994). Invariant points are shown by dots. Note that the mineral abbreviations used in text, this diagram and others are given in Appendix I (based on Kretz, 1983; Bucher & Frey, 1994).

(ASH). The stable assemblages in this system are linking petrographic, theoretical and experimental information. Such diagrams are widely used by petrologists to assess conditions of metamorphism, and graphically to depict the P–T histories of particular rocks based on textural and mineralogical observation. The P–T conditions of bounding reactions for key minerals and assemblages are being increasingly refined based on new experimental work and improved thermodynamic data sets. On such diagrams, an area between reaction lines, in which a specific assemblage is stable, is referred to as a **divariant field** (Fig. 1.2): that is, it has two degrees of freedom. As long as P–T conditions stay within the region defined by the field, P and T can change independently of each other without reactions occurring and a change in assemblage taking place. The reaction curves (lines) bounding divariant fields are termed **univariant curves** (Fig. 1.2). The line separating the divariant fields of kyanite stability and sillimanite stability is one such univariant curve, and at points precisely on this curve both kyanite and sillimanite are in stable coexistence. Because univariant curves have an extra phase that is at equilibrium (compared to divariant fields), in order to satisfy the phase rule they have one less degree of freedom (i.e. they are univariant). In this situation, to stay on the univariant curve, a change in T has to be matched by an appropriate change in P, so only one out of T and P can change independently. The position at which univariant curves intersect is known as an **invariant point** (Fig. 1.2). At such a position another phase is added to the equilibrium assemblage and thus the assemblage has no degrees of freedom, and defines a point in P–T space. Changing either of the independent variables, T and P, would change conditions off the invariant point and the assemblage would no longer be at equilibrium.

1.3.2 The energy of the system

Rocks, like all materials, have a certain energy content, referred to as the **free energy** of the system, linked to features such as atomic vibration and bonding. The energy (units = Joules) can be considered in various forms (thermal, chemical, mechanical and electrical). Of these, chemical and thermal energy are of greatest interest to the petrologist. There are two thermodynamic properties that are of fundamental importance when considering reactions. The first of these, **enthalpy** (symbol H, units J mol^{-1}), reflects the heat content of a phase or system ($H \equiv E + PV$, where E is the internal energy – a combination of heat and work – P is the pressure and V the volume). Of greatest interest to metamorphic petrologists is the change in enthalpy on reaction, i.e. the **enthalpy of reaction** $(\Delta H_r) = H_{products} - H_{reactants}$.

Where ΔH_r is positive, a reaction consumes energy, such that heat energy must be added to the system for the reaction to proceed. Prograde dehydration reactions in pelitic schists are good examples of this type of reaction, and are described as **endothermic**. The converse case is that of reactions that give off heat. These are said to be **exothermic**, and include retrograde hydration reactions. Whatever the case, the First Law of Thermodynamics states that *for any cyclical process, the work produced in the surroundings is equal to the heat removed from the surroundings* (i.e. conservation of energy in the system). The second property, **entropy** (symbol S, units J mol^{-1} K^{-1}), is a term representing the degree of randomness in a system, and reflects the thermal energy of the system. As well as variations in entropy between individual mineral phases, it is worth noting that for any given material (e.g. H$_2$O), the vapour phase has higher entropy than the liquid (water) phase, and the liquid phase a higher entropy than the solid (ice) phase.

The total energy of the rock system, referred to as the **Gibbs free energy** (G), is the sum total of the energy contributed by each phase in the system. The Gibbs free energy is defined as

$$G \equiv E + PV - TS.$$

Since

$$H = E + PV,$$

then

$$G \equiv H - TS.$$

Because the Gibbs free energy varies according to the amount of a given material, it is standard practice to normalise and speak in terms of molar Gibbs free energy. Values of molar Gibbs free energy at the standard state ($T = 298.15$K and $P = 10^5$ Pa (1 bar)) have been determined for most common rock-forming minerals, and are tabulated in publications such as those of Berman (1988) and Holland &

Powell (1990). Where minerals form simple two-component solid-solution series, such as in the case of plagioclase feldspar (from the sodic end-member, albite, to the calcic end-member, anorthite), the concept of chemical potential has to be considered. The **chemical potential** (μ) of a component i in solution (units J mol^{-1}) is defined as the rate of increase of G of a solution of constant composition, at constant P, T, when 1 mole of component i is added. More complicated solid solutions exist for minerals with several end-member components. Such minerals include the Ca–Mg–Fe pyroxenes (end-members diopside CaMgSi$_2$O$_6$, hedenbergite CaFeSi$_2$O$_6$, enstatite Mg$_2$Si$_2$O$_6$ and ferrosilite Fe$_2$Si$_2$O$_6$) and the Ca-amphiboles (end-members tremolite, hornblende (*sensu stricto*), edenite, pargasite and tschermakite). In such cases, where a number of end-member components (j components) are involved, the molar Gibbs free energy (of a given phase) is defined as

$$G_p = \left(\sum_{i=1}^{i=j} \mu_i n_i\right)/n_p, \quad (1.1)$$

where μ_i is the chemical potential of the ith component, n_i is the number of moles of the ith component and n_p is the number of moles of the phase. Thus for a system comprising m phases, the Gibbs free energy can be expressed as

$$G_s = \sum_{k=1}^{k=m} G_{p,k} n_{p,k}, \quad (1.2)$$

where $G_{p,k}$ is the molar Gibbs free energy of the kth phase and $n_{p,k}$ is the number of moles of this phase. As P–T conditions change, so the values of Gibbs free energy for individual phases change, and thus G_s changes. For equilibrium, G_s must be maintained at the lowest possible value for the P–T experienced. This is achieved via reactions, where phases with lower Gibbs free energy are formed at the expense of unstable phases with higher Gibbs free energy. The reaction with the lowest activation energy for nucleation will be that which occurs most rapidly.

At constant P, T, the enthalpy change (ΔH) associated with a reaction is comprised of a thermal energy component due to entropy change ($T\Delta S$) plus the change in free energy (ΔG). As a reaction proceeds, the net entropy of the system always increases (Second Law of Thermodynamics) and ΔG diminishes towards zero. A fundamental concept is that at equilibrium $\Delta G = 0$, and the relationship between entropy and enthalpy is given as $T\Delta S = \Delta H$ (see Philpotts, 1990, for further details). From this and the preceding discussion, it is clear that knowledge of some of the fundamental aspects of thermodynamics is crucial to understanding what drives metamorphic reactions, and consequently the petrographic features of metamorphic rocks.

1.3.3 Reaction types

Prograde metamorphism gives rise to extensive solid-state recrystallization accompanied by various metamorphic reactions, as some minerals become unstable and break down to form new minerals in equilibrium with the prevailing P–T conditions. In order to keep the energy of the system at a minimum for a given set of conditions, reactions will always proceed by the lowest energy route.

Most reactions are of a **univariant** or **discontinuous** character, and define a unique univariant curve in P–T space (which has one degree of freedom). For example, in the ASH system of Fig. 1.2, if we consider the case of progressive increase in T, while maintaining P constant at 6 kbar, then the stable assemblage will shift from Kln + Qtz to Prl + Qtz, to Ky + Qtz and ultimately to Sil + Qtz. Each of these changes involves breakdown of one aluminous phase and formation of another, at a precise point in P–T space defined by the various univariant reaction curves. Other reactions are **divariant** (also known as **continuous** or **sliding** reactions). In such cases, the reactants and products coexist over a range of P–T (divariant field) and the reaction proceeds by varying the composition and modal amounts of coexisting phases, until eventually one of the reactants is exhausted. This is typical for reactions involving phases that are solid solutions, and can vary their compositions by cation exchange with other phases in the system. The breakdown of chlorite and formation of garnet, by the reaction Chl + Ms ⇌ Grt + Bt + H_2O is a good example.

Metamorphic reactions can be classified into six main types:

(i) solid–solid reactions (*reactions involving solid phases only*);
(ii) dehydration reactions (*reactions that liberate H_2O*);
(iii) decarbonation reactions (*reactions that liberate CO_2*);
(iv) oxidation–reduction reactions (*reactions that change the valence state of Fe-oxide phases*);
(v) cation-exchange reactions (*e.g. Fe–Mg exchange between coexisting ferromagnesian phases*);
(vi) ionic reactions (*reactions that are balanced by inferring involvement of ionic species derived from the fluid phase*).

Types (i)–(iv) are single reactions involving the conversion of a mineral or set of minerals to another mineral or set of minerals (**net-transfer reactions**), the fifth type (**exchange reactions**) involve exchange of atoms between two or more coexisting phases without producing new minerals, and the final type (**ionic reactions**) involves ionic exchange between several different reaction sites within the system in order to facilitate a particular transformation.

Solid–solid reactions

Solid–solid reactions can be subdivided into: (i) those reactions that are **phase transformations**, where a single reactant produces a single product that is chemically identical, but has different crystallographic arrangement (e.g. Ky ⇌ Sil, Cal ⇌ Arg); and (ii) those reactions involving various chemically distinct reactants and products (e.g. Ab ⇌ Jd + Qtz). At equilibrium, solid–solid reactions are independent of the fluid phase and, because of this, are potentially useful indicators of P–T conditions. However, because ΔG is often small (e.g. Ky ⇌ Sil), it means that the kinetics of reaction are slow and metastable persistence of the reactant phase is commonplace. Thus some rocks may contain two Al_2SiO_5 phases, seemingly at equilibrium, but it is not safe to assume that they record peak conditions of the appropriate univariant curve, since conditions significantly beyond the curve may have been attained.

Univariant curves have slopes defined by the **Clausius–Clapeyron** equation:

$$dP/dT = \Delta H/T\Delta V = \Delta S/\Delta V, \quad (1.3)$$

where ΔH, ΔS and ΔV are changes in enthalpy, entropy and volume on reaction. Therefore, the slope can be expressed in terms of change in entropy and change in volume. Since the up-temperature side of reactions has higher entropy and usually increased volume, most univariant curves have a positive slope.

Dehydration reactions

There are many hydrous mineral phases, and consequently many prograde metamorphic reactions are dehydration reactions. Some dehydration reactions are univariant reactions (e.g. Prl ⇌ And + 3Qtz + H_2O), whereas others are divariant (e.g. Chl + Ms ⇌ Grt + Bt + H_2O). Most common dehydration reactions recorded by metamorphic rocks have positive ΔS and ΔV, and therefore have positive slopes. At pressures exceeding a few kilobars, most dehydration reactions are very steep; indeed, they are often near to isothermal due to very large ΔS and small (but positive) ΔV (see the reaction Prl $\rightleftharpoons$ Al$_2$SiO$_5$ + Qtz + H$_2$O in Fig.1.2). Dehydration reactions have much larger ΔS compared to solid–solid reactions, and consequently are always much steeper. At pressures exceeding 10 kbar, H$_2$O becomes highly compressed, such that dehydration reactions commonly give rise to negative ΔV and thus the slope of the univariant curve becomes negative.

Decarbonation reactions

This type of reaction (e.g. Cal + Qtz $\rightleftharpoons$ Wo + CO$_2$) involves the breakdown of carbonate minerals and liberation of CO$_2$. Like dehydration reactions, the univariant curves of decarbonation reactions over the normal range of P–T conditions usually have steep, positive slopes. The position of such a reaction in P–T space is strongly affected by the composition of the fluid phase (Fig. 1.3). In the wollastonite-forming reaction illustrated, higher XCO$_2$ is a consequence of the reaction and thus in turn inhibits the reaction, such that higher temperatures are required for the reaction to proceed to completion.

Oxidation–reduction reactions

Iron within stable Fe minerals can occur in two different oxidation states, namely ferrous (Fe^{2+}) or ferric (Fe^{3+}). *Ferrous* iron (Fe^{2+}) shares two electrons in forming bonds with other atoms (i.e. it is divalent), and silicate phases rich in ferrous ions are commonly green (e.g. olivine and hornblende). *Ferric* iron (Fe^{3+}) is trivalent – that is, it has three of its electrons bonded – and minerals rich in ferric iron are commonly red or orange (e.g. haematite).

Reactions that involve a change in valency of Fe–Ti phases in the system are known as *oxidation–reduction reactions*. If a reaction causes iron to increase its valency from Fe^{2+} to Fe^{3+}, it is termed an *oxidation* reaction, whereas in the reverse case, *reduction*, the valency decreases. Many metamorphic minerals are susceptible to *redox* (reduction–oxidation) reactions, but it is

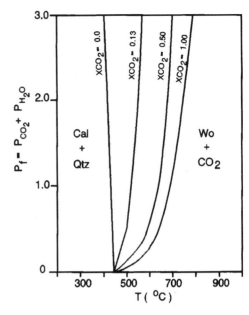

FIG. 1.3 A fluid pressure (P_f) – temperature (T) – composition (X) diagram, showing how the reaction Cal + Qtz $\rightleftharpoons$ Wo + CO$_2$ is strongly influenced by the fluid composition of the system. XCO$_2$ = 1.00 means 100% CO$_2$ fluid composition, whereas XCO$_2$ = 0 means 100% H$_2$O fluid composition (modified after Greenwood, 1967).

most significant in relation to Fe–Ti phases. In metamorphic systems the availability of oxygen is usually discussed in terms of *oxygen fugacity*, (fO_2), or else the partial pressure of oxygen, (PO_2). Oxygen fugacity is analogous to *activity* in describing the effective concentration of a given component in a non-ideal solution such as a melt or a metamorphic fluid. Since magnetite (Fe$_3$O$_4$) and ilmenite (FeTiO$_3$) tend to be the stable Fe–Ti oxides in most metamorphic rocks, whereas primary haematite is usually absent, reference to Fig. 1.4 indicates that reducing conditions are predicted under typical metamorphic temperature ranges. In certain environments of low-grade metamorphism oxidising conditions may prevail, but at deeper levels in the crust (higher grades of metamorphism), conditions tend towards moderate or strongly reducing.

Environments and processes of metamorphism

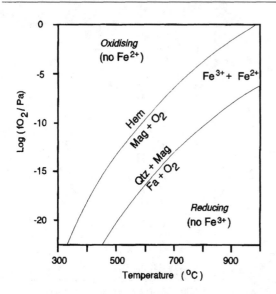

FIG. 1.4 The stability of different phases in the Fe–Si–O$_2$ system at different values of oxygen fugacity $f(O_2)$ over the temperature range 300–1000°C (modified after Gill, 1996). The top curve is generally referred to as the haematite–magnetite (HM) buffering reaction, and the lower curve as the quartz–fayalite–magnetite (QFM) buffer.

Cation-exchange reactions

This type of reaction involves ionic substitution between two or more phases in the system. Usually the cations are of the same or similar charge, and possessing similar ionic radius. In metamorphic rocks partitioning of Fe^{2+} and Mg^{2+} between ferromagnesian minerals such as garnet and biotite is a common example of such a reaction. Such changes occur in order to minimise the energy of the system and maintain equilibrium during retrograde processes, but they do not involve the breakdown or growth of minerals. Such reactions can be predicted based on theoretical/thermodynamic considerations, and while they cannot be identified petrographically (except perhaps with SEM back-scattered electron imaging), they can be determined by electron microprobe analysis.

Ionic reactions

In natural rock systems, the process of metamorphic reaction is complex, and may consist of one or more of the reaction types operating simultaneously and involving many if not all phases in the system, as the rock attempts to minimise the free energy of the system and maintain equilibrium. A classic example of this is the seemingly simple reaction concerning the breakdown of kyanite and formation of sillimanite, each with the composition Al$_2$SiO$_5$. It might be expected that with increasing temperature (Fig. 1.2), sillimanite would form by atom-for-atom replacement of kyanite. In reality, the evidence seen in thin section does not support this, since the normal observation at the sillimanite isograd is for the sillimanite to have nucleated and grown in an aggregate of Bt + Qtz crystals. Carmichael (1969) explained such observations by proposing that three intimately linked reactions take place (Fig. 1.5) at different sites within the rock matrix. He proposed that aluminium remains largely immobile, but that other components are transferred between reaction sites via an intergranular fluid. For further discussion of the complexity of this sillimanite-forming reaction, see Foster (1991). The seemingly straightforward retrograde pseudomorphing of an andalusite or kyanite porphyroblast by a fine-grained aggregate of white mica (sericite) is another example of a reaction in which components not present in the reactant(s) (i.e. K$^+$ and H$_2$O) must be derived from elsewhere in the system to make the product(s), or alternatively an external source for the components might be postulated, as in the case of metasomatic reactions.

1.3.4 Reaction rates

The various reactions described above will give rise to different microstructural and textural changes within a metamorphic rock, and these changes will be considered in subsequent chapters. The precise reaction sequence that occurs in a rock is a function of

Chemical processes of metamorphism

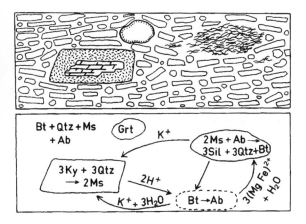

FIG. 1.5 A schematic illustration of a *Carmichael-type* ionic exchange reaction; in this case illustrating the complexity of the kyanite ⇌ sillimanite transformation (based on Carmichael, 1969). The upper half of the diagram shows a simplified representation of the interrelationships of the various phases in the assemblage, and the lower diagram summarises the various aspects of the ionic exchange reaction. In the upper diagram, aligned *rectangles and small 'circles'* represent the Bt + Qtz + Ms + Ab matrix; the large *stippled rectangle* represents the pseudomorphic replacement of kyanite by muscovite, with a *cross-hatched* remnant of kyanite in the core; the *large circle* with dots around the perimeter represents a garnet porphyroblast; and the *mesh of wispy lines* represents a sillimanite (fibrolite) aggregate mixed with biotite.

bulk rock chemistry, fluid chemistry and the relative rates at which reactions occur. For a given prograde reaction, once the product or products have nucleated, their growth is controlled by either diffusion or interface kinetics. The parameter with the slower rate is rate-controlling with respect to the reaction progress. As *P–T* conditions decline in association with uplift and erosion, the high-grade minerals move outside their stability fields. If a fluid is present, extensive retrograde metamorphism usually occurs (Section 7.1).

1.3.5 Diffusion

Diffusion is the process by which atoms, molecules or ions move from one position to another within a fluid or solid, under the influence of a chemical potential gradient. In solids this means the periodic jumping of atoms from one site in the structure to another. If effective, the process of diffusion will cause homogenisation of zoned solid-solution minerals such as garnet in an approach towards equilibrium (see Chapter 5 for further details). The rate of diffusion is measured as the **flux** (J) of atoms across a unit area down a concentration gradient over unit time. It is given by

$$J_x = -D \frac{dC}{dx}, \qquad (1.4)$$

which is known as **Fick's first law** of diffusion, where D is the diffusion coefficient (units $m^2 s^{-1}$) of the diffusing material, C is the concentration of the diffusing material and x is the direction of diffusion. Since the concentration of diffusing material will vary with time, a more useful equation is **Fick's second law** of diffusion, given as

$$\frac{dC}{dt} = \frac{d}{dx} D \frac{dC}{dx}. \qquad (1.5)$$

For most minerals over metamorphic temperatures, the diffusion of atoms within a crystal lattice is an extremely slow process, with values of D generally less than $10^{-18} m^2 s^{-1}$ (see Freer, 1981, for data on diffusion in silicate minerals). As well as depending on D and time, the process of diffusion within and between materials is also strongly temperature dependent. The higher the temperature, the more rapid diffusion in silicates becomes, such that at 300°C diffusion is fairly insignificant in metamorphic processes, but from 650°C upwards diffusion becomes increasingly important. Diffusion through the matrix of the rock via an intergranular fluid, while still a slow process, is many orders of magnitude faster than diffusion through the crystal lattice. Ildefonse & Gabis (1976) determined that the value of D for SiO_2 in an aqueous fluid at 550°C and 1 kbar (0.1 GPa) was on the order of $10^{-8} m^2 s^{-1}$.

1.3.6 Fluid phase

It is only in recent decades that petrologists have properly appreciated that it is not just the mineral being replaced and the obvious products of a texturally observed reaction that are actually involved in the reaction. More realistically, the final texture observed is the end result of a complex series of replacement and ion exchange reactions involving many phases in the system, and especially involving a grain-boundary fluid. This fluid is an efficient means of transfer of ions within a rock by diffusive processes. The presence of a fluid is essential for most metamorphic reactions to proceed, and its chemistry has a strong control over precisely which reactions occur. All things considered, in most cases it is fluid and temperature that have the greatest influence on most reactions and on reaction rates. The precise process of movement of 'volatile' and other ionic components in metamorphic rocks is something that still requires further study. Modification of chemical zonation profiles in porphyroblasts such as garnet gives evidence for solid-state diffusion within metamorphism, while the nature of pressure-solution cleavage gives ample evidence for diffusive mass transfer via the fluid phase. In the vicinity of major intrusions and hydrothermal systems there are often extensive metasomatic changes in associated metamorphic rocks, while the focusing of fluids into fault and thrust zones gives rise to extensive retrogression. Outside these environments metamorphic petrologists often prefer, for simplicity, to consider most rocks as essentially isochemical ('closed') systems. Although it is clear from the presence of veins that a certain degree of element migration (mass transfer) must occur, in most cases this does not appreciably modify the bulk rock chemistry. Bulk rock analyses of metamorphic rocks suggest that in the majority of cases their chemistry is not significantly different from the likely precursor sedimentary and igneous rocks (e.g. amphibolites generally have a basaltic or basaltic andesite chemistry).

1.4 Physical processes acting during metamorphism

1.4.1 Volume changes during reaction

In Section 1.3, the chemical processes involved in metamorphic reactions were introduced, but reactions also involve physical processes and physical change. Since minerals have different densities and many reactions involve liberation of a vapour phase, it is unlikely that volume will remain constant after a reaction. In many cases (e.g. reactions involving solids only), the change in volume (ΔV) is usually small, and may be considered negligible, but in other cases ΔV may be substantial. For example, most prograde devolatilisation reactions show large positive values for ΔV. If reaction rates are fast, rapid increases (or decreases) in volume can induce extensive fracturing and modification of the pre-existing microstructure, especially in low-pressure regimes. For example, the reaction Cal + Qtz → Wo + CO_2 involves a volume decrease of up to 35%, and serpentinisation of peridotite by breakdown of olivine may involve 35–45% volume increase.

1.4.2 Deformation processes on the macro- and microscale

At low P–T conditions of high crustal levels, rocks behave in a largely brittle manner, involving fracturing on all scales (cataclasis). In contrast, under moderate and high P–T conditions the behaviour is essentially ductile, involving power-law creep processes, such as dislocation creep (Chapter 8). This is considered to be the dominant mechanism for most metamorphic situations in the crust and upper mantle. At high temperatures and low strain rates in the upper amphibolite facies, and higher grades of metamorphism, rearrangement of grains by diffusive transfer of atoms (diffusional creep) can be a significant process facilitating ductile deformation.

Somewhere between the two end-member situations of brittle and ductile deformation there is a macroscopic 'brittle–ductile' transition, where rocks display features of both brittle and ductile conditions (Murrell, 1990). This change in behaviour varies considerably between rock types, and according to particular P, T and strain rate. The variation between rock types is due to the fact that the constituent minerals of each rock type have different mechanical properties. This means that, for a given sequence of rocks, some rocks may be experiencing broadly brittle deformation at a time at which others are experiencing broadly ductile deformation. On the scale of an individual rock, some minerals will display features of brittle deformation, whereas others are undergoing ductile deformation. A good example of this is the case of granitic mylonites. At low metamorphic grades the feldspars form porphyroclasts displaying brittle deformation, while quartz in the same assemblage experiences extensive grain-size reduction, and overall ductile deformation. At higher temperatures, both quartz and feldspar experience ductile deformation (Fig. 8.3). Other minerals such as hornblende also experience brittle deformation at low metamorphic grades, but behave in a ductile manner at high metamorphic grades. Even the same minerals in the same rock may show major differences in mechanical behaviour, since the heterogeneity of rock deformation will lead to strain partitioning on all scales, so that not only will high strain zones develop on a macroscale, but on the microscale there will be strain and strain rate variations. This will produce different textural and microstructural features in both monomineralic and polycrystalline assemblages.

1.4.3 Crystal defects and surface energy

When considering chemical processes in relation to mineral assemblages and reactions, a basic assumption in the analysis presented was that the minerals of each phase were perfect crystals. In the case of metamorphic rocks, it is clear from petrographic observation using standard microscopy, and especially from SEM/TEM studies, that metamorphic minerals are essentially imperfect crystals, and that their boundaries with each other are imperfect. To gain a more complete understanding of the processes involved in the microstructural development of metamorphic rocks, it is therefore essential to have an understanding of defects in crystals and the energy of crystal surfaces.

Atoms at the crystal edge do not have all their bonds satisfied, and consequently the surface of the crystal is less stable, and has excess energy, termed **surface energy** (or **interfacial energy**). The contribution of surface energy to the overall free energy of the system may seem small, but in rocks of fine grain size (e.g. mylonites), where surface area is obviously greater, the contribution becomes significant, and consequently fine-grained aggregates are generally more reactive. In all metamorphic rocks, the interfacial energy drives grain growth (coarsening) and causes modification of grain-boundary relationships and grain shape, in an approach towards the most stable configuration of grains. This is achieved by minimising the contribution of grain-boundary (interfacial) energy to the total free energy of the system. The process of coarsening is most conspicuous in monomineralic rocks, but is also apparent in many polymineralic assemblages.

As well as surface irregularities and imperfections causing increased free energy, natural crystals also have internal imperfections, or defects. These defects have the combined effect of increasing the energy of the crystal, and the higher the number of defects, the greater is the instability of the individual grain. A three-fold classification of crystal defects can be made, namely: (i) point defects (vacancies); (ii) line defects (dislocations); and (iii) surface (grain boundary) defects. Chapter 8 examines these in more detail, and introduces some of the characteristic deformation-related textural and microstructural features of deformed metamorphic rocks, and the processes responsible for such features. However, for more detailed information, the reader is referred to publications such as Poirier (1985) and the various papers contained within Barber & Meredith (1990) and Boland & Fitzgerald (1993).

1.5 Deformation–metamorphism interrelationships

The interrelationships between deformation and metamorphism are fundamental to the overall mineralogical and microstructural evolution of a given rock. Deformation processes accompanying metamorphism vary as a function of the prevailing temperature, the confining pressure, the strain rate and lithological factors (e.g. mineralogy, grain size, porosity and permeability). All of these variables affect the rheological behaviour of a given rock. At high crustal levels (i.e. low $P-T$) and moderate strain rates, grain-boundary sliding and diffusive mass transfer are the dominant processes, while at high strain rates disaggregation and brecciation occur. At middle to low crustal levels the significant increase in temperature means that rocks deform very differently. At these depths crystal–plastic processes associated with ductile deformation are most important, with dislocation creep being the dominant mechanism in operation.

The relationship between porphyroblastesis and deformation is an area of special interest, since it can often provide the key to interpreting the crustal evolution of a region. The interpretation of porphyroblast–foliation relationships has proved to be a controversial topic over the years. Much of the discussion has focused on inclusion trails of syntectonic garnets, and the interpretation of whether or not the porphyroblasts have rotated (Chapter 9). The controls on porphyroblast nucleation and growth within a metamorphic rock are many, and encompass a wide range of variables such as $P-T$ conditions, fluid chemistry, bulk rock chemistry, strain rate gradients and chemical potential gradients. In shear zones, bulk strain and strain rates have an important control over the processes that operate and the structures observed. Increased dislocations in crystals provide more sites to which fluids will be attracted and at which reactions may occur. Grain boundaries are shown to be crucial as sites of metamorphic reactions, as well as controlling most textural and microstructural changes. Despite this, there is still much to be learnt about the precise nature of grain-boundary conditions, and the processes that operate.

References

Barber, D.J. & Meredith, P.G. (eds) (1990) *Deformation processes in minerals, ceramics and rocks.* Mineralogical Society Monograph No. 1, Unwin Hyman, London, 423 pp.

Barker, A.J. (1994) Interpretation of porphyroblast inclusion fabrics: limitations imposed by growth kinetics and strain rates. *Journal of Metamorphic Geology*, 12, 681–694.

Bates, R.L. & Jackson, J.A. (1980) *Glossary of geology.* American Geological Institute, Falls Church, Virginia, 751 pp.

Berman, R.G. (1988) Internally-consistent thermodynamic data for stoichiometric minerals in the system $Na_2O-K_2O-CaO-MgO-FeO-Fe_2O_3-Al_2O_3-SiO_2-TiO_2-H_2O-CO_2$. *Journal of Petrology*, 29, 445–522.

Bischoff, A. & Stöffler, D. (1992) Shock metamorphism as a fundamental process in the evolution of planetary bodies: information from meteorites. *European Journal of Mineralogy*, 4, 707–755.

Boland, J.N. & FitzGerald, J.D. (eds) (1993) *Defects and processes in the solid state: geoscience applications.* Developments in Petrology No. 14 ('The McLaren Volume'), Elsevier, Amsterdam, 470 pp.

Bucher, K. & Frey, M. (1994) *Petrogenesis of metamorphic rocks.* Springer-Verlag, Berlin, 318 pp.

References

Carmichael, D.M. (1969) On the mechanism of prograde metamorphic reactions in quartz-bearing pelitic rocks. *Contributions to Mineralogy and Petrology*, 20, 244–267.

Coombs, D.S. (1961) Some recent work on the lower grade metamorphism. *Australian Journal of Science*, 24, 203–215.

Engelhardt, W.V. & Stöffler, D. (1968) Stages of shock metamorphism in crystalline rocks of the Ries Basin, Germany, in *Shock metamorphism of natural materials* (eds B.M. French & N.M. Short). Mono Book Corporation, Baltimore.

England, P.C. & Thompson, A.B. (1984) Pressure–temperature–time paths of regional metamorphism I. Heat transfer during the evolution of regions of thickened continental crust. *Journal of Petrology*, 25, 894–928.

Foster, C.T. (1991) The role of biotite as a catalyst in reaction mechanisms that form sillamanite. *Canadian Mineralogist*, 29, 943–963.

Freer, R. (1981) Diffusion in silicate minerals and glasses: a data digest and guide to the literature. *Contributions to Mineralogy and Petrology*, 76, 440–454.

Frey, M. (ed.) (1987) *Low temperature metamorphism*. Blackie, Glasgow, 351 pp.

Gill, R. (1996) *Chemical fundamentals of geology*, 2nd edn. Chapman & Hall, London, 320 pp.

Greenwood, H.J. (1967) Wollastonite: stability in H_2O–CO_2 mixtures and occurrences in a contact metamorphic aureole near Almo, British Columbia, Canada. *American Mineralogist*, 52, 1669–1680.

Grieve, R.A.F. (1987) Terrestrial impact structures. *Annual Review in Earth and Planetary Science*, 15, 245–270.

Holland, T.J.B. & Powell, R. (1990) An enlarged and updated internally consistent thermodynamic dataset with uncertainties and correlations: the system K_2O–Na_2O–CaO–MgO–MnO–FeO–Fe_2O_3–Al_2O_3–TiO_2–SiO_2–C–H–O_2. *Journal of Metamorphic Geology*, 8, 89–124.

Ildefonse, J.-P. & Gabis, V. (1976) Experimental study of silica diffusion during metasomatic reactions in the presence of water at 550°C and 1000 bars. *Geochimica et Cosmochimica Acta*, 40, 297–303.

Kerrick, D.M. (ed.) (1991) *Contact metamorphism*. Reviews in Mineralogy No. 26, Mineralogical Society of America, Washington, DC, 847 pp.

Kretz, R. (1983) Symbols for rock-forming minerals. *American Mineralogist*, 68, 277–279.

Kretz, R. (1994) *Metamorphic crystallization*. John Wiley, Chichester, 507 pp.

Miyashiro, A. (1973) *Metamorphism and metamorphic belts*. George Allen & Unwin, London, 492 pp.

Miyashiro, A. (1994) *Metamorphic petrology*. UCL Press, London, 404 pp.

Murrell, S.A.F. (1990) Brittle-to-ductile transitions in polycrystalline non-metallic materials, in *Deformation processes in minerals, ceramics and rocks*, (eds D.J. Barber & P.G. Meredith). Mineralogical Society Monograph No. 1, Unwin Hyman, London, Ch. 5, 109–137.

Philpotts, A.R. (1990) *Principles of igneous and metamorphic petrology*. Prentice Hall, Englewood Cliffs, New Jersey, 498 pp.

Poirier, J.-P. (1985) *Creep of crystals: high-temperature deformation processes in metals, ceramics and minerals*. Cambridge University Press, Cambridge, 260 pp.

Sandiford, M. & Powell, R. (1986) Deep crustal metamorphism during continental extension: modern and ancient examples. *Earth and Planetary Science Letters*, 79, 151–158.

Spear, F.S. (1993) *Metamorphic phase equilibria and pressure–temperature–time paths*. Mineralogical Society of America Monograph, Washington, DC, 799 pp.

Weber, K. (1984) Variscan events: early Paleozoic continental rift metamorphism and late Paleozoic crustal shortening, in *Variscan tectonics of the North Atlantic region* (eds D.H.W. Hutton & D.J. Sanderson). Geological Society of London Special Publication No. 14, Blackwell Scientific, Oxford, 3–22.

White, J.C. (1993) Shock-induced melting and silica polymorph formation, Vredefort Structure, South Africa, in *Defects and processes in the solid state: geoscience applications* (eds J.N. Boland & J.D. FitzGerald). Developments in Petrology No. 14 ('The McLaren Volume'), Elsevier, Amsterdam, 69–84.

Yardley, B.W.D. (1989) *An introduction to metamorphic petrology*. Longman, London, 248 pp.

Chapter two

Facies concept and petrogenetic grids

2.1 Metamorphic facies, grade and zones

Metamorphic rocks show variations in their mineralogical and microstructural features that are linked to variations in $P-T$ conditions of metamorphism. At low $P-T$ conditions, typical minerals are hydrous and generally of low density (e.g. Chl, Ms, Qtz and Pl). With increasing $P-T$, less hydrous and anhydrous minerals become dominant (e.g. Stt, Grt, Ky, Pyx and Hbl). Similarly, with increasing $P-T$, grain size generally increases. There are a number of ways of monitoring these changes (i.e. classification) to record changing conditions.

The zonal scheme was the first to be introduced, by Barrow (1893, 1912). In the Scottish Highlands, he established that metamorphic zones could be mapped out based on differences in the mineral assemblages of pelitic rocks. Zonal boundaries were defined by **isograds** representing the first appearance of key index minerals. From lowest to highest metamorphic grade, the order of these key zonal minerals, referred to as **Barrovian zones**, is chlorite, biotite, garnet (almanditic), staurolite, kyanite and sillimanite. It has subsequently proved possible to relate these changes directly to the $P-T$ conditions experienced by the rocks. The Barrovian zonal sequence has been recognised in metapelitic sequences from many areas of orogenic metamorphism, but it is not the only zonal sequence recognised. The Barrovian sequence characterises medium-pressure regimes, whereas in low-P, high-T regimes, kyanite and almandine garnet are absent, while andalusite and cordierite are important zonal minerals.

There are limitations to the assessment of metamorphic conditions based on single zonal minerals. Not least of these is the fact that particular minerals will only develop in rocks of a certain chemistry, and even within the range of rocks with appropriate chemistry, the first appearance of a particular mineral will not be synchronous. The use of mineral assemblages, rather than individual minerals, is a much more reliable approach to evaluating $P-T$ conditions in any detail.

TABLE 2.1 Differences between the characteristic mineral assemblages of metasediments in the metamorphic aureoles around intrusions of the Oslo region (Goldschmidt, 1911) and Orijärvi, Finland (Eskola, 1914, 1915).

Oslo (Goldschmidt, 1911)	Orijärvi (Eskola, 1915)
Kfs + And	Ms + Qtz
Kfs + Crd	Bt + Qtz
Kfs + Hy + An	Bt + Hbl
Hy	At

Facies concept and petrogenetic grids

The now firmly established concept of metamorphic facies was introduced by Eskola (1914, 1915) following his work in Orijärvi (Finland), and with due consideration of the observations of Goldschmidt (1911) from the Oslo area of Norway. Eskola had come to realise that in order to achieve or approach chemical equilibrium the mineral assemblages observed in metamorphic rocks changed as a function of the P–T conditions experienced. It was apparent to Eskola that for a given range of rock compositions and a particular range of P–T conditions certain characteristic mineral associations always occurred, and that as P–T conditions varied so these mineral associations varied. By comparison of the Orijärvi assemblages with those recorded by Goldschmidt (1911) from zones around intrusions in the Oslo region, Eskola noticed that there were consistent mineralogical differences between the two areas (Table 2.1). Those assemblages from Orijärvi are dominated by hydrous phases such as biotite and muscovite, while those from the Oslo region contain much K-feldspar. Since broadly comparable rock types were involved, Eskola concluded that although metamorphism in both areas was related to high-level intrusions, that at Orijärvi occurred at lower T than that at Oslo. This difference occurred due to the former being associated with granite intrusions, whereas the latter was in association with small mafic bodies, intruded at significantly higher temperature. With further work, Eskola developed the 'metamorphic facies' concept and, in a classic paper (Eskola, 1920), defined the term '**metamorphic facies**' to designate a group of rocks characterised by a definite set of minerals, which under the conditions that prevailed during their formation were in perfect equilibrium with each other. The mineral compositions in the rocks of a given facies vary gradually in correspondence with variations in the chemical compositions of the rocks. Tilley (1924) emphasized that a given metamorphic facies may include rocks of quite different whole-rock chemistry. Indeed, mineralogical studies of a range of rock types in a given area often allow greater certainty in assigning a particular metamorphic facies. Eskola (1920) proposed five metamorphic facies to define different conditions of metamorphism, with a further three facies added in 1939. The work of Coombs and co-workers in the early 1960s (e.g. Coombs, 1961) introduced the 'zeolite' and 'prehnite–pumpellyite' facies, such that by this time ten distinct metamorphic facies had been recognised. There then followed a period during which, rather than simplifying metamorphism using the 'facies' concept, the literature became cluttered and confused, as various workers began to subdivide the original 'facies' into two or more 'sub-facies' based on slight mineralogical differences. It was not always easy to assess whether these slight variations were solely P–T controlled, or whether they were controlled by some other factor such as fluid composition. In any case, the so-called 'sub-facies' were really 'facies' by definition, since they contained a distinctive set of minerals characterising certain conditions of formation.

There are problems with the facies concept as defined by Eskola. For example, he did not address the role of the fluid phase in any detail. Eskola, and others, either considered the fluid to be unimportant or assumed that all fluid was aqueous, and thus that $P_{total} = P_{fluid} = P_{H_2O}$. This is certainly not the case, and for many minerals the $XH_2O:XCO_2$ ratio, and fluid salinity, are crucial to their fields of stability in P–T space. For carbonate and calc-silicate rocks the facies approach has very limited use, because of the strong influence of fluid chemistry on the phase assemblage present. However, for metapelites and metabasites most geologists still use the facies concept and tentatively regard $P_f = P_{H_2O}$ (or close to it), accepting that in some instances there are variable degrees of dilution by CO_2, CH_4 and various salts. In addition, it must be recalled that zonal and facies approaches were based solely on hand specimen and basic petrographic analysis of thin sections, and only qualitative ideas of the link between mineralogical changes and P–T conditions. Since then, there have been major advances, such as the advent of the microprobe and computerised thermodynamic data sets. These advances have allowed stability fields of key minerals and assemblages to be defined ever more clearly. This means that the P–T ranges represented by the mineral assemblages defined by Eskola in developing the facies concept are now known in a much more quantitative basis. Accordingly, it is more common nowadays for petrologists to associate a facies with a range of P–T space rather than individual mineral associations.

In Fig. 2.1, P–T space has been divided into 12 facies, based on experimental work, empirical observations and theoretical stability fields of key minerals and assemblages based on thermodynamic data sets. Between the zeolite facies and the greenschist facies is an area referred to as the sub-greenschist facies. This term has been used to encompass the prehnite–pumpellyite, prehnite–actinolite and pumpellyite–actinolite facies that are referred to in some publications. The reason for preferring to amalgamate under the general heading 'sub-greenschist' facies is because recent work has shown that the previously used prehnite–pumpellyite, prehnite–actinolite and pumpellyite–actinolite facies show partial or complete overlap in P–T space as a function of even small whole-rock controls. The facies boundaries of Fig. 2.1 are of necessity, and for the sake of simplicity, drawn up for systems in which $P_{H_2O} = P_{total}$. In many cases the effects of CO_2 and other fluids on given reactions are still either totally unknown or poorly defined, so at the present state of understanding it would be difficult to draw them in any other fashion. It should be noted that metastable persistence of phases (Chapter 1) means that the facies boundaries are not always sharply defined, and that transitional assemblages are the norm.

Rather than adopting the facies concept, Winkler (1979) preferred simply to divide P–T space into large divisions of metamorphic grade based on key mineral reactions in common rocks. These divisions are designated as very low, low-, intermediate- and high-grade metamorphism. They are largely temperature-related divisions, since most of the reactions on which they are based are much more strongly temperature controlled, and in many cases nearly isothermal. Winkler emphasised that a qualitative estimate of pressure should also be stated (e.g. 'high-grade and high-pressure metamorphism' or 'high-grade and low-pressure metamorphism' etc.) in order to give a better idea of the environment of metamorphism. As such, this particular approach gives a convenient overview of metamorphic conditions, without having to be too specific. It provides a useful and straightforward first-order definition of metamorphic conditions based on initial field observations. Within the present text, the approach taken is to use 'grade' to make a generalised qualitative description of metamorphic P–T conditions, while using 'facies' when speaking more precisely. **Key mineral assemblages associated with major compositional groups of rocks for each metamorphic facies are listed in Appendix III.**

2.2 Petrogenetic grids

Experimental work has now reliably established the P–T stability fields of many minerals and mineral assemblages. Combining this information allows the construction of petrogenetic grids. On such grids, the intersection of univariant reaction curves at invariant points enables us to define the bounding conditions for particular equilibrium assemblages. From this, the metamorphic conditions experienced

Facies concept and petrogenetic grids

by a particular rock can be quantified in terms of a specific $P-T$ range. A few minerals – such as jadeite, kyanite and andalusite – allow reasonable estimation of maximum- or minimum-pressure conditions, but the majority of minerals and reactions are most sensitive to changes in temperature. For this reason, the accurate assessment of metamorphic temperatures based on individual minerals or assemblages is considerably easier than assessing pressures. Petrogenetic grids for the major chemical systems corresponding to metabasites, metapelites and other systems relevant to metamorphic rocks are comprehensively dealt with

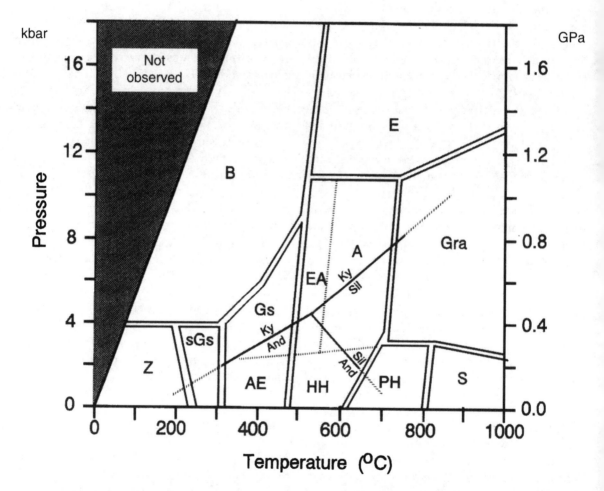

FIG. 2.1 A simplified diagram illustrating the positions of the different metamorphic facies in $P-T$ space. The facies abbreviations are as follows: A, amphibolite facies; AE, albite–epidote hornfels facies; B, blueschist facies; E, eclogite facies; EA, epidote–amphibolite facies; Gra, granulite facies; Gs, greenschist facies; HH, hornblende hornfels facies; PH, pyroxene hornfels facies; S, sanidinite facies; sGs, sub-greenschist facies; Z, zeolite facies. The boundaries between GS and AE, and between EA, A and HH, are not shown as solid lines because the distinction between these facies is not always clear. The lines radiating from a dot (invariant point) are the univariant curves defining the stability fields of Al_2SiO_5 polymorphs, based on Salje (1986). Note that key assemblages of major rock types for different metamorphic facies are given in Appendix III.

by Spear (1993) and will not be evaluated in this text (see Appendix III for key assemblages).

The greater the number of components in a system, the more restricted the stability field of an individual phase or phase assemblage becomes. As an example, one of the best known and most studied systems is the one-component Al_2SiO_5 system. The stability fields of the three Al_2SiO_5 polymorphs (andalusite, kyanite and sillimanite) have been of considerable interest to petrologists and have been extensively studied both on an experimental and theoretical basis (Kerrick, 1990). The stability fields for andalusite, kyanite and sillimanite shown in Fig. 1.2 are based on the thermodynamic data set of Berman (1988). Note how the addition of H_2O into the system prevents the formation of andalusite and kyanite at lower temperatures and gives rise to the formation of pyrophyllite instead. The addition of further components gives rise to further limitations and an increasingly complex system, more closely resembling pelitic rocks. Strens (1967) examined the effect of small amounts of Fe_2O_3 on the stability fields of the Al_2SiO_5 polymorphs, noting that the univariant lines separating the stability fields of different phases become divariant zones over which two of the polymorphs coexist. Bulk rock and fluid chemistry variations significantly influence whether or not a particular mineral or mineral association is present at a given metamorphic grade. In consequence, the absence of a phase such as staurolite in a garnet–mica schist does not necessarily preclude the possibility of staurolite grade conditions having been attained. As a general rule, it is inadvisable to place too much emphasis on the observations from a single thin section, but much more meaningful to draw conclusions from a group of thin sections of related rocks.

Originally, it was experimental data and empirical observation that aided the construction of petrogenetic grids, but over recent decades such grids are largely being defined on the basis of internally consistent thermodynamic data sets (e.g. Berman, 1988; Holland & Powell, 1990). This approach enables the calculation of all the univariant curves and divariant fields of interest for a particular chemical system, as long as reliable thermodynamic data are available for all phases of the system. For simple end-member systems the data are very good, and the grids are well defined. However, for complex systems such as those of metapelites and metabasites, involving phases with complex solid solution between end-members (e.g. chlorite, micas, garnet and amphiboles), and sliding reactions in natural systems with variable fluid compositions, it is less easy to construct a wholly reliable petrogenetic grid. Even so, important advances are being made, and some of the important grids constructed over the past ten years include those of Will *et al.* (1990) for ultramafic rocks; Connolly & Trommsdorff (1991) for metacarbonates; Frey *et al.* (1991) for metabasites; and for metapelites the grids of Spear & Cheney (1989), Powell & Holland (1990) and Symmes & Ferry (1992). For comprehensive treatment of metamorphic phase equilibria, the text of Spear (1993) is strongly recommended.

References

Barrow, G. (1893) On an intrusion of muscovite–biotite gneiss in the south-eastern Highlands of Scotland, and its accompanying metamorphism. *Quarterly Journal of the Geological Society*, **49**, 330–358.

Barrow, G. (1912) On the geology of lower Dee-side and the southern Highland Border. *Proceedings of the Geologists Association*, **23**, 274–290.

Berman, R.G. (1988) Internally-consistent thermodynamic data for minerals in the system Na_2O–K_2O–CaO–MgO–FeO–Fe_2O_3–Al_2O_3–SiO_2–TiO_2–H_2O–CO_2. *Journal of Petrology*, **29**, 445–522.

Connolly, J.A.D. & Trommsdorff, V. (1991) Petrogenetic grids for metacarbonate rocks: pressure–temperature phase-diagram projection for

mixed-volatile systems. *Contributions to Mineralogy and Petrology*, **108**, 93–105.
Coombs, D.S. (1961) Some recent works on the lower grades of metamorphism. *Australian Journal of Science*, **24**, 203–215.
Eskola, P. (1914) On the petrology of the Orijärvi region in southwestern Finland. *Bulletin de la Commission Geologique de Finlande*, 40.
Eskola, P. (1915) On the relations between the chemical and mineralogical composition in the metamorphic rocks of the Orijärvi region. *Bulletin de la Commission Geologique de Finlande*, 44.
Eskola, P. (1920) The mineral facies of rocks. *Norges Geologisk Tiddskrift*, **6**, 143–194.
Frey, M., de Capitani, C. & Liou, J.G. (1991) A new petrogenetic grid for low grade metabasites. *Journal of Metamorphic Geology*, **9**, 497–509.
Goldschmidt, V.M. (1911) Die kontaktmetamorphose im kristianiagebiet. *Vidensk. Skrifter. I. Mat.-Naturv. K.*, 11.
Holland, T.J.B. & Powell, R. (1990) An enlarged and updated internally consistent thermodynamic dataset with uncertainties and correlations: the system $K_2O-Na_2O-CaO-MgO-MnO-FeO-Fe_2O_3-Al_2O_3-TiO_2-SiO_2-C-H_2-O_2$. *Journal of Metamorphic Geology*, **8**, 89–124.
Kerrick, D.M. (1990) *The Al_2SiO_5 polymorphs*. Reviews in Mineralogy No. 22, Mineralogical Society of America, Washington DC, 406 pp.
Powell, R. & Holland, T.J.B. (1990) Calculated mineral equilibria in the pelite system, KFMASH ($K_2O-FeO-MgO-Al_2O_3-SiO_2-H_2O$). *American Mineralogist*, **75**, 367–380.
Salje, E. (1986) Heat capacities and entropies of andalusite and sillimanite: the influence of fibrolitization on the phase diagram of Al_2SiO_5 polymorphs. *American Mineralogist*, **71**, 1366–1371.
Spear, F.S. (1993) *Metamorphic phase equilibria and pressure–temperature–time paths*. Mineralogical Society of America Monograph, Washington, DC, 799 pp.
Spear, F.S. & Cheney, J.T. (1989) A petrogenetic grid for pelites in the system $SiO_2-Al_2O_3-FeO-MgO-K_2O-H_2O$. *Contributions to Mineralogy and Petrology*, **101**, 149–164.
Strens, R.G.J. (1967) Stability of Al_2SiO_5 solid solutions. *Mineralogical Magazine*, **31**, 839–849.
Symmes, G.H. & Ferry, J.M. (1992) The effect of whole-rock MnO content on the stability of garnet in pelitic schists during metamorphism. *Journal of Metamorphic Geology*, **10**, 221–237.
Tilley, C.E. (1924) The facies classification of metamorphic rocks. *Geological Magazine*, **61**, 167–171.
Will, T.M., Powell, R. & Holland, T.J.B. (1990) A calculated petrogenetic grid for ultramafic rocks in the system $CaO-FeO-MgO-Al_2O_3-SiO_2-CO_2-H_2O$ at low pressures. *Contributions to Mineralogy and Petrology*, **105**, 347–358.
Winkler, H.G.F. (1979) *Petrogenesis of metamorphic rocks*, 5th edn. Springer-Verlag, New York, 348 pp.

Chapter three

Compositional groups of metamorphic rocks

Metamorphic rocks display a wide range of chemical compositions and mineral assemblages, reflecting the variety of original rock types that become metamorphosed. Metasedimentary rocks can broadly be subdivided into three main categories; namely, metapelites, metacarbonates/calc-silicates and quartzofeldspathic metasediments. There is of course a complete spectrum of rocks between these end-members (Fig. 3.1). In addition to metasedimentary rocks, there are three main compositional groups of meta-igneous rocks; namely, metabasites, metagranitoids and meta-ultrabasics. The mineralogical and chemical characteristics of these various metamorphic rocks are now discussed, with emphasis on key minerals and assemblages. The mineral abbreviations used throughout this book are those suggested by Kretz (1983), and extended by Bucher & Frey (1994). For quick reference, they are tabulated in Appendix I.

3.1 Pelites

Rocks of this category represent metamorphosed argillaceous sediments (mudstones). As well as being moderately to highly siliceous, such sediments have a chemistry characterised by high Al_2O_3 and K_2O, variable FeO and MgO and much H_2O in hydrous minerals.

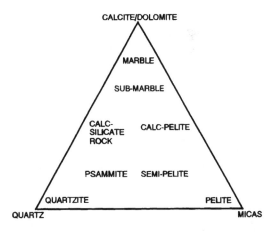

FIG. 3.1 A ternary diagram illustrating the nomenclature in common use for metasedimentary rocks based on the relative proportions of quartz, micas and carbonate minerals.

Accordingly, this system is known in petrologist's shorthand as KFMASH. In addition to the major components of the system, minor or trace amounts of CaO, Na_2O, Fe_2O_3, MnO and TiO_2 may also be present, and thus give rise to extra mineral phases other than those expected for the strict KFMASH system. The ratio of individual oxides varies from one rock to the next, and this complex chemistry gives rise to the development of a large range of minerals during metamorphism. The appearance of certain of these minerals can be very important for assessing metamorphic grade.

3.1.1 Medium-pressure 'Barrovian' metamorphism

Pelitic rocks initiate as mudstones, which in the diagenetic field are dominated by fine-grained clay minerals, quartz, chlorite and minor detrital feldspar. The dominant clay minerals have mixed-layer structure (e.g. illite–smectite). With increasing heating and burial, the mixed-layer clay minerals become ordered with the onset of 'metamorphic' conditions, indicated by the mineral illite at around 200°C. Over the approximate temperature range 200–280°C, there are recognisable changes in illite structure (crystallinity), as recorded by XRD and TEM methods. Ultimately, illite is replaced by muscovite, the 2M mica-in reaction curve being located at about 270–280°C.

Chlorite may occur either as authigenic or detrital crystals. During low-grade metamorphism it changes from a Type Ib to a Type IIb structure, this change occurring at temperatures of the order of 150–200°C. At around 300–325°C, there is a transition from sub-greenschist facies into the lower part of the true greenschist facies. At low greenschist facies conditions, although often fine-grained, the mineral assemblage is recognisable in thin section. For pelites, the characteristic assemblage is Chl + Ms (often phengitic) + Qtz + Ab (Appendix III). This assemblage typifies temperatures of the order of 325–425°C almost regardless of pressure. The muscovites present are usually phengitic, having significant Fe (or Mg) substitution for Al in octahedral sites. As the temperature rises these micas become more muscovitic, with the Al content increasing at the expense of Fe. A regular relationship of increasing Al substitution for Si in the octahedral site with increasing temperature and decreasing pressure has been noted, and calibrated for use as a geothermobarometer (Velde, 1965, 1967). More recently, Massone & Schreyer (1987) have extended Velde's work, and established that for the assemblage Phe + Phl + Kfs + Qtz there is a very strong, almost linear, increase of Si content per formula unit of phengite with respect to pressure, and a slight decrease in Si content per formula unit with increasing temperature. This relationship has also been calibrated for use as a geobarometer. In certain pelitic lithologies, minerals such as stilpnomelane, paragonite or pyrophyllite may occur.

The increase in temperature from low greenschist to upper greenschist facies conditions is marked by the incoming of **biotite**, by the reaction Chl + Kfs $\rightleftharpoons$ Ms + Bt. At 4 kbar (KFMASH system) this occurs at around 440° ± 20°C. This first appearance of biotite by the reaction given is not strictly in true pelites, but more correctly relates to greywackes with detrital K-feldspar. In true pelites the biotite-in isograd occurs at slightly higher temperatures, by a continuous reaction involving the breakdown of chlorite and the transformation of white mica from phengite to muscovite. The range of temperatures for first appearance of biotite is due to the continuous or divariant nature of the reaction. In rocks with a high Fe:Mg ratio (i.e. high XFeO) the reaction will proceed at 420°C, whereas for rocks with low XFeO temperatures > 450°C may be needed before biotite forms.

In particularly Mn-rich rocks, spessartine-rich garnets may occur in rocks of the lower greenschist facies, but for typical pelites of the KFMASH system, porphyroblasts of **almanditic garnets** do not occur until temperatures of the order of 460–500°C. The experimental work of Hsu (1968) has shown that pure almandine does not form until even higher T is attained. As with a significant spessartine component, a significant grossular component in almandine garnet has the effect of lowering the temperature of first formation. The bulk rock chemistry clearly has a strong influence on the composition of almanditic garnets that form, and as such the garnet-in isograd covers a broad range of temperatures and will vary from one area to the next. Although spessartine-rich garnets may occur in lower greenschist facies rocks, the presence of almanditic garnets is an indication of upper greenschist (epidote–amphibolite) facies conditions or higher. The continuous reaction Chl + Ms → Grt + Bt + H_2O is probably the most important reaction to form almanditic garnet. As temperature rises, the Mg:Fe ratio of garnet steadily increases as a result of a cation-exchange reaction with biotite.

At about the same P–T conditions as for the formation of almanditic garnet, the small proportion of plagioclase present in pelitic rocks becomes more calcic, changing from albite in greenschist facies rocks to oligoclase in epidote–amphibolite and amphibolite facies rock. In terms of anorthite component, the change is typically from An_{0-5} (albite) in low greenschist facies rocks, to An_{5-10} in upper greenschist facies rocks, and then a jump to An_{18-25} (oligoclase) in rocks of the epidote–amphibolite and low amphibolite facies. This jump marks the 'peristerite gap' corresponding to a structural change in plagioclase feldspar. This change in feldspar type is not recognisable by standard optical microscopy, but can be readily confirmed by electron-microprobe analytical work.

Another mineral characteristic of greenschist and epidote–amphibolite facies conditions is **chloritoid**. However, its occurrence is strongly influenced by bulk rock chemistry, and it has been shown that it only occurs in highly aluminous pelites with a very low calcium content. In the pure FASH system (not true pelites), it can start to appear at sub-greenschist facies conditions (c. 230°C) by the reaction Chl + Prl ⇌ Cld + Qtz (Fig. 3.2), but in average pelites of the KFMASH system chloritoid appears at temperatures of around 300°C, by reactions involving the breakdown of chlorite. Chloritoid frequently occurs as porphyroblasts, and exists over a broad range of pressure conditions. It is not uncommon in contact metamorphic aureoles (Plate 6(b)), as well as occurring in regionally metamorphosed pelites. Petrographic studies, experimental work and petrogenetic grids constructed on the basis of internally consistent thermodynamic data sets have established that at temperatures of the order of 520–560°C over the normal range of P–T conditions (3–10 kbar) during orogenic metamorphism (higher T with higher P) chloritoid reacts out to produce **staurolite**-bearing assemblages (Fig. 3.2). In view of this, the presence of chloritoid gives a good indication of the maximum temperature experienced by pelites, and the incoming of staurolite gives a good estimate of the minimum temperature conditions attained. Stable coexistence of the assemblage Grt + St + Cld, although only seen in certain aluminous pelites, is a good indication of temperature conditions of about 550°C (Bucher & Frey, 1994).

Over the temperature range 550–600°C, pelites in the KFMASH system with high alumina content and $X_{FeO} > 0.5$ have the assemblage Grt + Chl + St + Ms + Qtz. Those pelites with similar X_{FeO}, but lower alumina lack staurolite, but maintain the Chl + Bt association (i.e. Grt + Chl + Bt + Ms + Qtz). Above 600°C (almost irrespective of pressure), the association Grt + Chl is no longer stable and in Fe-rich pelites the assemblage Grt + St + Bt + Ms + Qtz occurs, while in pelites with a higher Mg content, the assemblage St + Bt + Chl + Ms + Qtz is seen. This St + Bt association is crucial, and marks the start of mid-amphibolite facies conditions. By a combination of field and thin-section study, it is an isograd that is quite easily defined. The Grt + St + Bt association is diagnostic of pelites of the mid-amphibolite facies (Ky + St + Bt [*no* Grt] in more aluminous pelites), but once upper amphibolite facies conditions in excess of 670°C are reached, staurolite is no

Compositional groups of metamorphic rocks

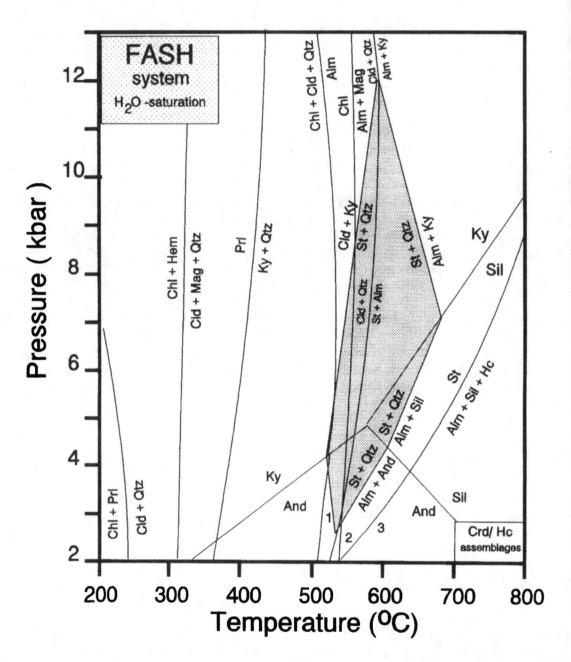

FIG. 3.2 A petrogenetic grid for the FeO–Al$_2$O$_3$–SiO$_2$–H$_2$O (FASH) system (modified after Bucher & Frey, 1994; constructed using data of Holland & Powell, 1990; details of reactions involving cordierite and hercynite not shown). Shaded field marks St + Qtz stability.

longer stable in pelites, and breaks down by the reaction St ⇌ Grt + Bt + Ky/Sil, to give the characteristic association Grt + Bt + Ky/Sil.

At mid-amphibolite facies conditions (P of the order of 5–10 kbar, T = 600–670°C), **kyanite** is the stable Al_2SiO_5 polymorph, whereas in the upper amphibolite facies, and passing into the granulite facies (typically P of the order of 5–10 kbar, T ≥ 700°C), it is **sillimanite** that is the stable Al_2SiO_5 phase present (Fig. 3.3). At temperatures around 650–660°C a certain degree of *in situ* partial melting ('anatexis') commences in metasedimentary sequences (H_2O-saturated conditions) to produce a 'granitoid' melt fraction. The melt fraction is usually represented by irregular pods, discontinuous layers and segregations parallel to the schistosity/gneissosity of the rock. At higher temperatures the melt fraction may be more extensive and may give rise to the development of migmatites (Fig. 4.11). For H_2O-saturated pelitic rocks with quartz in excess, an important reaction indicative of very high temperature conditions (typically T ≥ 700–750°C at P = 4–7 kbar) is the decomposition of muscovite and formation of **K-feldspar** by the reaction Ms + Qtz ⇌ Kfs + Al_2SiO_5 + H_2O (Fig. 3.3).

3.1.2 Low-pressure assemblages

The succession of assemblages and reactions observed in KFMASH pelites under lower-pressure conditions (T ≤ 4.5 kbar), while showing some similarities to those developed during medium-pressure 'Barrovian' metamorphism, also exhibits some notable differences. The incoming of biotite is one of the first clearly recognisable mineralogical changes when moving into a contact metamorphic aureole (or an area of regional high heat flow). The biotite-in isograd indicates temperatures of about 425°C or greater. At these temperatures and above, porphyroblasts of **andalusite**

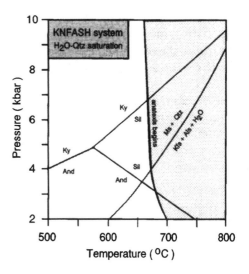

FIG. 3.3 The P–T diagram for the KNFASH (K_2O–Na_2O–FeO–Al_2O_3–SiO_2–H_2O) system, showing the stability fields of the Al_2SiO_5 phases, the curve for the Ms + Qtz ⇌ Kfs + Als + H_2O reaction and the curve defining the conditions under which anatexis commences. The reaction curves shown are based on the thermodynamic data set of Holland & Powell (1990). For simplicity, the numerous other reactions in this system are not depicted.

(commonly var. chiastolite; Plate 1(c)) are common in suitable pelitic lithologies. Andalusite is a very good indicator of low-pressure conditions. It is unstable above P = 4.5 kbar and most typical of P < 4 kbar (Fig. 3.3). Above 515–550°C (depending on P and bulk rock chemistry), porphyroblasts of **cordierite** form, and the assemblage And + Crd + Bt is common to many rocks. In other cases staurolite (Plate 1(a)) develops to give the common association of And + St + Bt. Precisely which assemblages develop is largely a function of bulk rock chemistry and pressure.

For a suite of rocks being metamorphosed at P = 3.5 kbar, the andalusite–sillimanite reaction line is crossed at about 650°C (Fig. 3.3). Commonly, andalusite can exist metastably even once this line has been crossed, but above temperatures of 660–670°C it is common to record fibrolitic sillimanite, as well as K-feldspar development from muscovite.

Prismatic crystals of sillimanite tend to be more characteristic of higher temperatures. At temperatures above 680°C, the assemblage Qtz + Pl + Kfs + Bt + Sil is most characteristic of pelitic rocks. In common with higher-pressure regional metamorphic regimes, the highest-temperature parts of contact aureoles or belts of regional high heat flow are dominated by rocks comprising a significant proportion of granitoid melt.

3.1.3 High-pressure assemblages

In the high-P/low-T blueschist facies terranes, pelites exhibit few reactions or assemblages of note. At medium pressures they are characterised by much the same phase assemblage as greenschist facies rocks. However, it is worth noting that the temperature conditions of the blueschist facies are generally too low for the formation of biotite. The brown, strongly pleochroic sheet-silicate phase sometimes mistakenly identified as biotite is in fact usually **stilpnomelane** (Figs 5.5 & 5.13(a)). Hence the characteristic phase assemblage of blueschist facies pelites is therefore Chl + white mica (often phengitic) + Ab + Qtz (+ Stp, Cal, Ep) + opaques (often Ilm or Py). While abundant in metabasites at this grade, sodic amphiboles are notable by their absence from pelites.

3.2 Metacarbonates and calc-silicate rocks

Sedimentary carbonates are chemically dominated by $CaCO_3$, $MgCO_3$ and SiO_2, with increasing amounts of Al_2O_3 and K_2O as they grade into pelites. The mineral assemblages produced during progressive metamorphism are dependent not only on P, T and bulk rock composition, but additionally fluid composition has a profound effect on the assemblages produced and the temperatures at which particular reactions occur. This is true of all metamorphic rocks, but is especially true of meta-carbonates and calc-silicate rocks. H_2O and CO_2 are the dominant components in typical metamorphic fluids, and the amount of each is commonly expressed as a value between 0.0 and 1.0 of total fluid. For example, a fluid with XH_2O = 0.8 and XCO_2 = 0.2 is a fluid that is 80% H_2O and 20% CO_2. The values of XH_2O and XCO_2 show considerable variation in fluids of carbonate and calc-silicate rocks, and act as an important control over the assemblages produced and the temperatures at which certain reactions occur (Trommsdorff & Evans, 1977; Walther & Helgeson, 1980). In general, high XCO_2 inhibits most reactions, so that for any given reaction higher T is required for it to occur (Fig. 1.3). The salinity of aqueous fluids also has an important effect. Generally speaking, the higher the salinity, the higher is the temperature required for a given reaction to proceed.

Metamorphism of a pure $CaCO_3$ limestone will give rise to recrystallisation of the calcite, to produce a fairly equigranular calcite marble, probably becoming coarser with increasing grade. Only under very high pressure conditions will any mineralogical change be seen, and under such conditions the polymorphic transformation of calcite to the denser polymorph, aragonite, will occur (Fig. 12.19).

In the case of siliceous dolomitic limestones ($CaO-MgO-SiO_2-H_2O-CO_2$; CMSHC system), the increased variety of original rock chemistry means that many reactions are possible during progressive metamorphism. In low-pressure orogenic metamorphism, at temperatures less than 500°C, in the presence of a H_2O-rich fluid, **talc** should be a diagnostic mineral, but it is often only seen as a retrograde phase, since on the prograde path it breaks down by the **tremolite**-forming reaction: 2Tlc + 3Cal $\rightleftharpoons$ Tr + Dol + H_2O + CO_2. For fluid compositions in the range XH_2O = 0.3–0.9, the tremolite-forming reaction in siliceous dolomitic limestones is 5Dol + 8Qtz + H_2O $\rightleftharpoons$ Tr + 3Cal + 7CO_2. Since this consumes H_2O and liberates CO_2, then in a closed, internally buffered system, the fluid composition is driven to increasingly higher XCO_2. At low- to mid-amphibolite facies conditions (T = 500–670°C), tremolite-bearing assemblages dominate, as long as XH_2O is ≥0.3. At upper amphibolite conditions and higher, **diopside** is the characteristic calc-silicate phase. At high XCO_2, it forms by a reaction involving the breakdown of Dol + Qtz, but at moderate to high XH_2O, it forms from the breakdown of calcite + tremolite.

Assemblages containing **forsterite + diopside** require high-T conditions and characterise metamorphosed siliceous dolomites from the innermost parts of contact metamorphic aureoles, and situations of very high T hydrothermal metamorphism. **Wollastonite** may be seen in certain metacarbonates and calc-silicate rocks near to major intrusions, but is not associated with orogenic metamorphism. It commonly forms by the decarbonation reaction Cal + Qtz = Wo + CO_2, but the temperature at which this reaction occurs is greatly influenced by the $H_2O : CO_2$ ratio of the fluid (Fig. 1.3). At total fluid pressure (P_f) in the range 1–3 kbar, for XCO_2 = 0.13 the reaction proceeds at about 520–550°C whereas for XCO_2 = 1.0 (i.e. pure CO_2 fluid) temperatures of 700–800°C are required (Greenwood, 1967).

Impure carbonates containing a significant pelite component will have a chemistry with a large proportion of Al_2O_3. When metamorphosed, Ca–Al silicates such as epidote, grossular, hornblende, idocrase and anorthite are commonly produced. The K_2O component of the original clays usually gives rise to muscovite or phlogopite in the calc-schists and calc-silicate rocks produced.

3.3 Quartzofeldspathic metasediments

When metamorphosed, clastic sediments such as sandstones and arkoses give rise to quartzitic and quartzofeldspathic metamorphic rocks. During metamorphism pure quartzites are extensively recrystallised to produce a more stable granoblastic-polygonal aggregate (Fig. 5.14(b)). In immature sandstones, detrital minerals give increased variety, although the metamorphic assemblage will usually be little more than Qtz + Pl + Kfs + mica. There is a complete gradation from quartzites via psammites and semi-pelitic schists to pelitic schists. However, since there are no standardised definitions for these rocks based on the relative proportions of the main end-member components, no definite boundaries have been marked on Fig. 3.1.

The simple mineralogy of metamorphosed quartzofeldspathic sediments generally makes them of little use in assessing metamorphic grade. However, during the metamorphism of such rocks at blueschist facies conditions, a crucial change in mineralogy does occur. With increasing pressure, albite becomes unstable and reacts out to form jadeitic pyroxene + quartz. In metaclastic rocks, jadeite is never pure but always contains at least 10% of other pyroxene components, such as diopside, hedenbergite and aegirine. The purity of the jadeitic pyroxene gives a good indication of the P conditions attained, with the purest jadeite being indicative of the highest P.

Although their mineralogy can hardly be described as exciting, the textural and microstructural changes observed during the metamorphism and deformation of quartzites and quartzofeldspathic rocks have attracted considerable interest over the decades. This is largely due to the differences in behaviour of quartz and feldspar during deformation, the simple slip systems of quartz and the simple mineralogy, which makes it easier to isolate and understand those processes in operation under different experimental and geological conditions (see Chapter 8 for more details).

3.4 Metabasites

Metabasic rocks represent metamorphosed basalts, dolerites, gabbros and basaltic

Compositional groups of metamorphic rocks

andesites. Their chemistry is typified by 45–55 wt% SiO_2 and dominant MgO, FeO, Fe_2O_3, CaO and Al_2O_3, with lesser amounts of Na_2O and TiO_2, and very little K_2O. Consequently, a varied assemblage forms during metamorphism of mafic rocks, dominated by silicates enriched in the major components listed. Like metapelites, metabasic rocks show mineralogical changes that have proved useful for assessing metamorphic conditions. At low metamorphic grades original igneous textures and microstructures may be recognisable, but with increasing metamorphism and deformation such features are usually lost. Metabasites from environments of orogenic metamorphism or subduction-zone metamorphism generally have a pronounced schistosity, while basic hornfelses and those rocks that have experienced granulite facies or eclogite facies metamorphism normally show a granoblastic structure.

Pressures of less than 3 kbar and temperatures of up to 300°C characterise areas of burial metamorphism, very low grade orogenic metamorphism, ocean-floor metamorphism and metamorphism in geothermal regions. At such grades it is typical to observe mixed disequilibrium assemblages containing high-T igneous minerals in association with low-T metamorphic minerals. This is because fluid access to the rock is very heterogeneous, depending strongly on the extent of fracturing and/or primary porosity, such as in lava flow tops. The zeolite group of minerals (Fig. 5.12(c)) can prove useful as a means of documenting changes in metamorphic conditions. However, these only form in association with high XH_2O fluids. If the fluids have moderate or high XCO_2, carbonate minerals form.

Various zeolite sequences have been documented in the basaltic lava piles of east Iceland, in various ophiolites worldwide and in the geothermally active volcanic regions of New Zealand. Such zonal sequences are variable from one area to another, and are dependent on variables other than temperature. These include porosity/permeability, fluid composition ($XH_2O : XCO_2$) and pressure, which at such high crustal levels may be at or close to $P_{hydrostatic}$, rather than $P_{lithostatic}$. Although complex, the general trend is one of most hydrous zeolites at lowest T being replaced by least hydrous zeolites at higher T. A combination of experimental work and improved thermodynamic data sets has enabled the stability fields for many of the key zeolites to be well defined with reference to simple chemical systems. For example, in the CASH system, heulandite–analcime represents temperatures of the order of 100–200°C. It is succeeded by laumontite, representative of about 200–275°C. The high-temperature zeolite, wairakite, is common in the active geothermal system (Taupo Volcanic Zone) of the North Island of New Zealand. It is indicative of temperatures of the order of 250–400°C in the CASH system, but is rarely encountered in other areas. In the complex chemical system of natural metabasites (NCMFASH), the equating of specific temperature stability ranges for the various zeolites is less assured. In many areas it is more usual to pass directly from laumontite assemblages into rocks bearing the minerals prehnite, pumpellyite and actinolite, along with ubiquitous quartz and albite. In simplified NMASH/NCMASH systems, it is possible to recognise distinctive facies that appear to be diagnostic of discrete P–T regimes. These have been referred to as prehnite–pumpellyite, prehnite–actinolite and pumpellyite–actinolite facies. However, in more complex systems (NCFMASH) and natural metabasites, such distinct 'facies' show substantial overlap. For this reason I use the term 'sub-greenschist facies' for P–T conditions represented by assemblages with prehnite and/or pumpellyite (Fig. 2.1), rather than attempting more detailed subdivision.

With increasing temperature, prehnite and pumpellyite become unstable and are replaced by epidote group minerals, in the transition to the greenschist facies. In the greenschist facies, (~325°C to 475°C), the characteristic assemblage of metabasites is Act + Chl + Qtz + Ab + Ep + Spn (Appendix III). The transition from greenschist facies to amphibolite facies via the epidote–amphibolite facies of Eskola (1939) is indistinct in most areas, and is marked by assemblages in which two Ca-amphiboles (actinolite and hornblende) coexist. In practice, this means that it is usually simpler to identify definite greenschist facies rocks and definite amphibolite facies assemblages, and to refer to intermediate assemblages as transitional. The change from greenschist to amphibolite facies is marked by a number of simultaneously occurring reactions. The most notable of these are: (i) the actinolite to hornblende reaction in the Ca-amphibole solid-solution series by Tschermak substitution (MgSi (or FeSi) = $Al^{IV}Al^{VI}$ exchange), coupled with chlorite breakdown, and (ii) the simultaneously occurring reaction involving breakdown of albite and epidote to form oligoclase, with Na released in this reaction incorporated in hornblende.

As well as the miscibility gap in the Ca-amphiboles there is also a miscibility gap in the case of the plagioclase (the 'peristerite gap'). The 'peristerite gap' in metabasic rocks at low P (2 kbar) has been studied in detail by Maruyama et al. (1982). They found that the transition zone consisting of 'peristerite pairs' + Ep + Chl + Ca-amph (usually Act + Hbl) + Qtz + Spn occurs between 370°C and 420°C. At higher T, this assemblage is succeeded by one in which actinolite is absent, chlorite persists and oligoclase is present ($Pl(An_{20-50})$+ Hbl + Chl + Spn + Ilm. At temperatures of about 500°C and moderate pressures of orogenic metamorphism, amphibolites have the characteristic assemblage Olg + Hbl + Ep + Rt (+ Qtz + Grt ± Bt). As upper amphibolite facies conditions are attained, garnet becomes modally more important, and plagioclase becomes increasingly calcic at the expense of epidote (which becomes increasingly scarce or absent). With increasing temperature of metamorphism, there is a tendency for hornblende to become more Ti-rich, and for the pleochroic scheme to change from blue green – green under epidote amphibolite facies conditions to green–brown in the upper amphibolite facies.

The boundary between hydrous amphibolite facies and anhydrous granulite facies is very much transitional, but as the temperature rises the last remaining hydrous phase (hornblende) reacts out to form pyroxenes. Although some hornblende persists in granulite facies rocks, most has usually gone by 750–800°C. When metabasites have been metamorphosed under granulite facies or pyroxene hornfels facies conditions, in addition to the characteristic granoblastic structure (Section 5.5), they also display a distinctive mineralogy consisting of orthopyroxene (hypersthene) + anorthitic Pl + Cpx (± Spl ± Grt). Under high-pressure granulite facies conditions (bordering eclogite facies) the anorthitic plagioclase becomes increasingly unstable and ultimately reacts out. The precise position of the plagioclase-out reaction in mafic and ultramafic rocks is strongly dependent on the composition of the parent rock. While it is generally absent from pyroxene hornfels facies rocks, garnet is common in granulite facies rocks, and becomes increasingly represented at higher P. Again, the extent of the P–T conditions suitable for garnets in metabasites/ultramafites of the granulite facies is strongly bulk-rock controlled. The change from spinel-bearing to garnet-bearing pyroxenites and lherzolites occurs within the top granulite facies through eclogite facies and reflects increased pressure. The exact position of this reaction varies according to bulk rock chemistry, but the presence of garnet in such rocks is a good indication of very high pressure.

In the eclogite facies the characteristic mineral association is Cpx(omphacite) + Grt, commonly in equal porportions. Accessory phases often present include quartz, rutile and kyanite. Green & Ringwood (1967) undertook a classic piece of experimental work examining the transformation of gabbros into eclogites with increasing pressure and under virtually anhydrous conditions. Although the precise boundaries for the eclogite facies are not well defined on the P–T grid, such rocks are clearly representative of high-P/high-T conditions.

Metabasites from the environment of blueschist facies metamorphism have a distinctive mineralogy dominated by sodic amphiboles such as glaucophane or crossite. Experimental work has shown that the phase assemblages of blueschist facies rocks are indicative of high-P/low-T conditions. In the lower-pressure part of the blueschist facies, with P of the order of 5–8 kbar and T = 200–350°C, the metabasite assemblage is typically Gln + Ep (or Lws) + Spn + Ab + Qtz (+ Chl + white mica + Stp + Cal). Higher-pressure blueschists may contain small amounts of jadeitic pyroxene in addition to glaucophane. Such rocks lack albite, and if a carbonate phase is present it is most likely to be aragonite rather than calcite. Many blueschist facies rocks are transitional into the greenschist facies (Section 12.3.5). In addition to Na-amphiboles (glaucophane–crossite), such higher-temperature blueschist facies rocks commonly contain garnet, and a second amphibole such as actinolite or the Na-rich sub-calcic hornblende known as barroisite.

3.5 Metamorphosed ultramafic rocks

Ultramafic rocks such as dunite, lherzolite, harzburgite and pyroxenite have an assemblage characterised by olivine and pyroxene, with the proportions varying according to rock type. Hydrated ultramafic rocks generally have a simpler chemistry compared to metapelites and metabasites, and are dominated by MgO, SiO_2, H_2O with CaO and Al_2O_3, and so can be considered as a MSH to CMASH system. For hydrated lherzolites the system is more strictly CFMASH. In cases in which the fluid has a significant CO_2 component, the systems become MS–HC, CMAS–HC and CFMAS–HC, and carbonate phases (e.g. magnesite) form part of the mineral assemblage.

Highly sheared and metamorphosed ultramafic rocks occur as discontinuous pods and lenses in fault and shear zones, as well as in thick units at the base of many ophiolite slabs. Rocks from these environments often show pronounced and pervasive ductile shear fabrics. Fluid flow through the shear zones readily leads to the retrogression of the primary igneous assemblage to give serpentinites. While serpentine is widely associated with metamorphosed ultramafic rocks, other Mg-rich minerals such as anthophyllite, talc, magnesite and clinochlore are also common in certain ultramafites. The Fe component of ultramafites is re-utilised in the new growth of Fe-oxides, Fe-bearing amphiboles and chlorites. None of these minerals is especially useful for tightly defining P–T conditions, especially if the fluid composition is not well known.

During metamorphism of simple MgO–SiO_2 ultramafites (harzburgites), it is well established that fluids rich in CO_2 generate entirely different mineralogies compared to H_2O-rich fluids. In mixed fluids the precise H_2O : CO_2 ratio significantly controls the temperature at which a given reaction occurs for given fluid pressure (Johannes, 1969). Serpentine group minerals are indicative of very aqueous fluids (i.e. very low XCO_2) and while often associated with metamorphism at temperatures less than 400°C, the extreme upper limit is shown to be at about 480°C/1 kbar to 590°C/7 kbar (Evans & Trommsdorff, 1970). Chrysotile is the stable serpentine group mineral below about 250°C, but at higher temperatures (greenschist–amphibolite facies) antigorite is the characteristic serpentine phase. The upper temperature limit for magnesite is somewhat similar to that for serpentine minerals, although magnesite can form over a very broad range of fluid composition from high H_2O to high CO_2.

Talc forms over a wide range of fluid and P–T conditions. The lower stability limit ranges from as low as about 310°C for highly aqueous fluids at 1 kbar, to as high as 670°C if the fluid is very H_2O-rich (see Johannes, 1969, for detailed reaction curves). The presence of anthophyllite, enstatite or forsterite in metamorphosed ultramafics signifies a high-grade assemblage. Anthophyllite is stable at temperatures between about 500°C and 760°C, depending on bulk rock chemistry, fluid composition and P_f. Forsterite and enstatite are commonly part of the original igneous assemblage. The reaction curves determined by Johannes (1969) suggest that the lower stability limit for these minerals is of the order of 500–550°C at 2 kbar.

CMASH (CFMASH) systems of metamorphosed lherzolites, as well as containing serpentine and talc, also have chlorite, Ca-amphibole (tremolite) and/or Ca-pyroxene (diopside) as key phases at greenschist and amphibolite facies. Passing into the granulite facies, spinel forms, while chlorite and amphibole disappear, to give a typical assemblage of Fo + En + Cpx + Spl at low to moderate pressure, and Fo + En + Cpx + Grt at high pressure, bordering eclogite facies.

3.6 Meta-granitoids

The characteristic mineral assemblage of deformed and/or metamorphosed 'granitoid' rocks is essentially the same as that for deformed and/or metamorphosed quartzofeldspathic sediments. At low metamorphic grades in granitoid rocks that have experienced only limited deformation, the original igneous texture and mineralogy is usually recognisable. However, the feldspars commonly show signs of alteration. They frequently show partial replacement or complete pseudomorphing by a fine aggregate of sericite, typically accompanied by epidote group minerals. Under similar conditions, clastic sediments will also retain much of their original texture, and thus it is easy to distinguish them from metagranitoid rocks. However, at higher temperatures the original rock microstructure becomes blurred, and ultimately lost, as the quartz–feldspar–mica assemblages recrystallise to a more stable arrangement. As temperature increases, the plagioclase becomes more calcic, and at high grades of metamorphism muscovite will ultimately break down to K-feldspar. At these high grades it is virtually impossible, from microscopic work alone, to distinguish rocks that were originally 'granites' from those that were quartzofeldspathic sediments.

Similarly, highly sheared or mylonitised granites can be difficult to distinguish from mylonitised arkosic sediments. However, a good understanding of the field relationships and associated lithologies to each side of the shear zone will normally resolve the problem.

References

Bucher, K. & Frey, M. (1994) *Petrogenesis of metamorphic rocks*. Springer-Verlag, Berlin, 318 pp.

Eskola, P. (1939) Die metamorphen Gesteine, in *Die Entstehung der Gesteine* (eds T.F.W. Barth, C.W. Correns & P. Eskola). Julius Springer, Berlin (reprinted, 1960, 1970), 263–407.

Evans, B.W. & Trommsdorff, V. (1970) Regional metamorphism of ultramafic rocks in the Central Alps; paragneisses in the system CaO–MgO–SiO_2–H_2O. *Schweizerische Mineralogische und Petrographische Mitteilungen*, 50, 481–492.

Green, D.H. & Ringwood, A.E. (1967) An experimental investigation of the gabbro to eclogite transformation and its petrological implications. *Geochemica et Cosmochimica Acta*, 31, 767–833.

Greenwood, H.J. (1967) Wollastonite: stability in H_2O–CO_2 mixtures and occurrence in a contact-metamorphic aureole near Salmo, British Columbia, Canada. *American Mineralogist*, 52, 1669–1680.

Holland, T.J.B. & Powell, R. (1990) An enlarged and updated internally consistent thermodynamic dataset with uncertainties and correlations: the system K_2O–Na_2O–CaO–MgO–MnO–FeO–Fe_2O_3–Al_2O_3–TiO_2–SiO_2–C–H_2–O_2. *Journal of Metamorphic Geology*, 8, 89–124.

Hsu, L.C. (1968) Selected phase relationships in the system Al–Mn–Fe–Si–O–H: a model for garnet equilibria. *Journal of Petrology*, **9**, 40–83.

Johannes, W. (1969) An experimental investigation of the system $MgO-SiO_2-H_2O-CO_2$. *American Journal of Science*, **267**, 1083–1104.

Kretz, R. (1983) Symbols for rock-forming minerals. *American Mineralogist*, **68**, 277–279.

Maruyama, S., Liou, J.G. & Suzuki, K. (1982) The peristerite gap in low-grade metamorphic rocks. *Contributions to Mineralogy and Petrology*, **81**, 268–276.

Massone, H.J. & Schreyer, W. (1987) Phengite geobarometry based on the limiting assemblage with K-feldspar, phlogopite, and quartz. *Contributions to Mineralogy and Petrology*, **96**, 212–224.

Trommsdorff, V. & Evans, B.W. (1977) Antigorite–ophicarbonates: contact metamorphism in Valmalenco, Italy. *Contributions to Mineralogy and Petrology*, **62**, 301–312.

Velde, B. (1965) Phengite micas: synthesis, stability and natural occurrence. *American Journal of Science*, **263**, 886–913.

Velde, B. (1967) Si^{+4} content of natural phengites. *Contributions to Mineralogy and Petrology*, **14**, 250–258.

Walther, J.V. & Helgeson, H.C. (1980) Description of metasomatic phase relations at high pressures and temperatures: 1. Equilibrium activities of ionic species in non-ideal mixtures of CO_2 and H_2O. *American Journal of Science*, **280**, 575–606.

Part B

Introduction to metamorphic textures and microstructures

The chapters in this part aim to give a broad introduction to the description and interpretation of the fundamental textures and microstructures of metamorphic rocks. The significance of each will be emphasised and, where relevant, current theories and controversies relating to their origin will be discussed. Before doing this, however, it is first necessary to define texture, microstructure and equilibrium in relation to metamorphic rocks.

Definition of texture and microstructure

In many previous texts dealing with metamorphic rocks, there has been a more or less synonymous usage of the terms TEXTURE and MICROSTRUCTURE to describe the shapes and arrangement of grains within the rock. However, following the reasoning of Vernon (1976), it is advocated here that the term MICROSTRUCTURE should be used to cover all aspects of the microscopic arrangements and interrelationships between grains, while restricting the term TEXTURE to those arrangements in which there is some preferred orientation. This more specific use of the term 'texture' is favoured for a number of reasons. In particular, since the modern approach to materials science groups all crystalline materials together in terms of basic processes, it is logical to use a common descriptive terminology. In synthetic materials (e.g. metals) 'texture' specifically relates to cases of preferred orientation. It therefore makes sense to use the term in a similar way when referring to rocks, rather than perpetuate a more general use of the term and promote possible confusion.

Equilibrium and equilibrium assemblages

Assessing the degree of equilibrium attained by a particular metamorphic assemblage may not always be easy. The strict definition of equilibrium is that state of a rock system in which the phases present are in the most stable, low-energy arrangement, and where all phases are compatible with the given P, T and fluid conditions. This concept has been introduced in Sections 1.3 & 1.4.

If the equilibration rate is rapid compared to the rate of change of P, T and/or fluid conditions, then the rock will maintain equilibrium. This may be true through certain parts of a rock's $P-T$ trajectory, but is seemingly not the case throughout the whole trajectory. The many high-grade metamorphic assemblages preserved in rocks at the Earth's surface provide a clear indication of this lack of equilibration. It therefore seems that the rate of equilibration is much slower than the rate of change in P–T during uplift, and that the assemblage observed records some earlier stage of the P–T evolution. In most cases it is taken to represent the assemblage at peak metamorphic conditions. How close to equilibrium this assemblage is depends as much as anything on how long the particular rock was held at or close to peak conditions, since the longer a particular set of conditions prevails, the more time there is to equilibrate. In many rocks there are signs of partial equilibration to lower P–T conditions in that some phases show signs of retrogression in the form of reaction rims or pseudomorphs, and yet other phases of the high-grade assemblage appear entirely fresh. In such cases the complete assemblage is very much a disequilibrium assemblage. It contains more phases than would satisfy the Phase Rule and it contains phases representing both high-grade and low-grade metamorphic conditions. The degree of microstructural equilibrium also depends on the rate of change of the different variables. The most stable arrangement of grains is a polygonal aggregate of unstrained crystals with flat faces, whereas aggregates of strained crystals with irregular boundaries show a much lower degree of equilibrium.

Reference

Vernon, R.H. (1976) *Metamorphic processes*. George Allen & Unwin, London, 247 pp.

Chapter four

Layering, banding and fabric development

4.1 Compositional layering

The presence of LAYERING or BANDING in metamorphic rocks can be either primary or secondary in origin, and can occur on the macro-, meso- or microstructural scale. In metamorphosed sedimentary and volcanic sequences, original COMPOSITIONAL LAYERING is often still recognisable. This is due to primary chemical differences that influence the nature of metamorphic mineral assemblages that can develop. Such differences are often recognisable in thin section by virtue of mineral assemblage or grain size variations (Fig. 4.1). In low- to medium-grade pelitic and semi-pelitic sequences it is often possible to recognise fining-upwards sequences and to determine 'way-up' based on grain size variations or subtle changes in the proportion of quartz and phyllosilicate minerals present (Fig. 4.2).

In high-grade gneisses and migmatites, original compositional layering still has a strong influence on nature of banding observed, but at such high metamorphic grades recrystallisation is so extensive that almost all primary features are obliterated. Diffusive mass transfer and *in situ* partial melting contribute to the enhancement of any primary compositional banding, as well as producing additional segregation of felsic and mafic minerals. The combination of processes described above, often synchronous with active shearing at elevated temperatures, produces the characteristic interbanding of quartzofeldspathic and mafic layers referred to as GNEISSOSE STRUCTURE (GNEISSOSITY) or MIGMATITIC BANDING. Such banding is generally coarse (centimetre to decimetre scale) and while recognisable in large hand specimens is not readily appreciated in thin section. AUGEN GNEISS is a special name applied to rocks with a structure consisting of 'augen' (eyes), usually of feldspar (typically K-feldspar), in a strongly foliated gneissic (typically Qtz–Bt–Ms) matrix (Fig. 4.3). This structure develops by a combination of deformation and recrystallisation of original coarse-grained quartz–feldspar–mica assemblages (e.g. metagranitoids) at low to moderate shear strains and moderate to high temperatures. At higher shear strains mylonitic rocks are produced (Chapter 8). In fact, there is a complete transition between the appearance of typical augen gneisses and granitic mylonites. Although augen gneiss is most commonly developed in metamorphic rocks of original

Layering, banding and fabric development

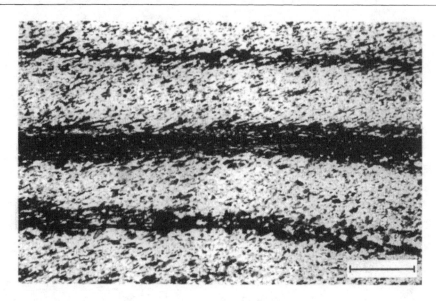

FIG. 4.1 Compositional layering in greenschist facies semi-pelitic schist from Sierra Leone. Alternating layers are richer and poorer in biotite with respect to quartz. Note also the weak schistosity trending oblique to the layering from the top right to the bottom left. Scale = 1 mm (PPL).

granitic to granodioritic composition, mafic gneisses may also display augen structure.

A further example of compositional (mineralogical) banding seen in metamorphic rocks is the case of zoned or banded SKARNS. These are often formed in contact metamorphic aureoles and other situations in which hydrothermal silica-saturated fluids have infiltrated carbonate (metacarbonate) rocks, or interbanded carbonate and siliceous rocks. Bands of calc-silicate minerals (skarns) form due to metasomatic reaction between carbonate and silicate systems in mutual contact. The banding develops by diffusive migration of ions both towards and away from the reaction front: some atoms such as Si are clearly sourced from the silicate system, while others such as Ca and Mg are likely to have originated from the marble. However, the origin of some elements present in skarn bands may be ambiguous, and it is quite likely that they have been introduced with an externally derived fluid rather than coming from immediately adjacent rocks.

4.2 Introduction to stress, strain and fabric development

The forces acting on a rock mass produce a set of stresses. Since force (unit = newton (N)) is a vector quantity, the stresses can be defined in terms of direction and magnitude. The unit of stress (pascal (Pa) = 1 N m^{-2}) is the same unit as that of pressure, and although many metamorphic petrologists talk of pressure in terms of bars and kilobars, the SI unit is strictly the pascal (1 bar = 10^5 Pa; 1 kbar = 0.1 GPa (100 MPa)).

When a body of rock (metamorphic or otherwise) is subjected to a system of superimposed stresses, it experiences a certain degree of strain reflecting the stress imposed. *Strain is defined as the change in size and shape of a body resulting from the action of an applied stress field* (Park, 1989). Strain can be recorded in terms of volume change, distortion or both. The shape change may be non-rotational (coaxial), as in the case of *pure shear* regimes, or

Planar fabrics in metamorphic rocks

FIG. 4.2 A field photograph of grading in pelitic/semi-pelitic rocks, Snake Creek, Queensland, Australia. The pencil is 12 cm long; the younging direction (left to right) is shown by the symbol in the lower left corner. Platy layers represent micaceous (originally fine-grained clay-rich) tops to cycles.

have a rotational component, where superimposed stresses form a couple, as in the case of *simple shear* regimes. Three mutually perpendicular principal strain axes, X, Y and Z, can be used to define a strain ellipsoid, where $X \geq Y \geq Z$ (i.e. X is the axis of maximum elongation, Z the principal shortening direction and Y the intermediate axis). Because rocks are highly heterogeneous in character, their deformation is a very uneven process. Different rock types have different rheological properties and therefore behave very differently under a given set of conditions. Even on the scale of an individual rock, certain domains within the rock will be weaker than others. As a consequence, rock deformation is heterogeneous on all scales and this leads to pronounced strain partitioning.

When considering strain experienced by a given metamorphic rock, we can examine aspects of bulk strain, but on closer inspection microdomains of different strain can be identified. In a garnet–mica schist for example, the garnet porphyroblasts typically have subrounded to euhedral form, and with the exception of examples showing brittle fracturing, most garnets show no sign of being deformed. By contrast, the quartz–mica matrix shows evidence of pronounced mineral alignment and deformation. It therefore follows that there must be a strain gradient between porphyroblast and matrix (Fig. 9.7). The strain recorded in a given metamorphic rock can be evaluated on the basis of textural and microstructural observation and measurement. Features of different stages in the evolution of the rock may be recognised, but the nature of the strain ellipsoid deduced is that of the bulk finite strain ellipsoid (i.e. the sum total of all the strain experienced).

4.3 Classification of planar fabrics in metamorphic rocks

As well as compositional layering or banding, most deformed metamorphic rocks exhibit some kind of preferred orientation of constituent grains. Such structures result from alignment of inequidimensional grains (i.e. GRAIN-SHAPE FABRIC) or else alignment of crystal structures (CRYSTALLOGRAPHIC PREFERRED ORIENTATION). In the vast majority of deformed metamorphic rocks,

43

Layering, banding and fabric development

FIG. 4.3 A hand specimen of augen gneiss (Baltic Shield, Norway), comprising large K-feldspar crystals in a biotite-rich matrix.

parallel alignment of elongate grains is visible in hand specimen and therefore visible in thin section. They may define a FOLIATION (planar structure) or a LINEATION (linear structure) (Fig. 4.4). Many metamorphic rocks (e.g. schists and mylonites) are L–S TECTONITES, and are comprised of a linear and planar component. The lineation generally lies within the plane of the foliation, and defines the maximum elongation direction (X-direction) of the finite strain ellipsoid. The planar element lies parallel or very close to parallel with the X–Y plane, and is perpendicular to the principal direction of bulk shortening or compression. The high strains associated with ductile shear zones generate strongly foliated rocks termed MYLONITES and PHYLLONITES. However, these rocks and their fabrics will not be dealt with in this chapter, since they are covered comprehensively in Chapter 8. The remainder of this chapter deals with the classification of cleavage, and the processes involved in cleavage and schistosity development, followed by a section on the processes responsible for banding and layering in gneisses and migmatites. For further illustration and evaluation of the various cleavage types and other rock fabrics, the reader is referred to Borradaile et al. (1982) and Passchier & Trouw (1996).

Parallel planes of preferred splitting in a rock are known as CLEAVAGE, and are widely developed under conditions of low and medium metamorphic grade. Various types of cleavage have been recognised and described by geologists, but by the mid-1970s it was apparent that existing classifications of rock cleavage were rather confused. This was a consequence of a plethora of terms being introduced into the literature, often poorly defined and used differently by different workers, and a mixture of terminology based on morphological and genetic considerations. In 1976 a Penrose Conference on Cleavage attempted to resolve the problem and standardise the terminology in use. Although several classifications were attempted, each had their problems. Nevertheless, a subsequent paper by Powell (1979) presented a purely morphological classification of rock cleavage, which has considerable merits. An important advantage is that it provides an objective description of all cleavage

Cleavage and schistosity development

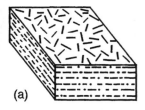 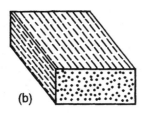

FIG. 4.4 Schematic block diagrams to illustrate the difference between S-tectonites (a) with a pronounced foliation (planar texture), and L-tectonites (b) with a pronounced lineation (linear texture). Metamorphic rocks such as schists and mylonites are generally L–S tectonites, and have both a linear and planar component.

types, without having any built-in genetic implications. Having established a clear description of cleavage type based on morphological features, any discussions of the cleavage-forming mechanism can follow, without fear of confusion about the type of cleavage under consideration.

Research has established that most cleaved rocks have a 'domainal' structure, comprising domains with strong mineral alignment (cleavage domains) separated by domains with a lesser degree of alignment (microlithons). Scanning electron microscopy has established that even the most finely cleaved slates are commonly domainal. Such cleavages have been termed SPACED CLEAVAGES, while those with apparently no domainal features are termed CONTINUOUS CLEAVAGES. This represents the most basic subdivision of cleavages on a morphological basis.

'Continuous' cleavages are those with a penetrative appearance in hand-specimen. They can be further subdivided into 'coarse' and 'fine', on the basis of mean grain size. Under the Powell classification, the term SCHISTOSITY (Fig. 4.5(a)) would be classified as a 'coarse continuous cleavage'. In sub-greenschist facies pelitic rocks of orogenic metamorphism, the fine-scale splitting referred to as SLATY CLEAVAGE is classed as a 'fine continuous cleavage' (Fig. 4.5(b)). For reviews of slaty cleavage development, reference should be made to Siddans (1972), Wood (1974) and Kisch (1991).

Spaced cleavages can be subdivided on the basis of whether or not there is a pre-existing planar anisotropy to the rock. Those cleavages post-dating an earlier planar anisotropy form CRENULATION CLEAVAGES, and are further divided into 'discrete' and 'zonal' types (Fig. 4.6). Spaced cleavages that have developed where there is no pre-existing anisotropy are termed DISJUNCTIVE CLEAVAGES. Under this heading come 'stylolitic', 'anastomosing', 'rough' and 'smooth' cleavages. In previous literature many of these types would have been loosely referred to as 'fracture cleavage', but this is not recommended, since this term has genetic implications. The spacing of cleavage domains varies from the scale of centimetres/decimetres for anastomosing and stylolitic cleavages, down to 0.1 mm – 1 cm for crenulation cleavages (Fig. 4.6(a)), and 0.01–0.1 mm for domainal 'slaty cleavage' (Fig. 4.7). Studies by scanning electron microscopy have shown that there is in fact a complete gradation from finely spaced cleavages into continuous cleavages, with domainal features still recognisable on the scale of 5–10 μm. At moderate and high grades of metamorphism where active deformation is involved, the preferential growth and alignment of medium- and coarse-grained inequant minerals (e.g. chlorite, micas and amphiboles) parallel to the X–Y plane of the strain ellipsoid leads to the development of a planar fabric termed **schistosity**.

4.4 Processes involved in cleavage and schistosity development

When argillaceous sediments are deposited, a primary fabric may develop sub-parallel to bedding due to the preferential alignment of minerals, especially phyllosilicates such as clays and detrital micas. When the sediment becomes

45

Layering, banding and fabric development

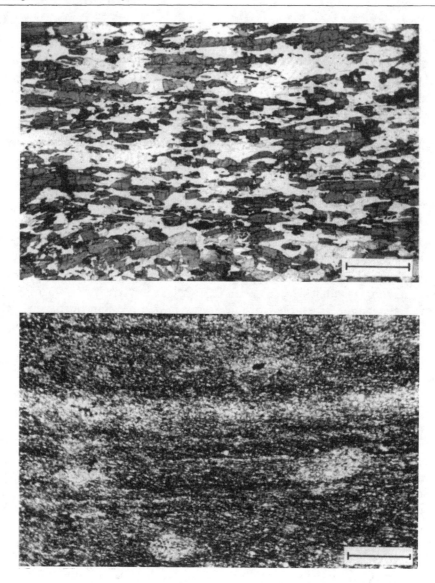

FIG. 4.5 (a) Schistosity, defined by aligned hornblende (dark crystals) plus feldspar and quartz (white). Schistose amphibolite, Norway. Scale = 1 mm (PPL). (b) A fine continuous cleavage, pervasively developed in the Skiddaw Slates. Contact aureole of the Skiddaw Granite, Lake District, England. Scale = 1 mm (PPL). Note the ghost-like oval pseudomorphs of cordierite, and the fact that the cleavage (defined by biotite, muscovite and elongate quartz) is slightly oblique (clockwise) to the horizontal compositional layering.

compacted and lithified during burial, this primary fabric may be enhanced. This accounts for the fissile nature of undeformed and unmetamorphosed mudrocks such as shales. However, for most metamorphic rocks, primary fabrics are insignificant and all major fabrics developed are secondary in origin, and caused principally by deformation of the rock. The relative importance of the main processes involved – namely, (i) mechanical rotation of

Cleavage and schistosity development

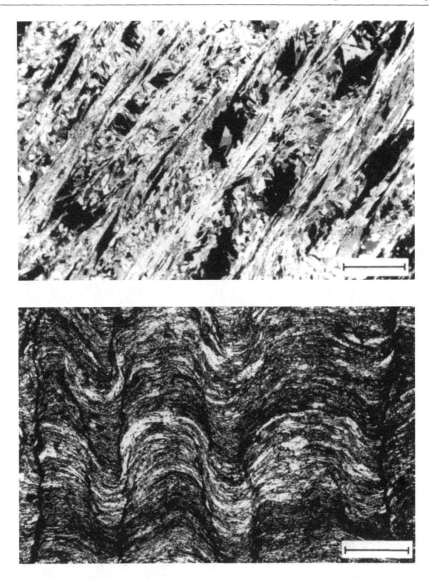

FIG. 4.6 (a) Zonal crenulation cleavage developed in a pelitic schist. The S_1 fabric trends top left to bottom right, and is overprinted by an S_2 crenulation cleavage at 90° to this. The separation into phyllosilicate-rich domains (P-domains), and quartz-rich domains (Q-domains) is clearly observed. Mica schist, Ox Mountains, Ireland. Scale = 1 mm (XPL). (b) Discrete crenulation cleavage developed in a fine-grained schist. Dark pressure solution seams defining S_2 trend top to bottom and overprint a crenulated S_1 fabric trending left to right. Semi-pelitic schist/phyllite, unknown locality. Scale = 0.5 mm (PPL).

pre-existing grains, (ii) deformation and ductile flow of individual crystals and (iii) dissolution and new mineral growth – has been and still is the focus of much debate with regard to cleavage formation. In many cases it is likely to be a combination of processes that operate (Knipe, 1981). The key factors controlling fabric development in metamorphic rocks are (i) rock composition, (ii) P–T conditions, (iii) stress orientation and magnitude, (iv) strain rate, (v)

Layering, banding and fabric development

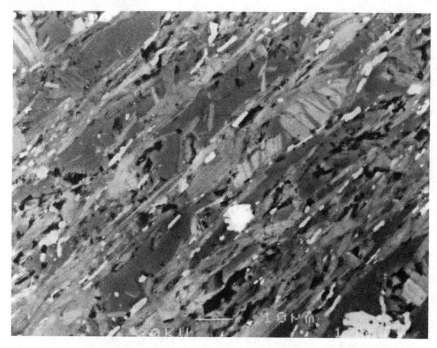

FIG. 4.7 An SEM photograph (back-scattered electron image) of domainal 'slaty cleavage' in slate (Luss, Scotland). Scale bar (lower centre) = 10 μm. Note the textural similarity to Fig. 4.6(a), but the difference in scale.

the amount of fluid present and (vi) fluid composition. In the model of Knipe (1981), it is envisaged that mechanical rotation dominates the initial stages of cleavage development, possibly accompanied by solution processes and grain-boundary sliding (Chapter 8). Later stages of cleavage development are interpreted as a more complex and heterogeneous interaction of deformation and metamorphic growth processes. The development of new phyllosilicates (e.g. phengite and chlorite) synchronous with deformation is an integral part of the cleavage-forming process.

To form a recognisable cleavage generally requires that rocks of appropriate composition have experienced at least 20–30% shortening. For a pronounced slaty cleavage, more substantial shortening is required. In many cases it has been estimated that typical shortening associated with a regionally pervasive slaty cleavage is around 60–75%. The fabric forms perpendicular to the principal shortening axis (Z) and thus defines the X–Y plane of the finite strain ellipsoid. As strain increases so the fabric intensifies, and where there is progressive increase in bulk strain the cleavage planes converge. The commonly observed phenomenon of cleavage refraction displays this feature well, especially in beds that show grading from psammite to pelite. In Fig. 4.8, note the change in orientation and curved nature of the main fabric as the cleavage passes from the less strained, more psammitic unit, into the more intensely deformed metapelitic layers in this metamorphosed sequence of clastic sediments. Assuming regional strain rates during cleavage formation of the order of 10^{-14} s^{-1}, it has been estimated by Paterson & Tobisch (1992) that formation of a regional cleavage would take 2–4 Ma. At faster strain rates, such as those operating in shear zones (e.g. 10^{-12} s^{-1}), an intense fabric could form in < 40 000 years.

Cleavage and schistosity development

As stated by Passchier & Trouw (1996), cleavage differentiation by solution transfer depends on a substantial amount of intergranular fluid in order to be effective, and consequently is most significant as a process at low metamorphic grades. During cleavage formation in sub-greenschist facies rocks there is good evidence (e.g. marker veins and oolites) of extensive dissolution and up to 50% volume loss (Fig. 4.9). The intensity of cleavage development will be a key factor, but for pressure solution cleavage development, 30–50% volume loss is a typical estimate given by many researchers. At higher metamorphic grades the evidence for substantial volume loss is less clear-cut, and wide-ranging estimates have been made. The majority of metamorphic petrologists have always viewed volume loss at higher metamorphic grades to be very low (or zero), because of limited fluid presence. However, authors such as Bell & Cuff (1989) have suggested that as much as 50% volume loss may occur during differentiated crenulation cleavage development by dissolution and solution transfer in phyllosilicate-rich rocks at greenschist and amphibolite facies. More recently, Mancktelow (1994) presents evidence from a range of classic areas to suggest that marked bulk volume change is not a prerequisite for the development of crenulation cleavage. Clearly, the topic remains a matter of current debate, with no unanimous view.

On the basis of detailed electron microprobe studies and SEM/TEM work, it is now clear that even in penetratively cleaved slates distinctly recognisable microdomains are often present. In crenulation cleavages, the existence of phyllosilicate-rich domains (P-domains), and quartz-rich domains (Q-domains) has long been known. The development of this distinctive 'domainal' or 'zonal' crenulation cleavage (Fig. 4.6(a)) commences with initial microfolding of an earlier formed cleavage or schistosity to give a series of gentle crenulations in the rock. Depending on the relative importance of pure shear and simple shear, the crenulations developed can range from upright and symmetrical to overturned and asymmetrical. The initial stage is dominated by kinking and bending of phyllosilicate minerals, but as cleavage development progresses, instability of the pre-existing phyllosilicates leads to the crystallisation of new

FIG. 4.8 Cleavage refraction in slates with varying quartz : phyllosilicate ratios (Bovisand Bay, Devon, England). The lens cap is 50 mm in diameter and located on one of the more pelitic (phyllosilicate-rich) layers.

phyllosilicates parallel to the crenulation axial surfaces. This becomes the dominant process of fabric development, whereas the significance of mechanical rotation of old grains is greatly diminished. The alignment of new phyllosilicate grains is the first clear sign of the newly developing crenulation cleavage fabric. Further increase in strain tightens the hinge-angle of the crenulations and leads to an intensification of the new fabric by preferential nucleation of phyllosilicates in the limbs (P-domains). At this stage of development, significant strain and chemical potential gradients exist between 'hinge' and 'limb' regions of the crenulations (the limbs becoming sites of higher strain and chemical potential compared to the hinges). This leads to soluble minerals of the limbs (e.g. quartz and calcite) preferentially entering into solution, and being transferred down chemical potential gradients, via grain-boundary fluid, to sites of deposition in the hinge regions. Gray & Durney (1979) established a mineralogical order in which, in terms of decreasing mobility by solution transfer, Cal > Qtz > Feld > Chl > Bt > Ms > opaques. Although the bulk rock chemistry and mineralogy may remain largely the same, this redistribution of material by solution transfer leads to significant changes on the domain scale. In other cases bulk rock chemistry may be significantly modified by transfer of more soluble material (e.g. silica) out of the local system by pressure solution along cleavage surfaces. In such cases, rather than distinct P-domains developing, thin dark cleavage 'seams' form (Fig. 4.9). These seams contain insoluble carbonaceous (graphitic) material and phyllosilicates (Gray, 1979; Gray & Durney, 1979). Such spaced cleavage is known as 'discrete' crenulation cleavage (Fig. 4.6(b)). Displacement of pre-existing compositional markers and veins gives the impression of microfaulting, but this is not the case, the offset simply representing major volume loss of material. Compared to 'zonal' crenulation cleavages which are common in both greenschist and amphibolite facies rocks, the development of 'discrete' crenulation cleavage is usually restricted to greenschist and sub-greenschist rocks. During polyphase deformation and metamorphism, an early fabric may be virtually obliterated, or

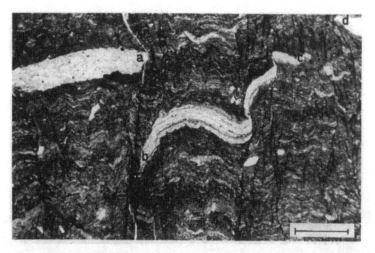

FIG. 4.9 Displacement and partial dissolution of marker horizons in crenulated phyllite. A quartz-rich horizon (pale) defining the original lamination (left to right) is cut by vertical pressure solution seams (thin black lines) related to a secondary crenulation cleavage. Area a–b shows displacement and partial dissolution, whereas c–d shows more pronounced dissolution. Phyllite, unknown location, UK. Scale = 1 mm (PPL).

completely transposed by a later intense fabric. In such cases, the only remaining evidence of the early fabric may be in the form of inclusion trails trapped in porphyroblasts, or in less intensely deformed areas in the porphyroblast strain shadows. The full approach to interpretation of porphyroblast–foliation relationships and deciphering of polydeformed metamorphic rocks is dealt with in Chapters 9 & 12.

Although mechanical rotation of grains occurs to some extent in rocks of low metamorphic grade, it is not an important process in the development of schistose fabrics of higher-grade rocks. At these deeper levels in the crust, with elevated temperatures (and pressures), crystal–plastic processes associated with ductile deformation become dominant in the modification of rock fabric. Dislocation creep gives rise to grain boundary migration and intragranular movement along certain crystallographically controlled slip planes. The general process of recrystallisation, both during and after deformation, plays an important role in the development of both grain-shape fabrics and crystallographic preferred orientations. Continual recrystallisation and equilibration of matrix minerals during regional metamorphism will tend to enhance the dimensional preferred orientation. During grain coarsening from slate to schist, those crystals most favourably oriented with respect to the prevailing stress will grow more rapidly than those with less favourable orientation, and in so doing will generate a pronounced alignment of grains to define a schistosity. The supply of material for nucleation and growth of new grains involves diffusive mass transfer from reactants via a grain boundary fluid. The process of MIMETIC GROWTH, which involves growth of new minerals in the direction of a pre-existing fabric, further enhances the fabric of the rock, and is common at moderate and high temperatures as a late-stage process in orogenic metamorphism. Further details on nucleation and growth are given in Chapter 5.

4.5 Processes involved in formation of layering in gneisses and migmatites

4.5.1 The nature and origin of gneissose banding

Gneisses and migmatites both reflect high-grade metamorphism. Gneisses are characteristically coarse-grained and comprise alternating felsic and mafic layers on a scale of centimetres to metres. They are strongly recrystallised, and display a granoblastic texture. Compared to the pronounced schistosity of schists, the fabric of most gneisses is often weaker (Fig. 4.10), especially in gneisses with a low mica content. Gneisses that had a sedimentary protolith are referred to as *paragneiss*, whereas those that were originally igneous rocks are referred to as *orthogneiss*.

The origin of gneissose banding has been a topic of considerable debate over the decades. Even the distinction between schists and gneisses is not always clear, since schist sequences often show compositional interlayering on the centimetre to metre scale, and gneisses may have a schistose fabric (alignment of inequant minerals) as well as displaying gneissic banding. An interlayered sequence of pelitic schists, semi-pelitic schists and psammites is commonly seen to pass gradationally up metamorphic grade to take on a progressively more gneissic or migmatitic appearance. So what are the key processes that give rise to the development of gneissose banding? There will of course be some original anisotropy, probably some original compositional layering in the rock. Added to this, metamorphic processes such as diffusional creep become significant as temperature increases. Compared to migmatites (discussed below), which largely involve *in situ* partial melting, the transformations leading to the development of gneissic banding are for the most part melt-absent solid-state processes. This sounds simple enough, but in practice, as with the distinction

Layering, banding and fabric development

between schist and gneiss, the distinction between migmatite and gneiss is not always clear, because there is a transition in both the appearance and the processes operating. For many gneisses, it has been suggested that metamorphic differentiation in the solid state is one of the key processes (e.g. Robin, 1979). This is achieved by reaction and diffusion of material down chemical potential gradients. However, others (e.g. Myers, 1978; Krøner et al., 1994) would argue that diffusion is a much less important process in the development of gneissic banding, and that in many cases the banding results from intense shear of original compositional layering and early oblique dykes and veins. In some – or possibly many – cases, gneissic banding may represent sheared migmatites!

Lucas & St-Onge (1995) studied Precambrian granulite facies banded rocks of the Ungava Peninsula, Canadian Shield. They interpreted the cm/dm-scale banding of tonalites, quartz diorites and monzogranitic rocks in terms of broadly layer-parallel emplacement of externally derived monzogranite veins into previously layered and strongly foliated host rock. Lucas & St-Onge (1995) argue for 'granite' vein emplacement into layer-parallel extension fractures, rather than rotation of pre-existing oblique veins into parallelism, on the basis that the layer-parallel 'granitic' veins are considerably less strained than the surrouding host rock. However, they also point out that the 'gneissose' banding is subsequently enhanced by later deformation.

From the above discussion, it is apparent that there are many possible models for the origin of banded gneisses. Original composition layering is often an important element, and intense shearing at high temperatures usually plays a part in enhancing the banding. However, to advocate a single model to account for all banded gneisses would be unwise, since other processes such metamorphic differentiation and emplacement of melt may also be involved to some degree. Therefore, before deciding on the likely origin of a particular banded gneiss it is vital that the gneiss in question has been examined carefully in both outcrop and thin section to assess the available evidence for and against the various interpretations.

FIG. 4.10 A weak fabric of gneisses. Hbl–Bt gneiss, Ghana. Scale = 1 mm (XPL).

4.5.2 The nature and origin of layered migmatites

A simple but useful summary of the main processes that may operate during migmatisation is provided in Table 4.1 (after Ashworth, 1985). One or more of these processes may operate, and all have been suggested as processes responsible for the formation of migmatites. Evidence suggests that *in situ* partial melting (anatexis) is responsible for the formation of most migmatites, or at least that is the interpretation currently favoured by most petrologists. On the basis that most (though not all) migmatites involve melt, they can largely be defined as coarse-grained heterogeneous rocks, characteristically with irregular and discontinuous interleaving of melt-derived leucocratic granitoid material (*leucosome*) and residual high-grade metamorphic rock (*restite*), also referred to as *mesosome*. The leucosome originates by partial melting of the high-grade metamorphic rock, and in many migmatites an accompanying dark coloured component (*melanosome*) is also present. Of the many structural types of migmatites recognised in the classic work of Mehnert (1968), it is *stromatic* migmatites and those with *schlieren* texture that are relevant to this chapter on layering, banding and fabrics in metamorphic rocks. Stromatic migmatites are those that have a pronounced layering (e.g. Maaløe, 1992). These are common in migmatites that have experienced only moderate degrees of partial melting (*metatexites*). Schlieren texture is the term used for migmatites that possess streaks or elongate segregations of non-leucosome material (usually biotite-rich) in leucosome (Fig. 4.11). This is especially common in migmatites that have experienced extensive partial melting and are leucosome dominated (*diatexites*). The schlieren represent entrained restite that has not been entirely separated from the melt.

The presence of biotite-rich melanosome selvages around layers and lenses of quartzo-feldspathic leucosome is a characteristic feature of certain migmatites (Fig. 4.12). The boundary between such melanosome selvages and the mesosome is commonly gradational. The development of these selvages results from melt segregation within migmatites, and is not seen in gneisses and lower-grade metamorphic rocks. This relationship can be viewed on the scale of an individual thin section, but most of the key migmatite relationships are best determined by field studies of individual exposures, followed by thin-section studies of specific igneous and metamorphic components of the migmatite. In the presence of an aqueous fluid, melting of high-grade schists and gneisses may commence in favourable (quartz–feldspar–mica) rocks at temperatures as low as 640°C, but extensive melting and migmatization typically occurs at 670–750°C (upper amphibolite facies and the innermost part of contact aureoles). Johannes & Gupta (1982) and Johannes (1988) describe the migmatisation of a layered paragneiss by progressive melting of individual layers. Because of the high viscosity of the granitoid melt, in the absence of open fractures, the melt has difficulty migrating away from the area in which it formed. It tends to accumulate at boundaries between layers, but will also migrate into localised areas of lower pressure such as boudin necks and shear zones. In most cases the leucosome is represented as discontinuous layers and lenses. Once the melt has migrated to some degree, the boundary between leucosome and other portions of the migmatite is usually clear, but in areas of melt generation the distinction between the leucosome and the refractory residue (the 'mesosome') often appears much more diffuse and nebulous. Because it is a rock formed from a melt, the granitoid leucosome often lacks any significant alignment of minerals to define a fabric. It should also display a random arrangement of constituent mineral phases, with little or no tendency to segregation

Layering, banding and fabric development

FIG. 4.11 A field photograph of schlieren-texture migmatite (St. Jacut, Brittany, France). The coin is 22 mm in diameter.

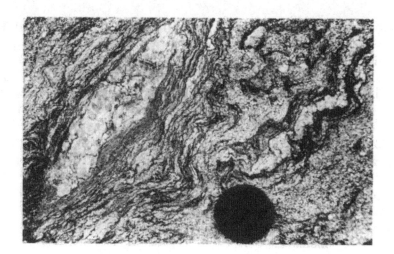

FIG. 4.12 A field photograph showing Bt-rich melanosome selvages in metatexitic migmatite (St. Jacut, Brittany, France). The lens cap is 50 mm in diameter.

TABLE 4.1 The processes involved in migmatisation (after Ashworth, 1985).

	Process requires open system	*Process does not require open system*
Process requires presence of a melt	Magmatic injection	Anatexis
Process does not require presence of a melt	Metasomatism	Metamorphic differentiation

or clustering of particular mineral phases (Ashworth & McLellan, 1985). This is true of well-preserved migmatites, but in a large number of cases the original migmatite textures have been modified by subsequent deformation and metamorphism. In such cases, the leucosome may also display a strong fabric, and may show mineral segregation. Many migmatites also display intense and irregular folding. There is often a lack of continuity of folds, with many appearing intrafolial (rootless) or dismembered. Intense and irregular veining, including ptygmatic veins, is another feature of many migmatites. For further insight into migmatites and the processes responsible for their formation, the text edited by Ashworth (1985) is strongly recommended.

References

Ashworth, J.R. (ed.) (1985) *Migmatites*. Blackie, Glasgow, 302 pp.

Ashworth, J.R. & McLellan, E.L. (1985) Textures, in *Migmatites* (ed. J.R. Ashworth). Blackie, Glasgow, Ch. 5, 180–203.

Bell, T.H. & Cuff, C. (1989) Dissolution, solution-transfer, diffusion versus fluid flow and volume loss during deformation/metamorphism. *Journal of Metamorphic Geology*, 7, 425–447.

Borradaile, G.J., Bayly, M.B. & Powell, C.McA. (eds) (1982) *Atlas of deformational and metamorphic rock fabrics*. Springer-Verlag, Berlin, 550 pp.

Gray, D.R. (1979) Microstructure of crenulation cleavages: an indicator of cleavage origin. *American Journal of Science*, 279, 97–128.

Gray, D.R. & Durney, D.W. (1979) Crenulation cleavage differentiation: implications of solution–deposition processes. *Journal of Structural Geology*, 1, 73–80.

Johannes, W. (1988) What controls partial melting in migmatites? *Journal of Metamorphic Geology*, 6, 451–465.

Johannes, W. & Gupta, L. (1982) Origin and evolution of a migmatite. *Contributions to Mineralogy and Petrology*, 79, 14–23.

Kisch, H.J. (1991) Development of slaty cleavage and degree of very low-grade metamorphism: a review. *Journal of Metamorphic Geology*, 9, 735–750.

Knipe, R.J. (1981) The interaction of deformation and metamorphism in slates. *Tectonophysics*, 78, 249–272.

Krøner, A., Kehelpannala, K.V.W. & Kriegsman, L.M. (1994) Origin of compositional layering and mechanism of crustal thickening in the high-grade gneiss terrain of Sri Lanka. *Precambrian Research*, 66, 21–37.

Lucas, S.B. & St-Onge, M.R. (1995) Syn-tectonic magmatism and the development of compositional layering, Ungava Orogen (northern Quebec, Canada). *Journal of Structural Geology*, 17, 475–491.

Maaløe, S. (1992) Melting and diffusion processes in closed-system migmatization. *Journal of Metamorphic Geology*, 10, 503–516.

Mancktelow, N.S. (1994) On volume change and mass transport during the development of crenulation cleavage. *Journal of Structural Geology*, 16, 1217–1231.

Mehnert, K.R. (1968) *Migmatites and the origin of granitic rocks*. Elsevier, Amsterdam.

Myers, J.S. (1978) Formation of banded gneisses by deformation of igneous rocks. *Precambrian Research*, 6, 43–64.

Park, R.G. (1989) *Foundations of Structural Geology*, 2nd edn. Blackie, Glasgow, 148 pp.

Passchier, C.W. & Trouw, R.A.J. (1996) *Microtectonics*. Springer-Verlag, Berlin, 289 pp.

Paterson, S.R. & Tobisch, O.T. (1992) Rates of processes in magmatic arcs: implications for the timing and nature of pluton emplacement and wall rock deformation. *Journal of Structural Geology*, 14, 291–300.

Powell, C.McA. (1979) A morphological classification of rock cleavage. *Tectonophysics*, 58, 21–34.

Robin, P.-Y.F. (1979) Theory of metamorphic segregation and related processes. *Geochemica et Cosmochimica Acta*, 43, 1587–1600.

Siddans, A.W.D. (1972) Slaty cleavage, a review of research since 1815. *Earth Science Reviews*, 8, 205–232.

Wood, D.S. (1974) Current views of the development of slaty cleavage. *Annual Review of Earth and Planetary Sciences*, 2, 369–401.

Chapter five

Crystal nucleation and growth

5.1 Nucleation

Crystallisation of metamorphic rocks in response to changing P–T conditions requires crystallisation of minerals by nucleation and growth. *Homogeneous nucleation*, involving spherical nuclei with uniform surface energy, randomly distributed throughout the host in which they develop, may be relevant in chemistry and metallurgy, but is an inappropriate description of nucleation in heterogeneous polycrystalline metamorphic rocks. *Heterogeneous nucleation* describes non-random nucleation on some pre-existing substrate, such as new crystals preferentially nucleating at pre-existing grain boundaries. Even the purest monomineralic quartzite or marble has heterogeneities such as microfractures or small detrital grains. Because of this there will always be some places where new crystals are more likely to nucleate, and other areas where nucleation is less likely. In view of this, heterogeneous nucleation is the most relevant way to consider nucleation in metamorphic rocks. Bulk rock chemistry also plays a key role in influencing the number and size of porphyroblasts that develop. An example from the aureole of the late Caledonian Corvock granite (Co. Mayo, Ireland) is shown in Fig. 5.1. In this case, cordierite is preferentially developed as larger

FIG. 5.1 Metasediments in the contact aureole of the Corvock Granite, Mayo, Ireland, showing widespread cordierite development. Due to bulk rock chemical controls on nucleation and growth, porphyroblasts have preferentially developed to a larger size in the pelitic layers compared to the semi-pelitic layers. The lens cap is 45 mm in diameter.

Crystal nucleation and growth

porphyroblasts in pelitic compared to semi-pelitic horizons.

With the exception of crystallisation of minerals from a melt (e.g. migmatites) and development of minerals in fluid-filled fractures or cavities, nucleation and growth of minerals in metamorphic rocks essentially occurs in the solid state. Classical nucleation theory, first proposed by Gibbs (1878), has been developed in more detail by various workers and was neatly summarised by Kretz (1994). It is outside the scope of this book to give a detailed theoretical treatment of the topic but, instead, the key aspects of nucleation will be examined.

If we consider a simple solid-state phase transformation in which phase A reacts to form phase B, the classical nucleation theory proposes that once the temperature of the reaction has been exceeded, microdomains (*embryos*) of phase B will start to nucleate in phase A. If it is assumed that the embryos are spherical, the free energy of formation of an embryo can be written as

$$\Delta G_e = \left(\frac{4}{3}\right) \pi r^3 \Delta G_v + 4\pi r^2 \sigma, \quad (5.1)$$

where r is the radius of the embryo, ΔG_v is the Gibbs energy change (per unit volume of B) for the reaction A → B (always < 0, because the reaction always proceeds to lower G) and σ (> 0) is the surface (interfacial) energy (per unit area) of the A–B interface, which exists because of surface tension between two phases in contact. Because ΔG_v is always negative, the first term on the right is always negative, and because σ is always positive the second term on the right is always positive. This means that, depending on the magnitude of ΔG_v, there will be some critical size (r_c), corresponding to maximum ΔG_e (Fig. 5.2), above which further growth of the embryo results in a progressively more stable state. ΔG_e at r_c is the activation energy of nucleation (ΔG^*) or, in other words, the energy that must be supplied to the system for a nucleus to form. An additional source of energy, not considered in (5.1) is *strain energy* (ε) stored in elastically strained crystals. Some or all of this energy may be released during reactions and recrystallisation, and thus reactions may commence earlier and proceed more rapidly in aggregates in which strained crystals are present.

Thermal fluctuations will cause embryos to change in size, but once r_c is exceeded, a *nucleus* is formed and from this a crystal can grow. Authors such as Christian (1975) have developed the classical theory further. Considering the activation energy of formation of a nucleus (ΔG^*) in terms of ΔG_v and σ, the relationship $r_c = -2\sigma/\Delta G_v$ is obtained and, when substituted into (5.1), gives

$$\Delta G^* = \frac{16\pi}{3} \frac{\sigma^3}{(\Delta G_v)^2}. \quad (5.2)$$

Activation energy (ΔG^*) represents an energy barrier that must be surpassed before atoms

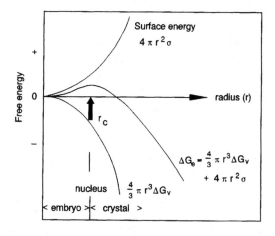

FIG. 5.2 The contributions of the Gibbs energy change per unit volume (ΔG_v) of B for the reaction A → B (lower curve), and the surface (interfacial) energy (σ) per unit volume of the A–B interface (upper curve), to the energy of formation of a spherical embryo ($r < r_c$), a nucleus ($r = r_c$) and a crystal ($r > r_c$). ΔG is the Gibbs free energy change, and ΔG^* is the activation energy that must be overcome for the reaction to proceed (modified after Kretz, 1994).

will transfer freely across the A–B interface and thus promote growth of B. One important implication of the relationships expressed in (5.2) is that the more the equilibrium position of transformation is overstepped (i.e. the more negative ΔG_v becomes), the smaller r_c and ΔG^* become, so that (other things being equal) the easier it will be for a nucleus to form. Correspondingly, no nucleation will be possible *at equilibrium*, since at that point $\Delta G_v = 0$ and thus ΔG^* will be infinite.

Because of the heterogeneous nature of polycrystal aggregates that constitute metamorphic rocks, nucleation of new phases is unlikely to show an even distribution. Preferential sites for nucleation include grain boundaries (Fig. 5.3), vein margins (Plate 8(e)), and on previously strained crystals. Nuclei preferentially develop at such sites because there are more loose bonds available to attach atoms, and the elevated energy of these disordered areas facilitates nucleation. The increased dislocation density of strained crystals, particularly at their margins, gives rise to an increase in surface energy and stored elastic strain energy. The reason for preferential nucleation at sites such as grain boundaries can be understood in terms of a simple extension of the homogeneous nucleation theory described above. If we take a spherical embryo of B formed at an A–A grain boundary that is being consumed, the surface energy term of (5.1) is modified to

$$4\pi r^2 \sigma^{A-B} - \pi r^2 \sigma^{A-A} \qquad (5.3)$$

This modification is valid because the sum of energy obtained from the A–A boundary being destroyed contributes to the formation of the embryo and can thus be subtracted from the term for homogeneous nucleation. This means that for heterogeneous nucleation the surface energy term is lower than for homogeneous nucleation and consequently ΔG_c is lowered, and thus grain-boundary areas are more favourable sites for nucleation. Grain edges (triple-junctions) and grain corners are even more favourable, as are strained grain boundaries, which have the the added component of stored elastic strain energy associated with

FIG. 5.3 Fibrolitic sillimanite preferentially nucleating at quartz grain boundaries. Connemara, Ireland. Scale = 0.1 mm (XPL).

dislocations. Coherent nucleation (e.g. oriented intergrowths and topotactic replacement) is more favourable than incoherent nucleation, since lattice-matching across the grain boundary reduces σ, and thus the surface energy term in (5.3) is diminished and ΔG^* is likewise lowered.

Nucleation theory predicts that for the nucleation of phase B from the breakdown of phase A, a certain amount of temperature (or pressure) overstepping of the A–B equilibrium boundary is necessary before significant nucleation rates occur (as discussed above in relation to (5.2)). Once a certain finite amount of overstepping has been achieved, nucleation of phase B will usually start abruptly. Temperature is the main control, since it provides energy in the form of heat to drive reactions. However, the precise amount of overstepping varies from one case to the next. A combination of experimental observation and theoretical calculations suggests that for dehydration reactions the amount of overstepping required is normally < 10°C, but for solid–solid reactions the value in some instances may be as much as 100°C. Significant modification of local fluid chemistry, either by local reaction or by infiltration of some externally derived fluid, can also induce rapid nucleation.

The above discussion has concentrated on the nucleation of a single product (B) from a single reactant (A), but the case of nucleation in polyphase aggregates is understandably more complex. If reaction rates are slow relative to atomic mobility, the product(s) may not form at the site of the reactant(s). This is commonly seen in the case of the polymorphic transformation of andalusite to sillimanite, or of kyanite to sillimanite (Section 1.3.3). Rather than the sillimanite nucleating on the precursor Al_2SiO_5 phase, it is commonly seen to nucleate at Qtz–Bt, and Qtz–Qtz boundaries (Fig. 5.3), at some distance from the reactant.

In the case of isothermal nucleation, the nucleation rate could be considered constant, in which case an even distribution of crystal sizes might be expected for the new phase. However, for some dehydration reactions the nucleation rate diminishes and can be attributed to a progressive depletion of favourable nucleation sites. A variation on this theme, proposed by Carlson (1989), is that growing porphyroblasts develop a diffusion halo, within which nucleation is inhibited. Consequently, the potential for nucleation of new porphyroblastic phases during the later stages of porphyroblastesis is limited.

5.2 Growth of crystals

From the development of a stable nucleus, further addition of atoms marks the start of growth of a crystal. From this point onwards, nucleation and growth become competing processes. Crystals formed in a fluid environment, such as a melt or a fluid-filled fracture or cavity, often display perfectly regular (planar) crystal faces. One of the principal mechanisms by which these faces advance relates to the emergence of screw dislocations (Frank, 1949; Griffin, 1950). These are a type of line defect that displaces part of a plane of atoms and causes a step in the crystal face, as distinct from edge defects, which mark the termination of a plane of atoms (Section 8.2). From these screw dislocations a crystal growth spiral develops. In detail, the growth of natural crystals is more complex than this, with surface imperfections and surface diffusive processes undoubtedly having a role. In solid-state transformations, growth is largely controlled by interface processes, diffusion processes or a combination of these. In the interface model (Christian, 1975), the rate of advance of the A–B interface in the transformation A → B is largely a function of the Gibbs energy change (ΔG) and activation energy (ΔG^*). At equilibrium, or close to it, the growth rate is approximately linear with respect to ΔG (i.e. as ΔG increases, the growth rate increases).

In cases in which the observed reaction microstructure exhibits reaction products that are chemically different from the reactants, it is likely that diffusion of atoms to and from the growth surface plays an important role. Indeed, it may be the rate-limiting process for crystal growth in such cases. This type of process is crucial during exsolution (Section 6.2) and in the formation of symplectites (Section 6.4). However, during many metamorphic reactions it is likely that a combination of interface processes and diffusive processes are involved in growth. Since metamorphic reactions often proceed during rising or falling temperature, it is possible that the rate-limiting process may change with time. Fisher (1978) gave a detailed evaluation of the rate-controlling mechanisms for crystal growth over the full range of metamorphic conditions. It was concluded that 'spherical' porphyroblasts pass from an initial reaction-controlled stage, through an intermediate grain-boundary diffusion-controlled stage and finally to a heat-flow-controlled stage. Computer simulations by Sempels & Raymond (1980) and simulations compared with natural samples (Carlson, 1989, 1991) have provided additional insight into the microstructural and textural features developed during crystal (porphyroblast) nucleation and growth in polyphase aggregates. The influence of deformation on nucleation and growth of minerals is considered in Section 8.3. For further details on the kinetics of heterogeneous reactions, especially in relation to contact metamorphism, the review by Kerrick et al. (1991) is highly recommended.

5.3 Size of crystals

It is important when describing any rock or texture to note the overall grain size and variations in relative size between the constituent mineral phases. The precise subdivisions of fine-, medium- and coarse-grained metamorphic rocks vary slightly from one author to the next. The classification that we shall adopt here considers rocks with matrix grain size <0.1 mm as fine-grained, those with grain size 0.1–1.0 mm as medium-grained and those with matrix grain size >1.0 mm as coarse-grained.

During initial prograde metamorphism, reaction products are generally fine-grained, and much of the original microstructure of the rock remains clear (e.g. original clastic features and delicate compositional layering (bedding) in sediments, phenocryst outlines and relationships in igneous rocks). However, with time, and in response to increasing P–T conditions, the rock recrystallises further. This leads to an overall coarsening of the matrix, and the original microstructural and mineralogical features of the rock are largely or completely obliterated. This includes the loss of all compositional layering (laminations) on a scale smaller than final matrix grain size.

Although monomineralic rocks such as marble and quartzite are typically equigranular, as are high-temperature rocks such as hornfels and granulite, other rocks such as pelitic schists and amphibolites commonly have a structure in which some minerals have grown to a much larger size than those of the matrix. Such minerals are known as PORPHYROBLASTS, and the structure is termed PORPHYROBLASTIC (Fig. 5.4). This is analogous to 'phenocrysts' and 'porphyritic structure', used to describe a similar feature in igneous rocks. Many factors contribute to the formation of this structure. They largely relate to the variable nucleation and growth rates of constituent minerals of the rock. These in turn are dependent on factors such as P, T, fluid, rock chemistry and a critical activation energy that needs to be overcome for nucleation and growth to occur. Since some minerals nucleate and grow more easily than others, this gives rise to an heterogeneous porphyroblastic structure. In particular minerals such as garnet and staurolite are almost always porphyroblasts,

Crystal nucleation and growth

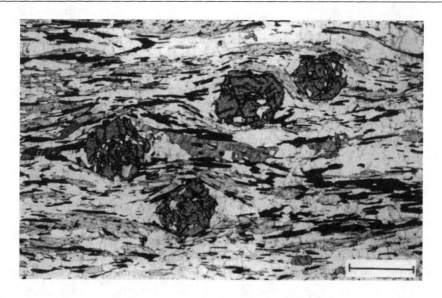

FIG. 5.4 Porphyroblastic structure. Porphyroblasts of garnet enveloped by a schistose fabric comprised of Chl + Ms + Bt + Qtz + Ilm. Garnet–mica schist, Norway. Scale = 1 mm (PPL).

whereas minerals such as quartz are always matrix phases.

In single-phase aggregates (e.g. metals, quartzite and marble) it has long been recognised that crystals coarsen with both time and increased temperature. This coarsening process, also known as 'Ostwald ripening', occurs in order to decrease the Gibbs free energy of the system and thus produce a more stable configuration, closer to equilibrium. This is facilitated by reducing the total grain boundary surface area and thus reducing the contribution of surface (= interfacial) energy to the total energy of the system. The coarsening is achieved by elimination or amalgamation of small grains by grain-boundary migration. Although coarsening increases with both time and temperature, it has been shown both experimentally and theoretically that the *rate* of coarsening diminishes as time and temperature increase. Joesten (1991) gives a comprehensive summary of the theory of grain coarsening and examples of natural and experimental studies in contact metamorphism, including the study by Buntebarth & Voll (1991) of quartz coarsening in quartzites within the Ballachulish contact metamorphic aureole, Scotland.

While coarsening is readily apparent in monomineralic aggregates, it does nevertheless also occur in bimineralic and polymineralic assemblages. However, in such situations the processes that occur and the overall kinetics of coarsening are considerably more complex. In pelites for example, the lowest grade, and most fine-grained, rocks are slates. These are transformed into phyllites and fine- to medium-grained schists at greenschist facies conditions. In turn, these are converted into medium- and coarse-grained schists and gneisses at higher metamorphic grades.

Factors such as the metamorphic fluid, diffusion rates, the grain-boundary energy (which affects the rate at which a boundary migrates) and the rate of change of temperature will all be significant in controlling the amount and rate of crystal growth. If the nucleation rate is high relative to the growth rate, numerous nucleation sites are utilised at an early stage following the onset of reaction. This results in a fine-grained reaction product disseminated throughout the rock (Fig. 5.5), and may result in site saturation. The converse of this is when the ratio of nucleation rate to growth rate is small. This results in rapid growth of just a few early nuclei, and often involves the consumption or inclusion of many small grains, thus eliminating many potential nucleation sites in the form of grain boundaries, corners, and so on. The development of a limited number of large porphyroblasts can be considered in this way (Fig. 5.4).

Once nucleated, the ultimate size of a given porphyroblast will be a function of the growth rate and the time available for growth. The growth rate is strongly influenced by the rate of diffusive transfer of required elements/ions to and from the reaction site(s), as well as the rate at which ions can be incorporated into the growing phase. The rate of interface migration (i.e. the rate at which the crystal faces can advance), can also be rate-controlling. If the growing porphyroblasts exhaust the supply of reactants, growth will terminate either permanently or until such a time as the matrix of the rock is replenished in relevant ions. This replenishment may occur by external fluid input or else by ionic release into the matrix system as a result of some other reaction (Fig. 5.6). Incomplete diffusion of reactants in the matrix may give a reaction halo around the porphyroblast (Fig. 5.7).

For a given rock, it might seem reasonable to expect that following some time interval after porphyroblast nucleation and growth, the earliest formed porphyroblasts (X) would be the largest, and the most newly formed porphyroblasts (Y) the smallest (Fig. 5.8(a)). However, this assumes constant growth rates

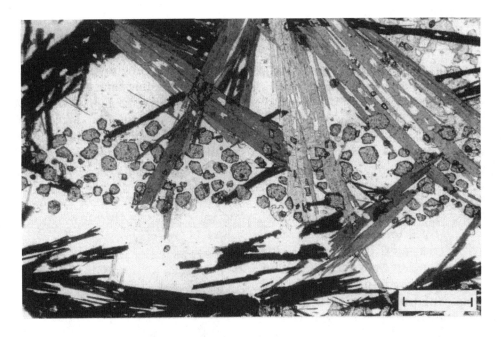

FIG. 5.5 A high nucleation rate relative to the growth rate has produced thin layers with numerous small spessartine garnets in this blueschist facies meta-ironstone. The acicular and lath-like crystals with variable orientation are of stilpnomelane, developed as a late-stage overprint. The white background is quartz. Blueschist facies meta-ironstone, Laytonville Quarry, California. Scale = 0.5 mm (PPL).

Crystal nucleation and growth

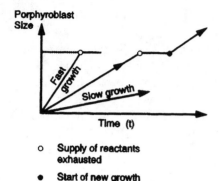

○ Supply of reactants exhausted
● Start of new growth

FIG. 5.6 A schematic illustration to show the influence of growth rate, time and supply of reactants on porphyroblast size. For simplicity, constant growth rates are shown, but in practice many porphyroblasts experience an initial phase of rapid growth followed by a period of slower growth.

FIG. 5.7 A schematic illustration (width approximately 1.5 mm) of reaction halo developed around a porphyroblast. This occurs due to incomplete diffusion within the rock, the material required for porphyroblast growth only being derived from the local environment immediately around the porphyroblast, rather than being diffused through the whole rock.

for all porphyroblasts of the same phase, and need not necessarily be true. For example, the most favourable site for nucleation may not necessarily be the most favoured site for growth. If diffusion rates are variable within the rock, perhaps as a result of small-scale compositional variations, then early formed porphyroblasts may grow more slowly than later formed porphyroblasts, as a direct result of variations in the rate of supply and removal of elements involved in the reaction. This being the case, after a certain time, by virtue of the faster growth rate, the later formed porphyroblast will become larger than the earlier porphyroblast (Fig. 5.8(b)).

In a comparison of grain size variations in basaltic hornfels and garnet–mica schists over a range of metamorphic grades, Cashman & Ferry (1988) established that the rocks from contact aureoles have linear crystal size distributions (i.e. the smallest sized crystals were most numerous, the largest crystals least numerous and there was a linear relationship inbetween). Such a relationship is indicative of continuous nucleation and growth of crystals. By contrast, regionally metamorphosed pelites have 'bell-shaped' crystal size distributions. In other words, the most numerous crystals are not the smallest grain size or the largest grain size, but have some intermediate value. The data from rocks of the chlorite zone through to the sillimanite zone all show this relationship, but rather than being a normal distribution are all skewed towards the smaller sizes. This distribution of grain sizes in regional metamorphic rocks is interpreted in terms of initial continuous nucleation and growth of crystals followed by a period of annealing after the cessation of nucleation. This annealing by Ostwald ripening leads to amalgamation of smaller crystals and hence modification of the initial linear crystal size distribution. The observed differences are readily explained in terms of the differences in

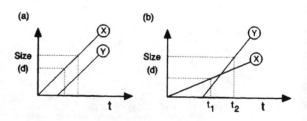

FIG. 5.8 A schematic illustration of the relationships between growth rate, time and porphyroblast size: (a) variation in porphyroblast sizes at a constant growth rate; (b) variation in porphyroblast sizes at different growth rates.

thermal history of the two regimes. Contact metamorphism involves high temperatures for a short period of time, while regional metamorphism involves high temperature followed by a prolonged period of cooling. In the latter case, the rock is held at moderate to high P–T for some considerable time (several tens of Ma). A final point to be aware of is that the presence of certain phases may inhibit Ostwald ripening of other phases (e.g. graphite inhibits muscovite growth), so the relationships outlined above are not always as simple as they might seem.

5.4 Absolute growth times

By substituting what they considered to be reasonable values for equilibrium temperature, growth rates and heating rates, Cashman & Ferry established that garnet growth times (garnets 0.1–1.2 mm in diameter) were in the range <100–40 000 years (i.e. geologically very rapid), and that nucleation and growth occurred at small ΔT. This compares favourably with the simplified theoretical models of Walther & Wood (1984), which suggest that porphyroblast growth times during regional metamorphism were probably of the order of 10^4–10^5 years. A more recent evaluation by Paterson & Tobisch (1992), utilising estimates by Ridley (1986) for metamorphic mineral growth (2×10^{-5} cm yr^{-1}) suggests that porphyroblasts formed during regional metamorphism could attain lengths of 5 cm in 250 000 years.

Radiometric dating provides an additional means of estimating porphyroblast growth times. Christensen et al. (1989) undertook Rb–Sr core–rim dating of 3 cm diameter garnets from Vermont (USA) and determined growth times of 6–10 Ma, indicating a mean radial increase of 1.1–1.7 mm Ma^{-1}. Sm–Nd, U–Pb and Rb–Sr dating of 1.0–1.5 mm garnets in Caledonian schists from Norway (Burton & O'Nions, 1991) demonstrated small age differences (approximately 1–2 Ma) between cores and rims. However, on the basis of error bar overlap, they conceded that their results were unable to resolve precisely the small time interval indicated.

Considering the various studies of absolute growth times for porphyroblasts, Barker (1994) concluded that for garnets <1.5 mm diameter, in situations of orogenic metamorphism, growth times of <1 Ma and possibly <0.1 Ma would seem a reasonable estimate. For larger (e.g. 1–3 cm diameter) garnets, available estimates for growth time vary from <1 Ma to as much as 5–10 Ma. Such conclusions relating to porphyroblast growth times have important implications for the interpretation of porphyroblast–foliation relationships (Chapter 9). For further details regarding rate and time controls on metamorphic and tectonic processes based on garnet chronometry, see Vance (1995).

5.5 Shape and form of crystals

The form of the constituent minerals of metamorphic rocks can be described as EUHEDRAL, SUBHEDRAL or ANHEDRAL. EUHEDRAL crystals are those with good crystal form and well developed crystal faces (Fig. 5.9(a)), while SUBHEDRAL crystals are less well formed but have some well developed faces (Fig. 5.9(b)) and ANHEDRAL crystals have irregular form with no well developed crystal faces (Fig. 5.9(c)). The controlling factors on whether a given mineral has euhedral or anhedral form are many, but the influence of growth kinetics is paramount. The development of euhedral porphyroblasts is most favoured by conditions of unimpeded slow growth in an anisotropic medium, whereas anhedral crystals commonly reflect rapid growth.

Considering euhedral crystals in a little more detail, the shape and number of faces depends on certain properties specific to the given

Crystal nucleation and growth

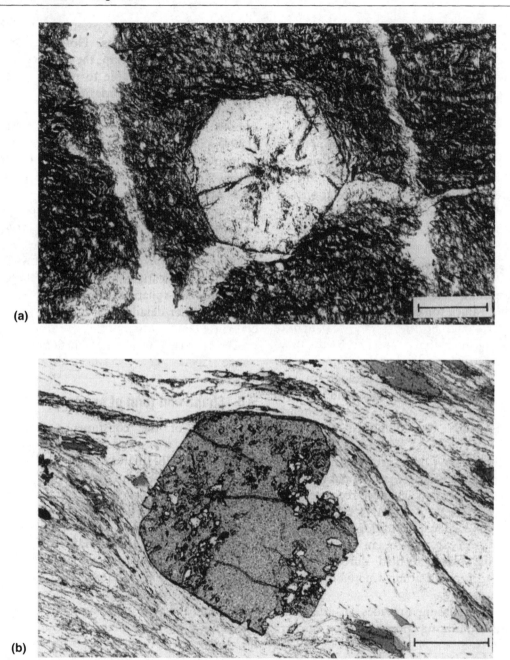

FIG. 5.9 (a) EUHEDRAL (or IDIOBLASTIC) porphyroblast of spessartine garnet. Porphyroblastesis was a late event in this rock, the garnet clearly overgrowing Qtz–Chl–Serc veinlets, and a slaty matrix with fine-scale crenulations. The rock is a pelite that has experienced low-grade regional metamorphism overprinted by later contact metamorphism. Isle of Man, England. Scale = 0.5 mm (PPL). Note the star-like arrangement of inclusions in the porphyroblast due to their concentration at interfacial boundaries (for further explanation, see Section 6.1, Fig. 6.4 & Plate 1(d)). (b) A SUBHEDRAL porphyroblast of almandine garnet, enveloped by a well-defined regional schistosity. Garnet–mica schist, Norway. Scale = 0.5 mm (PPL).

mineral phase, and on the growth process involved. The mineral will attempt to maintain the lowest-energy form, and this is controlled by the inherent surface energy, as well as the lattice energy of the crystal. If all faces had the same surface energy, then numerous faces would develop, and the crystal would approximate to a sphere. However, in the ideal form of all minerals, certain faces have lower surface energy than others, and these preferentially develop in the equilibrium shape, despite giving a greater surface area per unit volume than a sphere. A flat face has a lower energy than an irregular surface because the number of disturbed bonds is less, and of flat faces, those with the greatest density of atoms usually have lower surface energy than those with least density. Those faces with highest surface energy advance most rapidly, and form a proportionately smaller part of the crystal surface. Curie (1885) established that for a given crystal there was a constant relationship between the distance from a given crystal face to the crystal centre (d) divided by the surface free energy (γ) of that given face:

$$\frac{d_1}{\gamma_1} = \frac{d_2}{\gamma_2} = \ldots = \frac{d_n}{\gamma_n} = \text{constant.} \quad (5.4)$$

This expression is now generally referred to as *Wulff's theorem*. In the schematic example shown in Fig. 5.10, face B grows more rapidly than face A, and it can be seen that the slow-growing lower-energy faces predominate. Surface defects increase the surface energy of a given face such that a low-energy (slow-growing) face may grow more quickly than expected. Following original work by Harker (1939), metamorphic petrologists have recognised that certain minerals have a greater tendency to develop euhedral form than others, and are often porphyroblastic. It has been established that a given mineral develops good crystal faces when in contact with certain phases, but not when in contact with others.

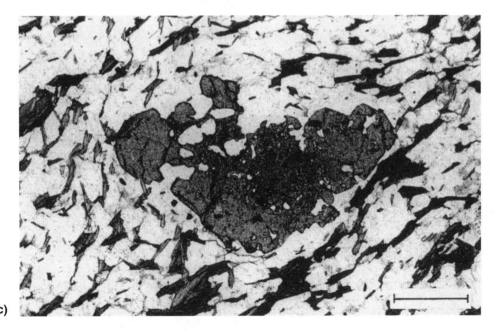

(c)

FIG. 5.9 (*contd*) (c) An ANHEDRAL porphyroblast of garnet in a biotite–quartz matrix, with poorly developed schistosity. Garnet—mica schist, Connemara, Ireland. Scale = 0.5 mm (PPL).

Crystal nucleation and growth

From this, minerals have been arranged in a sequence termed the *crystalloblastic series* (Table 5.1), with those at the top (e.g. pyrite, garnet and staurolite) having the greatest tendency towards euhedral form. The sequence reflects decreasing surface energy, and minerals higher in the sequence will always have a tendency to form euhedral faces against minerals lower in the sequence. Therefore, in a garnet–mica schist, the garnet will have a tendency towards euhedral form against both mica and quartz, whereas mica will develop euhedral form against quartz, but is not expected to develop euhedral form against garnet, except when favourably oriented with respect to the garnet boundaries.

Crystals with SKELETAL, and more rarely DENDRITIC, form are sometimes encountered in metamorphic rocks. Both develop as a result of very rapid growth around a limited number of nucleii. True dendritic crystals are uncommon in geological examples, although crystallites of dendritic olivine in volcanic glasses are good examples. In metamorphic rocks they are especially rare because of their high surface energy and thermodynamic instability. However, records of dendritic calcite developed during the epitaxial replacement of aragonite have been noted. Skeletal crystals form by rapid mineral growth along intergranular boundaries under circumstances of unfavourable nucleation. Truly skeletal porphyroblasts are not especially common in metamorphic rocks, but are most frequently observed in quartz-rich lithologies or segregations (Fig. 5.11).

ACICULAR (needle-like) crystals (Figs 5.12(a)–(c)), FASCICULAR bundles (Fig. 5.13(a)), BOW-TIE arrangements (Fig. 5.13(b)) and SPHERULITIC aggregates all result from predominance of growth over nucleation. Acicular crystals develop from a single nucleus, and may occur as scattered individual crystals or clusters throughout the rock. They are often concentrated in specific areas of favourable chemistry or nucleation (e.g. grain boundaries; Fig. 5.3), but may also occur as radiating aggregates (e.g. zeolite minerals in amygdaloidal basalts (Fig. 5.12(c)) and tourmaline in certain hornfelses). A fascicular growth consisting of a bundle of rods or needles initiates from a single nucleus, but then branches slightly. Further divergence of needles will give rise to a 'bow-tie' structure. This is common in amphiboles of *'garbenschiefer'*, calc-schists and meta-volcanic rocks (Fig. 5.13(b)).

Having considered free growth forms,

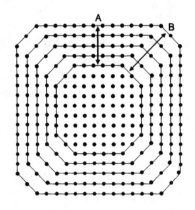

FIG. 5.10 A schematic illustration of how certain faces of a crystal grow more rapidly than others. In this example B grows more rapidly than A, and forms a smaller proportion of the crystal surface.

TABLE 5.1 The crystalloblastic series of minerals (modified after Harker, 1939; Philpotts, 1990). The sequence reflects decreasing surface energy, and those minerals higher in the sequence will always have a tendency to form euhedral faces against minerals lower in the sequence.

Magnetite, rutile, sphene, pyrite, ilmenite
Sillimanite, kyanite, garnet, staurolite, chloritoid, tourmaline
Andalusite, epidote, zoisite, forsterite, lawsonite
Amphibole, pyroxene, wollastonite
Muscovite, biotite, chlorite, talc, prehnite, stilpnomelane
Calcite, dolomite, vesuvianite (idocrase)
Cordierite, feldspar, scapolite
Quartz

Shape and form of crystals

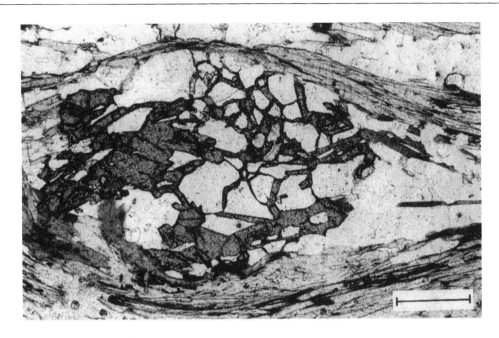

FIG. 5.11 A "skeletal" porphyroblast of garnet formed by growth between quartz grain boundaries. Garnet–mica schist, Norway. Scale = 0.5 mm (PPL).

largely in connection with porphyroblastesis, let us now consider the various aspects of mutual growth forms, in which adjacent crystals impinge on each other. This is mostly concerned with the process of matrix recrystallisation and adjustment towards equilibrium, but also relates to impingement of growing porphyroblasts having nucleated close together.

The lowest-energy system is the most stable one, and in order to maintain the lowest chemical free energy, mineral phases will tend to react in response to changing P, T and fluid conditions, and hence produce a more stable assemblage. During or after these mineralogical changes, the shapes of grains and grain-boundary arrangements will often become modified in an attempt to reduce grain-boundary energy. This energy is considerably smaller than the chemical free energy, but is nevertheless significant and, throughout the microstructural changes, the grain boundaries will become modified and rearranged to minimise this energy. Unless a grain has experienced appreciable internal strain, its boundaries against other grains are generally more atomically disordered than its internal structure. In order to become more stable there is a tendency to reduce the total area of grain boundaries and to develop more stable (regular) crystal faces. The degree to which an aggregate of grains in a metamorphic rock approaches stability is dependent on the time available for adjustments, and on prevailing conditions.

Non-equilibrium impingement structures consist of a range of grain sizes, irregular and variable grain shapes, curved and irregular boundaries and multiple junctions. The total surface energy in low-grade schists is often considerably higher than in higher-grade rocks because of the irregular and often curved nature of many of the grain boundaries (Fig. 5.14(a)). The minimisation of surface area during recrystallisation, and the development of GRANOBLASTIC–POLYGONAL (or

Crystal nucleation and growth

FIG. 5.12 (a) ACICULAR (needle-like) crystals of stilpnomelane and deerite developed in a quartzitic layer within a blueschist facies meta-ironstone sequence. Laytonville Quarry, California. Scale = 0.1 mm (PPL). (b) RODDED and ACICULAR sillimanite crystals. Sillimanite gneiss, Broken Hill, Australia. Scale = 0.1 mm (PPL).

MOSAIC) microstructure comprising many nearly planar grain boundaries (Fig. 5.14(b)), is indicative of a high degree of stability, and is especially common at higher metamorphic grades. It is commonly observed in quartzites (Fig. 5.14(b)), hornfelses and granulites, and the same microstructure is also recognised in metamorphosed massive sulphide ores (e.g. Craig & Vaughan, 1994). Such a microstructure could be considered in terms of random nucleation and growth during metamorphic crystallisation, but the fact that concentrations or clusters of certain minerals occur in many cases suggests that, even at high grades of metamorphism, pre-existing microstructural and mineralogical heterogeneities may have an important influence on the location of favourable and unfavourable nucleation sites for particular minerals.

In polygonal aggregates, interfaces generally meet at triple-points with interfacial angles of 120° (±10°). The interfaces are largely regular and planar, the regularity being achieved by surface (interfacial) energy-driven grain-boundary migration. This arrangement minimizes surface area and thus the contribution of surface energy to the total free energy of the system. However, at a triple-point, there will only be three angles of exactly 120° if the surface energies on all boundaries are the same. This will be approximately true if the three grains meeting at the triple-point are all of the same mineral, but where different minerals are involved, it will not. In a study of granulites from Quebec, Kretz (1966) established that the dihedral angle (Θ) for clinopyroxene against two scapolite grains was 128° (s.d.=13°), and clinopyroxene against two plagioclase grains was 109° (s.d.=16°). Similar studies have also been made for other phases. Another point to note in relation to polygonal aggregates – and for simplicity let us consider a monomineralic

(c)

FIG. 5.12 (*contd*) (c) Radiating acicular crystals of natrolite infilling a vesicle within altered basalt. Antrim, Northern Ireland. Scale = 0.1 mm (XPL).

Crystal nucleation and growth

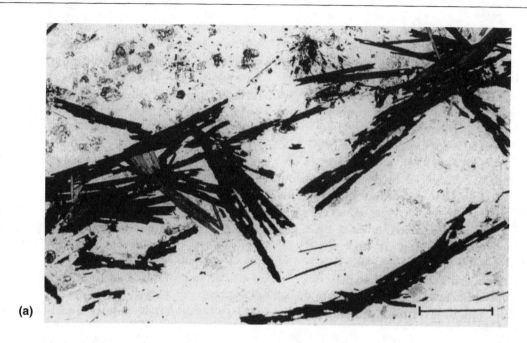

FIG. 5.13 (a) A fascicular bundle of stilpnomelane crystals developed in a quartzitic layer within a blueschist facies meta-ironstone sequence. Laytonville Quarry, California. Scale = 0.5 mm (PPL). (b) A bow-tie arrangement of actinolitic hornblendes on the schistosity plane of a hornblendic schist hand specimen from Troms, Norway.

Shape and form of crystals

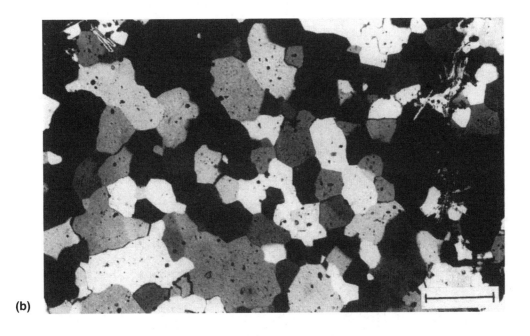

FIG. 5.14 (a) Irregular quartz grain boundaries in greenschist facies semi-pelitic schist. Loch Leven, Scotland. Scale = 0.5 mm (XPL). (b) A granoblastic–polygonal aggregate of quartz in a blueschist facies quartzitic rock. California. Scale = 0.5 mm (XPL). Note the straight grain boundaries and 120° triple-junctions between grains, indicating a very stable arrangement.

Crystal nucleation and growth

aggregate such as quartz in a quartzite – is that if all grains are hexagonal in cross-section, 120° triple-junctions are easily satisfied with straight crystal boundaries, but if the aggregate contains grains that vary from three- to eight-sided in cross-section, 120° triple-junctions can only be maintained with curved boundaries (Fig. 5.15). Grains with less than six sides have convex-outward boundaries, and phases with more than six sides have convex-inward boundaries. Curved grain boundaries are also required in polymineralic equant aggregates. For example, in a bimineralic aggregate of hexagonal grains, where phase A is much more abundant than phase B, grains of phase B entirely surrounded by phase A will be convex-outward if $\Theta_{ABA} > 120°$ and convex-inward if $\Theta_{ABA} < 120°$.

Certain aggregates of minerals develop a DECUSSATE structure (Fig. 5.16). This is a special type of granoblastic structure in which crystals are subhedral, prismatic or flaky, and randomly oriented, and have a strong crystal anisotropy (i.e. the surface energies of the different faces are very different). It is generally associated with monomineralic vein assemblages, monomineralic amphibole- or mica-hornfelses

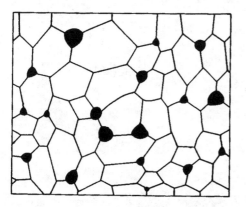

FIG. 5.15 A schematic drawing of a recrystallised polygonal aggregate of grains, to show that in order to maintain 120° triple-junctions in three- to eight-sided crystals, a certain number of curved boundaries is required.

FIG. 5.16 Decussate structure exhibited by axinite crystals in a vein from the contact aureole of the Bodmin Granite, Cornwall, England. Scale = 0.5 mm (XPL).

or granofelses, and monomineralic amphibole or mica domains within polymineralic hornfelses and granofelses. Following impingement, grain growth proceeds by grain-boundary migration and engulfment of small grains by large ones, to produce a rational low surface energy aggregate of subhedral crystals. Equal-angle (120°) triple-junctions between grains tend not to occur.

In bi-mineralic and polycrystalline aggregates, equilibrium textures and microstructures are more complex. In quartz–feldspar–mica aggregates (e.g. psammites and semi-pelitic schists), anisotropic minerals such as mica will tend to dominate the microstructure. The planar 001 surfaces of micas are very stable, and consequently do not become adjusted by quartz or feldspar impingement during growth. This results in quartz/quartz interfaces generally meeting mica(001)/quartz interfaces at 90° (Figs 5.17(a) & (b)). By restricting grain growth of quartz in certain directions, layer silicates will influence the size and shape of the quartz grains (i.e. quartz will tend to be elongate parallel to aligned micas; see Figs 5.17(a) & (b)).

5.6 Twinning

5.6.1 Introduction

Twinned crystals are comprised of two or more portions of the same crystal species in high-angle

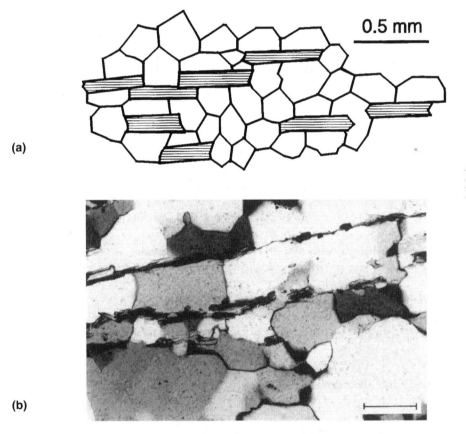

FIG. 5.17 (a) A schematic illustration showing how micas influence the shape of quartz grains during recrystallisation. (b) A natural example showing how the 001 faces of mica crystals have influenced the shape of quartz crystals, to make them sub-rectangular. Semi-pelitic schist, Norway. Scale = 0.1 mm (XPL).

contact with each other, and with a rational symmetry according to specific laws. They are easily identified in thin section because, due to their different optical orientation, different portions of the twinned crystal go into extinction at different times, or show different interference colour. Since twin boundaries represent high-angle contacts, they are usually sharply defined. SIMPLE TWINS are those comprised of two units (Plate 2(c)), while MULTIPLE or POLYSYNTHETIC TWINS (Plate 2(e)) consist of many. It is important to distinguish between PRIMARY TWINS, which represent twins formed at the time of crystal growth, and SECONDARY TWINS such as those formed due to crystallographic inversion or in response to stress during subsequent deformation.

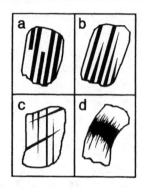

FIG. 5.18 A schematic illustration of the differences between (a) enclosed primary twins with abrupt or stepped terminations, and (b) enclosed deformation twins with tapered terminations. (c) Conjugate sets of deformation twins. (d) An example of deformation twins concentrated at a bend in a crystal.

5.6.2 Primary twins

Primary twins (or growth twins) commonly develop in minerals such as amphiboles, pyroxenes and feldspars. They can be further subdivided into 'layer growth twins' and 'twinned nuclei'. For further details on the various types of feldspar twinning, the reader is referred to the classic reference on feldspars by Smith & Brown (1988).

LAYER GROWTH TWINS (e.g. albite twins in feldspar) are those involving the addition of successive layers to the growing crystal faces in either the 'normal' or the 'twin' position. Partial or enclosed primary twins have abrupt or stepped terminations (Fig. 5.18(a)), in contrast to deformation twins (see below), which generally have tapered terminations (Figs 5.18(b)–(d) & Plate 3(b)). Feldspar, amphibole and pyroxene all commonly show primary twins in igneous rocks, but in metamorphic rocks growth twinning in these minerals is less common. This can be explained in terms of ongoing deformation at the time of growth, inhibiting twin development. Plagioclase feldspars, for example, are common in metabasic and pelitic rocks of the greenschist and amphibolite facies but are mostly untwinned. Where twinning is present, it is usually a simple twin on the Albite Law (Plate 2(c)), which suggests that twinning is difficult to produce under the temperatures and stresses associated with low- and medium-grade metamorphism. The effect of stress as a restraint on growth twins can be argued with reference to cordierite. In this case, sector twinning (or 'sector trilling') in cordierite (see below, and Fig. 5.19(a)) is commonly encountered in contact metamorphosed rocks where stresses are very low, but is uncommon in cordierite schists and gneisses from terrains of orogenic metamorphism, where stresses at the time of growth would be appreciably higher.

TWINNED NUCLEI represent a separate type of growth twin, and develop when crystals nucleate in the twinned state. Examples of this type of twin are the cruciform twins commonly observed in staurolite and chloritoid (Plate 2(d)). ANNEALING TWINS represent a further type of growth twin, and form during recrystallisation. They are produced when a migrating grain boundary of one crystal meets an adjacent crystal with its lattice suitably oriented. A twin boundary is of lower energy than a normal boundary and will be developed in preference.

5.6.3 Secondary twins

Two main types of secondary twin (those formed after crystal growth) can be recognised in minerals of metamorphic rocks. These are INVERSION (or TRANSFORMATION) TWINS and DEFORMATION TWINS.

Inversion twins occur in certain minerals when changing metamorphic conditions give rise to instability of the original crystal structure. This ultimately leads to a change in crystal habit. CROSS-HATCHED TWINNING of microcline (Fig. 5.19(b)) is one such example of this type of secondary twinning. For igneous rocks, the traditional interpretation of cross-hatched twinning in microcline has been in terms of inversion from higher-temperature monoclinic sanidine or orthoclase to lower-temperature triclinic microcline. In order to accommodate this crystallographic change from monoclinic to triclinic brought about by increasing degrees of Al, Si ordering, a complex network of albite twins (± pericline twins) develops, to give the characteristic cross-hatched twinning of microcline. However, in the case of metamorphic rocks, microcline has been shown to increase at the expense of orthoclase in rocks showing greatest deformation. Therefore, in addition to the control of temperature, this suggests that superimposed shear stress is another key factor that will cause orthoclase or sanidine to invert to microcline.

SECTOR TRILLING (also referred to as 'sector twinning' or 'sector zoning') in cordierite (Fig. 5.19(a)) has received considerable attention in the literature (e.g. Kitamura & Yamada, 1987). It is not a growth twinning feature, but is now widely accepted as developing in response to a transformation from metastable high-temperature hexagonal cordierite to stable low-temperature orthorhombic cordierite. While frequently observed in hornfelses of contact metamorphic aureoles, sector trilling is rarely reported from orogenic metamorphic terrains. However, interpenetrant sector twinning certainly occurs in cordierite from some high-grade gneiss terrains. The reason for the rather limited occurrence of sector trilling in cordierite from orogenic metamorphic terrains is probably related to the active shearing prevalent in such environments at the time of cordierite growth. For details relating to TEM imaging of transformation-induced microstructures, the reader is referred to the excellent review by Nord (1992).

DEFORMATION TWINS are the other dominant type of secondary twin seen in metamorphic minerals. They are extremely common in calcite (Plate 3(a)) and dolomite, but frequently occur in certain other minerals (e.g. plagioclase (Plate 3(b)), and pyroxenes). Deformation twins are occasionally simple, but are most commonly polysynthetic twins or in conjugate sets (Fig. 5.18 & Plate 3). Unlike 'enclosed' primary twins which have abrupt terminations, 'enclosed' deformation twins have tapered ends (Figs 5.18(b)–(d)). Other than this, it may be difficult in minerals such as plagioclase to decide whether the twins observed are 'primary' or 'secondary'.

The mechanism responsible for deformation twinning has similarities with that required for translation 'gliding' (or 'slip'), in that both processes involve displacement of a layer in the crystal lattice relative to its neighbour. However, there is a clear difference in terms of how this is achieved (Fig. 5.20). Translation gliding (or 'slip') involves individual layers of the lattice slipping past each other but coming to rest with crystal portions on each side of the slip plane being similarly oriented both before and after (Fig. 5.20(a)). This means that unless some marker (e.g. a primary twin) has been offset, slip planes are virtually impossible to detect optically. Deformation twins, on the other hand, involve (in simple terms) shear of part (or parts) of the crystal structure (Fig. 5.20(b)) by

Crystal nucleation and growth

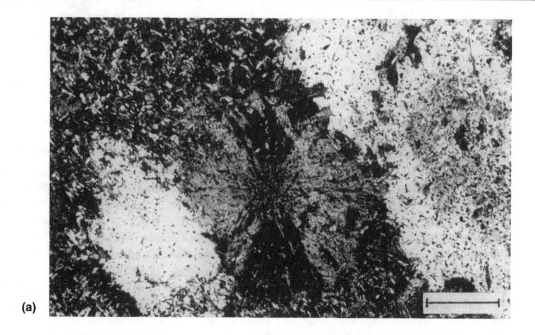

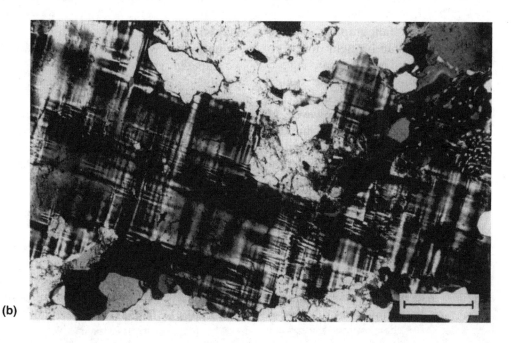

FIG. 5.19 (a) Sector trilling (or sector twinning) in cordierite. This results from the inversion of high-temperature hexagonal cordierite (indialite) to low-temperature orthorhombic cordierite during cooling. Cordierite Hornfels, Skiddaw, England. Scale = 0.5 mm (XPL). (b) Cross-hatched twinning of microcline, caused by the inversion from monoclinic sanidine or orthoclase to triclinic microcline. High-grade gneiss, Ghana. Scale = 0.5 mm (XPL).

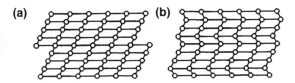

FIG. 5.20 A schematic illustration of the processes of lattice rearrangement in crystals undergoing deformation: (a) translation gliding; (b) deformation twinning.

twin-glide rather than offset and slip. This results in a reoriented (adjusted) structure that mirrors the original pattern (i.e. is in the twinned position). Deformation twinning gives homogeneous deformation in the twinned region, but the distribution of twins within a grain is generally heterogeneous (Plate 3(a)). In a polycrystalline aggregate, differences in the crystallographic orientation of individual grains relative to the superimposed shear stress will lead to some grains developing twins more readily than others. Nucleation occurs at positions of strain concentration such as dislocations and irregularities on grain boundaries. Twins will propagate along the favoured plane(s) (e.g. $\{01\bar{1}2\}$ in calcite, to form a continuous (and usually thin) twinned layer (Plate 3(a)). At temperatures above 200°C it is common to see two or three deformation twin sets in a single calcite grain. A section in Shelley (1993) gives a useful summary of the principal twin sets that develop in calcite and dolomite, and how they can be measured using a Universal Stage ('U-stage'). Recent work by Ferrill (1991) and Burkhard (1993) suggests that the geometry of deformation twins in calcite may have potential for indicating the temperature conditions at the time of deformation.

Experimental studies have established that deformation twinning is easier in coarse-grained than in fine-grained rocks. In carbonate lithologies it has been shown that there is a strong relationship between the orientation of deformation twins in calcite, and the orientation of the principal stress axes. In calcite, while twinning occurs over the full range of metamorphic temperatures, it will only develop when the direction of shortening is at a high angle to the c-axis of the crystal. Experimental work by Rowe & Rutter (1990) has examined the possibility of assessing the magnitude of palaeostresses based on the volume fraction and twinning incidence within calcite. They record that at constant differential stress there is a linear increase in both twinning incidence and volume fraction in relation to original grain size. For a rock of given grain size the volume fraction of twinning is found to increase non-linearly with differential stress, while the incidence of twinning increases linearly with stress. These factors are seemingly independent of temperature and strain rate. Bearing in mind grain-size considerations, these results suggest that there is considerable potential for the assessment of palaeostresses in naturally deformed carbonate rocks based on the abundance and orientation of deformation twins.

5.7 Zoning

A crystal growing by incremental addition to its outer surfaces will be uniform if the units added are of constant composition and orientation. However, if there is a change in either the source reservoir or in the element partitioning between the crystal and matrix minerals (due to changing P and/or T) the material added will probably change in composition. This will lead to a crystal the structure of which persists from core to rim, but which shows gradual or abrupt, continuous or cyclic compositional zoning (Tracy, 1982). Zoned crystals in igneous rocks (e.g. plagioclase and pyroxene) can be modelled in terms of incomplete reaction with a residual magma of changing composition. Zoning in metamorphic minerals relates to changing composition of the matrix and fluid, which in the case of a closed system is directly related to the availability of reactants and the

Crystal nucleation and growth

degree of completion of a reaction. At any given time it is only the rim of the zoned mineral that is at equilibrium with the matrix. The preservation of a zoned profile demonstrates that the interior of the crystal has not equilibrated under the changing conditions. As such, this gives considerable potential for gaining an insight into the early stages of the metamorphic history. This has been exploited by Spear & Selverstone (1983) and others, who have used variations in the chemistry of zoned porphyroblasts and inclusions trapped within them to give detailed evaluation of P–T–t trajectories of individual rocks. A crucial assumption to this approach is that the trapped phases were in local equilibrium at the time of trapping, and that post-trapping diffusive processes have not modified the chemistry of the inclusion or host. The whole approach is invalidated if such assumptions are not true, so a cautious approach must always be exercised. Prior to any interpretation of zoned minerals (e.g. plagioclase) in rocks of low metamorphic grade, the first step should always be to establish that they are not detrital phases.

Although not usually recognisable by optical means, one of the most widespread and extensively studied chemically zoned minerals of metamorphic rocks is almandine garnet. It largely develops from the breakdown of chlorite and, on the basis of electron microprobe studies, SEM back-scatter imaging and X-ray element mapping, it is found that rocks metamorphosed in the approximate range 450–625°C commonly have garnets that show Mn–Fe zonation from core to rim. The characteristic zonation ('Mn-bell' profile) is comprised of an Mn-rich core and an Mn-depleted rim, with a converse pattern for Fe (Figs 5.21 & 5.22). At higher metamorphic temperatures (approximately > 625°C) such primary chemical zonation is wiped out by diffusion throughout the crystal lattice (see below). Even before the advent of the electron microprobe, chemical zonation was realised by virtue of colour zoning in certain minerals

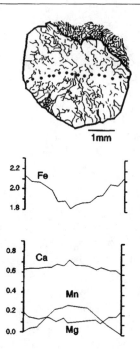

FIG. 5.21 Chemical profiles (based on electron microprobe analyses) for a typical almanditic garnet in a low amphibolite facies garnet–mica schist, from the Scandinavian Caledonides of arctic Norway. The substitution of Fe for Mn from core to rim has given a chemical zonation pattern typical of many low amphibolite facies garnets, with a characteristic Mn-bell profile. Dots on the drawing of garnet analysed represent probed points. Chemical variations are given in formula amounts of a particular element, on the basis O = 12 (see Fig. 5.22 for an SEM back-scatter image and X-ray element maps for the same garnet).

(e.g. pyroxenes (Plate 3(c)), tourmalines (Plate 3(d)) and amphiboles). Intracrystalline refractive index variation in minerals such as garnet had also enabled compositional zonation to be recognised.

The 'growth zoning' process introduced above involves reaction partitioning to give a continuous or discontinuous change in composition of material supplied to the growing crystal surface. It also requires slow volume diffusion in order that the interior of the crystal is isolated and does not equilibrate to new conditions. A second type of zoning, known as

Zoning

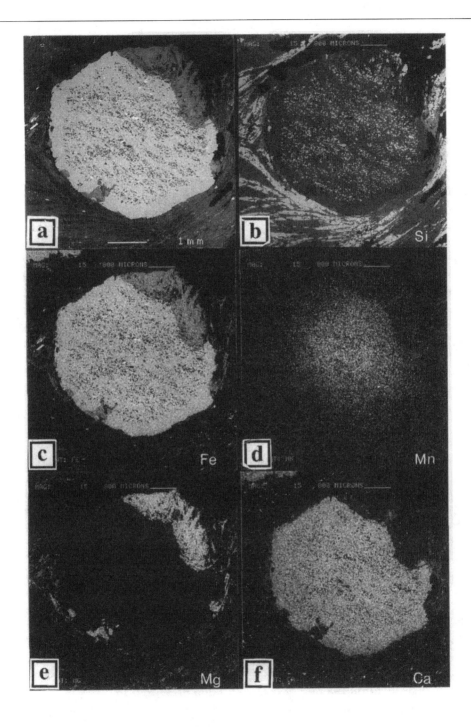

FIG. 5.22 (a) An SEM (back-scatter image) of the garnet shown in Fig. 5.21. (b–f) SEM X-ray element maps of the same garnet, showing chemical variations in the garnet and the matrix. Mn-enrichment in the garnet core shows up particularly well in (d), and the chlorite alteration (pale grey/white) at the garnet margins shows up well in (e).

Crystal nucleation and growth

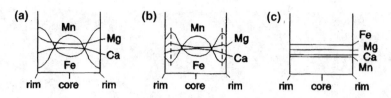

FIG. 5.23 Chemical zonation in garnet porphyroblasts. (a) A typical Mn-bell profile of garnet porphyroblasts that have grown largely at the expense of chlorite. (b) Garnet with an Mn-enriched rim: this commonly occurs as the result of late-stage diffusion zoning or resorption during retrogression. (c) Garnets that grow under high-grade metamorphic conditions, or that later experience high-grade metamorphic conditions, are largely unzoned: such garnets are said to be homogenized and have flat profiles– this is the result of complete intracrystalline diffusion.

'diffusion zoning', may occur in some metamorphic minerals, given the right circumstances. This is particularly common in garnet porphyroblasts of many high-grade metamorphic rocks, and has only been recognised following detailed microprobe work. The process involved is one of intra-crystalline diffusion driven by disequilibrium and reaction between the crystal (porphyroblast) surface and prevailing matrix conditions. Diffusion zoning is typically imposed on pre-existing crystals rather than being associated with growth. The crystal may or may not have been zoned to start with, and while normally associated with waning P–T conditions and retrogression it can also develop during 're-heating'. Diffusion zoning involves redistribution of atoms in the crystal structure and change in the relative abundance of certain atoms. It is usually recognised by sharp apparently 'reverse' zonation in the outer parts of crystals. In the case of garnets, Mn-enrichment at the margins is a characteristic diffusion zoning feature (Fig. 5.23(b)). This is commonly interpreted in terms of resorption of the garnet edge during retrogression, associated with a biotite- or cordierite-producing reaction. The extent of the resorption is dependent on the rate of cooling and on diffusion rates. Since the degree of resorption is often small, cooling rates must be fast relative to diffusion. Diffusion rates vary from one element to the next under given conditions, but below a certain temperature diffusion for all elements becomes so imperceptibly slow that zonation patterns do not equilibrate. At high temperatures, where diffusion is much faster, complete intracrystalline diffusion ('volume diffusion') occurs, and gives rise to homogenisation of earlier growth zoning profiles. The 'flat' chemical profiles characteristic of garnet porphyroblasts of the upper amphibolite facies are related to this process (Fig. 5.23(c)). In complex situations associated with polymetamorphism, discontinuous zonation profiles may be recognised, which result from resorption followed by new overgrowths. Within the leucosome component of migmatites, euhedral (idiomorphic) zoning in plagioclase is sometimes encountered. This is interpreted in terms of unimpeded growth zonation in the melt prior to solidification and impingement with surrounding grains (Ashworth, 1985).

Oscillatory zoning is recorded in some solid-solution metamorphic minerals, and is best known from pyroxenes and garnets of skarns (Plate 3(c)). While the significance of such zoning in igneous phenocrysts and sedimentary carbonate cements is well documented, its significance in metamorphic rocks has received limited attention. In an SEM study using back-scattered electron imaging, Yardley *et al.* (1991) document several examples of oscillatory zoning in metamorphic minerals. They interpret such zoning as diagnostic of open-system behaviour during metamorphism, and suggest that it may provide useful evidence for metasomatic mineral growth in metamorphic rocks. The fact that such zoning is preserved indicates sluggish intracrystalline diffusion through the minerals concerned.

As well as chemical zonation, textural zonation may be recognised in many porphyroblasts. For example, garnet porphyroblasts often have a heavily included core, but an inclusion-free rim (e.g. Figs 5.22(a) & (b)). This may represent two growth stages or, alternatively, may be interpreted as rapid initial growth followed by a slowing of the growth rate. More complex textural zonation occurs during polyphase metamorphism and deformation, and is discussed more fully in Chapters 9 & 12. In many porphyroblasts, breaks in chemical zonation commonly coincide with textural breaks, but in other cases there may be no clear relationship.

References

Ashworth, J.R. (1985) *Migmatites*. Blackie, Glasgow, 302 pp.

Barker, A.J. (1994) Interpretation of porphyroblast inclusion trails: limitations imposed by growth kinetics and strain rates. *Journal of Metamorphic Geology*, 12, 681–694.

Buntebarth, G. & Voll, G. (1991) Quartz grain coarsening by collective crystallization in contact quartzites, in *Equilibrium and kinetics in contact metamorphism* (eds G. Voll, J. Töpel, D.R.M. Pattison & F. Seifert). Springer-Verlag, Berlin.

Burkhard, M. (1993) Calcite-twins, their geometry, appearance and significance as stress-strain markers and indicators of tectonic regime: a review. *Journal of Structural Geology*, 15, 351–368.

Burton, K.V. & O'Nions, R.K. (1991) High-resolution garnet chronometry and the rates of metamorphic processes. *Earth and Planetary Science Letters*, 107, 649–671.

Carlson, W.D. (1989) The significance of intergranular diffusion to the mechanisms and kinetics of porphyroblast crystallization. *Contributions to Mineralogy and Petrology*, 103, 1–24.

Carlson, W.D. (1991) Competitive diffusion-controlled growth of porphyroblasts. *Mineralogical Magazine*, 55, 317–330.

Cashman, K.V. & Ferry, J.M. (1988) Crystal size distribution (CSD) in rocks and the kinetics and dynamics of crystallization: III. Metamorphic crystallization. *Contributions to Mineralogy and Petrology*, 99, 401–415.

Christensen, J.N., Rosenfeld, J.L. & De Paulo, D.J. (1989) Rates of tectonometamorphic processes from rubidium and strontium isotopes in garnet. *Science*, 244, 1465–1468.

Christian, J.W. (1975) *The theory of transformations in metals and alloys*. Pergamon Press, Oxford.

Craig, J.R. & Vaughan, D.J. (1994) *Ore microscopy and ore petrography*, 2nd edn. John Wiley, New York, 434 pp.

Curie, P. (1885) Sur la formation des cristaux et sur les constantes capillaires de leur différent faces. *Soc. Minéral. France Bull.*, 8, 145–150.

Ferrill, D.A. (1991) Calcite twin widths and intensities as metamorphic indicators in natural low-temperature deformation of limestone. *Journal of Structural Geology*, 13, 667–676.

Fisher, G.W. (1978) Rate laws in metamorphism. *Geochimica et Cosmochimica Acta*, 42, 1035–1050.

Frank, F.C. (1949) The influence of dislocations on crystal growth. *Discussions of the Faraday Society*, No. 5, 48–54.

Gibbs, J.W. (1878) On the equilibrium of heterogeneous substances. *Transactions of the Connecticut Academy III*, 1875–1876, 108–248; 1877–1878, 343–524.

Griffin, L.J. (1950) Observation of unimolecular growth steps on crystal surfaces. *Philosophical Magazine*, 41, 196–199.

Harker, A. (1939) *Metamorphism – a study of the transformations of rock masses*. Methuen, London, 362 pp.

Joesten, R.L. (1991) Kinetics of coarsening and diffusion controlled mineral growth, in *Contact metamorphism* (ed. D.M. Kerrick). Mineralogical Society of America, Reviews in Mineralogy No. 26, 507–582.

Kerrick, D.M., Lasaga, A.C. & Raeburn, S.P. (1991) Kinetics of heterogeneous reactions, in *Contact metamorphism* (ed. D.M. Kerrick). Mineralogical Society of America, Reviews in Mineralogy No.26, 583–671.

Kitamura, M. & Yamada, H. (1987) Origin of sector trilling in cordierite in Diamonji hornfels, Kyoto, Japan. *Contributions to Mineralogy and Petrology*, 97, 1–6.

Kretz, R. (1966) Interpretation of the shape of mineral grains in metamorphic rocks. *Journal of Petrology*, 7, 68–94.

Kretz, R. (1994) *Metamorphic crystallization*. John Wiley, Chichester, 507 pp.

Nord, G.L. Jr. (1992) Imaging transformation-induced microstructures, in *Minerals and reactions at the atomic scale: transmission electron microscopy* (ed.

P.R. Buseck). Mineralogical Society of America, Reviews in Mineralogy No. 27, 455–508.

Paterson, S.R. & Tobisch, O.T. (1992) Rates and processes in magmatic arcs: implications for the timing and nature of pluton emplacement and wall rock deformation. *Journal of Structural Geology*, **14**, 291–300.

Philpotts, A.R. (1990) *Principles of igneous and metamorphic petrology*. Prentice-Hall, Englewood Cliffs, New Jersey, 498 pp.

Ridley, J. (1986) Modeling of the relations between reaction enthalpy and the buffering of reaction progress in metamorphism. *Mineralogical Magazine*, **50**, 375–384.

Rowe, K.J. & Rutter, E.H. (1990) Palaeostress estimation using calcite twinning: experimental calibration and application to nature. *Journal of Structural Geology*, **12**, 1–17.

Sempels, J.-M. & Raymond, J. (1980) Mathematical simulation of the growth of single crystals. *Computers and Geosciences*, **6**, 211–226.

Shelley, D. (1993) *Igneous and metamorphic rocks under the microscope*. Chapman & Hall, London, 445 pp.

Smith, J.V. & Brown, W.L. (1988) Feldspar minerals. Springer-Verlag, Berlin.

Spear, F.S. & Selverstone, J. (1983) Quantitative *P–T* paths from zoned minerals: theory and tectonic applications. *Contributions to Mineralogy and Petrology*, **83**, 348–357.

Tracy, R.J. (1982) Compositional zoning and inclusions in metamorphic minerals, in Characterization of metamorphism through mineral equilibria (ed. J.M. Ferry). *Mineralogical Society of America, Reviews in Mineralogy*, **10**, 355–397.

Vance, D. (1995) Rate and time controls on metamorphic processes. *Geological Journal*, **30**, 241–259.

Walther, J.V. & Wood, B.J. (1984) Rate and mechanism in prograde metamorphism. *Contributions to Mineralogy and Petrology*, **88**, 246–259.

Yardley, B.W.D., Rochelle, C.A., Barnicoat, A.C. & Lloyd, G.E. (1991) Oscillatory zoning in metamorphic minerals: an indicator of infiltration metasomatism. *Mineralogical Magazine*, **55**, 357–365.

Chapter six

Mineral inclusions, intergrowths and coronas

Small solid-phase inclusions are commonly observed in metamorphic minerals. They can develop either by exsolution of a solute phase during cooling, by enclosure of residual foreign phases during porphyroblast growth, or due to incomplete pseudomorphing of an early phase by a later one. In certain high-grade rocks two or more phases may crystallise simultaneously to give rise to distinctive symplectic intergrowths. In granulite facies rocks, the development of concentric coronas of one or more phases around a core of another phase is a commonly observed feature. The characteristics of these various microstructures and the processes involved in their development will now be examined.

6.1 Growth of porphyroblasts to enclose residual foreign phases

Many porphyroblasts have rather cloudy cores due to very fine inclusions, but in many cases (e.g. in garnet, cordierite, staurolite and hornblende) the host porphyroblast contains abundant clearly discernible inclusions (e.g. quartz and opaques) giving the host crystal a 'spongy' appearance (Plate 1(a)). This is known as POIKILOBLASTIC structure, or sometimes termed 'sieve' structure. It is analogous to poikilitic structure of igneous rocks, with which it should not be confused. The inclusions may either show no preferred orientation, may be arranged with respect to the internal structure of the crystal (see below) or may be oriented in relation to some pre-existing fabric of the rock.

Inclusions increase the total free energy, because they give rise to a greater surface area in poikiloblastic crystals (depending on the number and size of inclusions) and because of the inherent problem of fitting inclusions of a different phase into the host porphyroblast. For this reason, poikiloblasts are less stable compared to inclusion-free porphyroblasts of the same phase. The development of euhedral porphyroblasts with few inclusions probably reflects slow growth, whilst the formation of anhedral poikiloblasts is favoured by rapid growth.

The included phases in porphyroblasts may represent an 'inert' phase not involved in the porphyroblast-forming reaction, but enveloped as the porphyroblast grows (e.g. zircon in garnet). Alternatively, they may represent

Mineral inclusions, intergrowths and coronas

residual excess of a phase involved in the porphyroblast-forming reaction but overtaken by the growth 'front' before it could be completely utilised (e.g. quartz in garnet, cordierite, andalusite, and so on). This occurs because diffusion rates in the host are too slow with respect to growth rate. In order to minimise the surface (interfacial) energy, inclusions tend towards shapes with minimal surface area per unit volume. This is limited in some cases by their crystallographic structure. Quartz, for example, commonly occurs as fairly rounded to elliptical inclusions in most minerals, while phases such as sillimanite, apatite and rutile form elongate inclusions with rounded ends (Fig. 6.1). Vernon (1976) emphasises that these 'ideal' inclusion shapes may not necessarily be attained under low-grade metamorphic conditions, and that growth under significant stress can give strong preferred orientation in a given direction.

Inclusions of zircon occur in many minerals, but may be overlooked because of their small size (typically < 50 µm). However, in micas, chlorites and certain other minerals, the presence of zircon inclusions may be recognised on account of the pleochroic haloes that surround them (Fig. 6.2). These haloes result from the radioactive decay of small amounts of U and Th contained within zircon, which affects the structure of the host mineral. The size of the pleochroic halo is proportional to the size of the zircon inclusion, but is commonly 10–150 µm diameter. Inclusions of allanite and monazite may also show such haloes.

Because of the two processes of ADSORPTION and ABSORPTION by which ions, molecules or minerals may become attached to the growing front of a crystal, the presence of inclusions need not necessarily mean a high energy or unstable situation. ADSORPTION is the loose physical bonding to the crystal surface of foreign material. The material is located at surface defects rather than forming

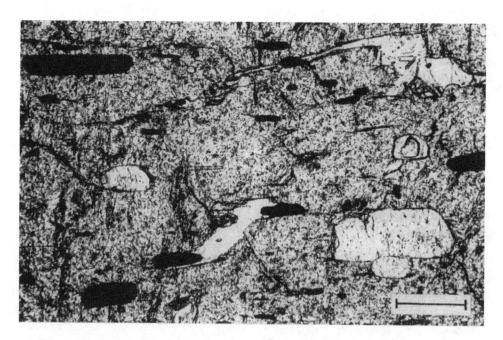

FIG. 6.1 Elongate rutile (dark) and tourmaline (light) inclusions in kyanite porphyroblast from kyanite schist. Ross of Mull, Scotland. Scale = 0.1 mm (PPL).

Growth of porphyroblasts

FIG. 6.2 Pleochroic haloes developed around zircon inclusions in biotite, as a result of radioactive decay of small amounts of U and Th in zircon affecting the structure of the host mineral. Pelitic schist, Snake Creek, Queensland, Australia. Scale bar = 200 μm (PPL).

an integral part of the structure. The adsorbed material may either be enclosed as the crystal grows or else accumulate in front of the advancing crystal face. Graphite (or amorphous carbon) has a strong tendency to become adsorbed to porphyroblasts and is commonly seen concentrated at the edges of andalusite, garnet, chloritoid and staurolite (Fig. 6.3). Harvey et al. (1977), Ferguson et al. (1980) and more recently Rice & Mitchell (1991) discuss the development of CLEAVAGE DOMES at the faces of idioblastic porphyroblasts. These domes, comprising graphite, muscovite and a low proportion of quartz compared to the adjacent matrix, are interpreted in terms of displacement of insoluble matrix graphite and muscovite during porphyroblast growth in a bulk hydrostatic stress field. Such domes, though rarely recorded, are the only reliable indicator of porphyroblast growth by di*splacement (*Rice & Mitchell, 1991). In most other situations porphyroblasts, are considered to grow by matrix re*placement.*

ABSORPTION involves chemical bonding and integration of foreign material into the crystal structure. If the impurity has or develops a low-energy interface with the growing porphyroblast, then during conditions of rapid growth (which is probably most usual) the activation energy of attachment is easily surpassed and the porphyroblast will build up around, and eventually enclose, the material as an inclusion. Because it is bounded by a low-energy immobile interface, the force of rejection will be small. It appears that certain crystals (e.g. quartz) are more readily included than others (e.g. micas). This suggests that porphyroblast–quartz interfacial energies are lower compared to porphyroblast–mica interfacial energies. Alternatively, it may indicate that mica is being consumed in the porphroblast-forming reactions and simply that there is an

Mineral inclusions, intergrowths and coronas

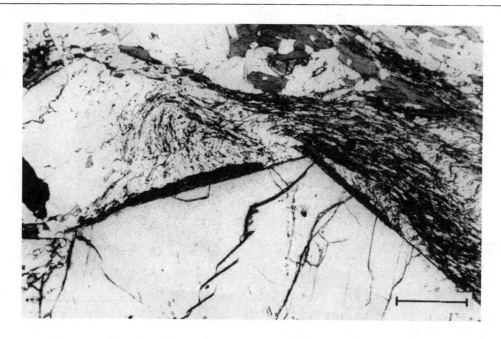

FIG. 6.3 Fine-grained graphite ADSORBED to the surface of a staurolite porphyroblast. Graphitic staurolite schist, Ghana. Scale = 0.5 mm (PPL).

excess of quartz in the system, such that even though quartz may be involved in the porphyroblast-forming reaction, small remnant inclusions of quartz are inevitable. Within a sequence of schists it is more than coincidence that the greatest density of quartz inclusions generally occur in porphyroblasts from those lithologies that are most quartz-rich.

The HOUR-GLASS structure frequently observed in chloritoid porphyroblasts (Plate 1(b)) and the characteristic cross-like inclusion arrangement of CHIASTOLITE (a textural variety of andalusite) from contact metamorphosed pelites (Plate 1(c)) represent regular geometrical patterns of inclusions arranged in relation to the host structure of the crystal (Kerrick, 1990, pp. 302–310). Staurolite has similarly been shown to have regular arrangement of inclusions in relation to crystal structure in some instances. Frondel (1934) advanced the idea that the regular arrangement of inclusions seen in crystals such as chloritoid and chiastolite could be explained in terms of preferential adsorption of impurities at certain crystal faces. This general model has been advocated by subsequent workers, and is reiterated here. Andalusite is an orthorhombic crystal with $(110)\wedge(1\bar{1}0)$ of 89°. It forms prismatic crystals elongate parallel to C. Depending on the thin-section cut, the basal sections of andalusite (var. chiastolite) show various inclusion patterns (Fig. 6.4 & Plate 1(c)), the inclusions usually being of fine carbonaceous material and quartz.

Although less frequently encountered, rhombdodecahedral garnets with inclusions of quartz concentrating at the interfacial boundaries are also recorded (e.g. Powell, 1966; Atherton & Brenchley, 1972; Anderson, 1984; Burton, 1986; Rice & Mitchell, 1991). These give spectacular examples of the same type of feature (Plate 1(d) & Figs 5.9(a) & 6.4). Two distinct types of inclusions can be recognised, namely, Type 1 inclusions of quartz, Fe–Ti oxides and graphite,

Exsolution textures

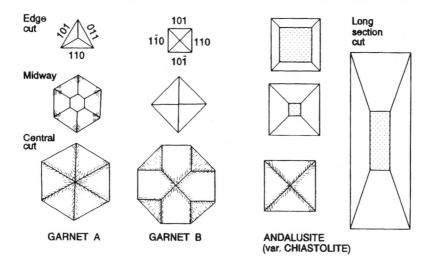

FIG. 6.4 The influence of the thin-section cut in relation to crystal orientation on the patterns seen in porphyroblasts where inclusions are concentrated at interfacial boundaries. (a) and (b) are for different cuts of garnet, and (c) is for different cuts of andalusite (var. chiastolite).

which were derived from the matrix and are preferentially located along the interfacial (sector) boundaries, and Type 2 inclusions, which are elongate or rodded quartz inclusions arranged perpendicular to the crystal faces. These Type 2 inclusions have tubular, or rodded form, and are not really inclusions, but intergrowths formed simultaneously with garnet growth. In all probability, growth was relatively slow in order to develop such a well defined crystallographically controlled arrangement.

In minerals which possess a strong crystal cleavage (e.g. amphiboles and micas) it is common for solid-phase inclusions to be preferentially incorporated and aligned within the cleavage planes. When studying thin sections, care should be taken not to misinterpret such well aligned inclusions as clear evidence for overgrowth of some earlier rock fabric, when in fact they result from a crystallographic control.

6.2 Exsolution textures

In igneous and metamorphic rocks, a number of minerals (e.g. feldspars, pyroxenes and amphiboles) representing non-ideal solid solutions (i.e. in which the enthalpy of mixing, $\Delta H_{mix} \neq 0$) show unmixing as temperature decreases. This immiscibility between solute and solvent gives rise to the development of an interphase boundary and recognisable inclusions of the solute within the host mineral. The degree of ordering of solute atoms is dependent on the rate of temperature reduction. If the drop is relatively slow it allows increased ionic mobility, and the solute ions will become increasingly organized before finally separating as discrete inclusions to give an EXSOLUTION texture within the host phase (Fig. 6.5). This unmixing occurs in order to minimise the Gibbs free energy of the system by producing an exsolved phase and a chemically changed host phase that have a combined energy contribution that is less than that of the original phase. On a T–X diagram (T = temperature; X = composition between two end-member components of a solution series), the *solvus* (Fig. 6.6) is a curve that separates the single phase field (above) from the immiscible two-phase field (below). If a horizontal line of

Mineral inclusions, intergrowths and coronas

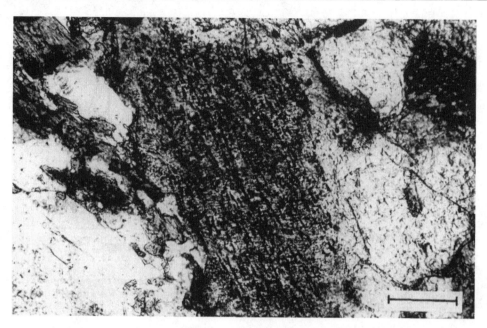

FIG. 6.5 Exsolved iron oxides along cleavage planes in hornblende. Epidiorite, Norway. Scale = 1 mm (PPL).

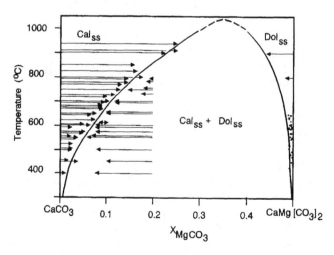

FIG. 6.6 The Cal–Dol solvus fitted to experimental data (modified after Anovitz & Essene, 1987). Arrowheads show the final compositions, and the direction of the arrow gives the direction of the composition shift. Experimental pressures are corrected to 2 kbar. The points near the dolomite limb represent natural dolomites coexisting with calcite.

constant temperature is drawn across the diagram, the two points of intercept with the solvus give the equilibrium compositions of the phases resulting from phase separation at that temperature. It is apparent from Fig. 6.6 that, as temperature decreases, the compositional difference between the two phases (calcite and dolomite) will steadily increase. The example of calcite–dolomite intergrowths is well documented by Kretz (1988). Other minerals that commonly show exsolution include Ca,Mg,Fe-pyroxene, Ca,Mg,Fe-amphiboles, feldspars and spinels.

On the basis of theory, experiment and empirical observation, *T–X* diagrams are well established for many solution series relevant to metamorphic phases. The systematic relationships observed have often been developed as geothermometers (e.g. Cal–Dol; Anovitz & Essene, 1987). In natural situations, exsolution usually requires high diffusion rates and sufficient time for nucleation and growth of the exsolved phases. Since these conditions are not always satisfied, some phases formed above the solvus will persist metastably at lower temperatures. However, certain mineral phases seem to have compositions that are so unstable below the solvus that even with rapid uplift they will exsolve.

On the basis of exsolution intergrowths observed in metals and alloys, Christian (1965) discussed different modes of precipitation from solid solutions. Most common mineral exsolution textures, particularly oriented intergrowths (e.g. perthites) are considered to form by a process of 'continuous precipitation', involving continuous draining of the solute throughout the host crystal's lattice. This is a distinctly different process from the model of simultaneous crystallization of two or more phases to give a symplectic texture.

The variations in size and shape of the exsolved phase are a function of ionic mobility (in turn dependent on temperature) and the interfacial energy between the phases. The exsolved phase often exists as abundant small blebs (Fig. 6.5). These commonly show an even distribution, suggesting homogeneous nucleation and minimal change in lattice spacing within the host. In other cases, concentration at the margin of the host crystal is observed, suggesting heterogeneous nucleation at grain boundaries. Other exsolution textures have an oriented lamellar form. This is frequently observed in feldspars and pyroxenes of certain igneous and high-grade metamorphic rocks. The development of such a texture must either have involved some primary ordering of the impurity atoms in the host, which became more ordered during cooling or, alternatively, may indicate restricted nucleation and growth along specific lattice planes. Throughout the exsolution process the framework of the host remains largely undisturbed, and the host retains its original shape and size. There is usually a strong relationship between the orientation of the exsolved inclusions and the atomic structure of the host (Fig. 6.5).

Unmixing of alkali feldspar during cooling from high temperatures gives rise to exsolution intergrowths of plagioclase (albite) and K-feldspar. This gives the characteristic PERTHITIC texture common to feldspars of many plutonic igneous and high-grade metamorphic rocks (Figs 6.7 & 6.8). The term PERTHITE strictly refers to feldspar intergrowth, in which the dominant phase is K-feldspar with inclusions of plagioclase.

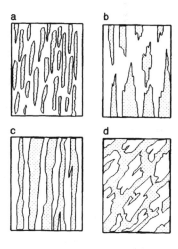

FIG. 6.7 Some common forms of perthite: (a) rodded or string; (b) flame-like; (c) banded or lamellar; (d) braided or interlocking. The width of the box is roughly 0.2 mm in each case.

Mineral inclusions, intergrowths and coronas

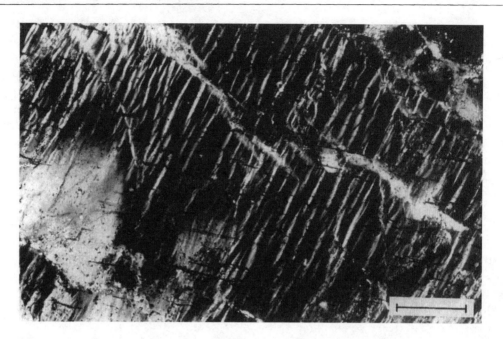

FIG. 6.8 A natural example of string perthite. Migmatite, Ghana. Scale = 0.1 mm (XPL).

ANTIPERTHITE is the converse and consists of K-feldspar enclosed in plagioclase (typically oligoclase–andesine). The term MESO-PERTHITE is given to exsolved feldspars with subequal volumes of intergrown K-feldspar and plagioclase. Although sometimes visible in hand specimen, most perthitic and antiperthitic intergrowths occur on the microscopic and sub-microscopic scale. Alling (1938) classified perthites in igneous rocks largely on the basis of shape. Many of these types are recognised in high-grade metamorphic rocks, and a selection of commonly observed types is shown in Fig. 6.7. Various processes have been suggested for the formation of perthites (and antiperthites). These are reviewed by Smith (1974), who concluded that the relative importance of 'replacement' and 'exsolution' are very much a function of host rock and of the crystallographic and chemical properties of the feldspar itself. In cases in which separate grains in the same sample show a constant proportion of host to inclusion, the 'exsolution' interpretation is favoured as the dominant mechanism relative to 'replacement', since the latter would give varying proportions (Hubbard, 1965). However, this is undoubtedly not the favoured interpretation in all examples. A recent study by Pryer & Robin (1995) examined the origin of flame perthite in deforming granites in the Grenville Front Tectonic Zone, Canada. They reviewed the arguments for and against exsolution and replacement and, in their particular example, concluded that flame perthite had developed due to retrograde metamorphic reactions involving the replacement of K-feldspar by albite. In another study of flame perthites, Passchier (1985) documented examples in which the origin is attributed to stress variations and the degree of deformation of K-feldspar grains. This certainly appears to be an attractive interpretation in the case of deforming high-grade gneisses, so it appears likely that more than one process is responsible for the formation of perthite in metamorphic rocks.

6.3 Inclusions representing incomplete replacement

Of the many examples in which porphyroblasts are packed with numerous small inclusions (poikiloblastic), there are certain cases in which the inclusions do not represent overgrown matrix minerals or remnants from a porphyroblast-forming reaction, but in fact represent incomplete replacement of the host phase by the small included phase(s). A good example of this type of reaction microstructure is the '*gefüllte plagioklas*' (= stuffed plagioclase) commonly developed in Alpine metagranodiorites and metatonalites (Angel, 1930; Ackermand & Karl, 1972). The microstructure comprises subhedral to euhedral plagioclase packed with numerous tiny, euhedral/subhedral and randomly oriented clinozoisite and muscovite inclusions (Fig. 6.9; see also Yardley *et al.*, 1990, p. 77). This feature can be interpreted in terms of original (moderately An-rich) plagioclase of the igneous protolith reacting with K-feldspar and H_2O to form more Ab-rich plagioclase and numerous inclusions of clinozoisite, muscovite and quartz, during Alpine greenschist/amphibolite facies orogenic metamorphism. In effect, the An-component of plagioclase has reacted with K-feldspar:

$4CaAl_2Si_2O_8 + KAlSi_3O_8 + 2H_2O =$
$2Ca_2Al_3Si_3O_{12}(OH) + KAl_3Si_3O_{10}(OH)_2 + 2SiO_2,$

An + Kfs + H_2O = Czo/Zo + Ms + Qtz.   (6.1)

A key feature of this type of inclusion is that some chemical components have to be able to diffuse through the host grains. Another example in which included phases are directly related to a complex reaction involving the host phase is the case of minute blebs of spinel giving a 'clouding' in plagioclase (e.g. Nockolds *et al.*, 1978, p. 355), as a result of the reaction

$CaAl_2Si_2O_8 + 2(Mg,Fe)_2SiO_4 = Ca(Mg,Fe)Si_2O_6 +$
$2(Mg,Fe)SiO_3 + MgAl_2O_4,$

An + Ol = Di + Hy + Spl   (6.2)

in metagabbros, caused by instability of Pl + Ol due to increasing *P* or decreasing *T*. In this case, the original labradoritic plagioclase shape is preserved, but the labradorite is transformed to andesine with a 'clouding' of minute blebs of spinel. Another case involving plagioclase (illustrated by Shelley, 1993, p.159) involves incoherent incomplete replacement of a large plagioclase (in a basic metatuff) by amphibole and quartz. A final example of a host phase containing numerous inclusions relating to an incomplete prograde reaction is the paramorphic replacement of andalusite by sillimanite. In this case, Kerrick & Woodsworth (1989) illustrate examples of abundant new crystals of sillimanite enclosed within single crystals of andalusite as a result of the incomplete transformation. Rosenfeld (1969) and Vernon (1987) similarly describe excellent examples of the paramorphic replacement of andalusite by sillimanite.

6.4 Symplectites

Two or more mineral phases frequently crystallise simultaneously in a metamorphic rock and yet remain largely separated. For example, during the prograde metamorphism of pelites, garnet and staurolite may develop contemporaneously as porphyroblasts and yet show a wide distribution throughout the rock because of abundant and dispersed sites of nucleation. In other cases, however, two or more phases can show intimate intergrowth textures. These can form regular oriented arrangements, but more usually occur as irregular and complex intergrowths. The most frequently encountered types in metamorphic rocks consist of irregular fine-grained mineral intergrowths, and are termed SYMPLECTITES, or SYMPLECTIC INTERGROWTHS.

Mineral inclusions, intergrowths and coronas

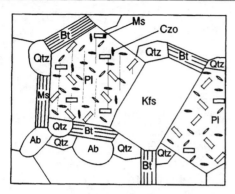

FIG. 6.9 A schematic illustration of plagioclase packed with Czo + Ms inclusions as a result of the prograde reaction 4An + Kfs + 2H$_2$O = 2Zo/Czo + Ms + 2Qtz (based on Angel, 1930; Ãckermand & Karl, 1972; Droop, pers. comm., 1997) in relation to metatonalites from the Hohe Tauern, Austria. The field of view is approximately 6 mm.

Most symplectites are associated with high-grade metamorphism, and are especially common in high-grade gneisses and granulites. Although some tri-mineralic intergrowths occur, most symplectites consist of intimate intergrowths of mineral pairs. In many cases the development of symplectites is considered the direct result of decompression during uplift (Section 6.5). This gives rise to instability and retrogression of the high-grade granulite assemblages. Commonly encountered symplectites include:

(i) Crd + Qtz (or Crd + Opx + Qtz) as a retrograde breakdown of Grt;
(ii) Opx + Pl symplectites as retrogression of Hbl + Grt, or developed due to instability of Cpx + Grt + Qtz (Fig. 6.10(a) is based partly on a diagram of De Waard (1967), which shows a symplectic intergrowth of orthopyroxene (C) and plagioclase (D), formed due to the instability of Grt(A) + Cpx(B) + Qtz);
(iii) Opx + Spl symplectites formed by reaction of Ol + Pl.

Symplectites generally occur either at boundaries between reacting minerals (Fig. 6.10(a)) (often referred to as 'kelyphytic rims'), or associated with the replacement of a primary phase by a pair of secondary phases (Figs 6.10(b) & (c)). In kelyphitic rims the intergrowths occur on a range of scales from those visible by standard microscopy down to ultra-fine intergrowths only visible by high-magnification SEM work (Fig. 6.11).

Diffusion (especially grain-boundary diffusion) is the main process involved in the formation of symplectites. Diffusion proceeds down the chemical potential gradient, so the direction and rate of diffusion is different for each element. In symplectites, the intergrowths comprise elongate minerals and small blebs arranged roughly perpendicular to the reacting interface. Slow diffusion of elements relative to the rate of progress of the reacting interface is held responsible for the formation of these symplectites. In many cases it is slow diffusion of Al and Si that are probably responsible, but many other elements are also involved, some diffusing towards the interface, and others moving in the opposite direction. The spacing of the rods relates to the range of the slowest diffusing species.

Kretz *et al.* (1989), in a study of corona and symplectite textures developed between olivine and plagioclase in metagabbros from Quebec, Canada, emphasised the complexity of diffusion pathways (A–E in Fig. 6.12) and the fact that the reaction zone is not entirely isochemical: that is, some elements involved in the reaction have diffused to the reaction area from outside. The study of olivine–plagioclase coronas in metagabbros of the Adirondacks, USA, by Johnson & Carlson (1990), similarly concluded that to achieve a mass balance for the reactions involved diffusional transport of material into and out of the reaction bands occurred. Since the width of a symplectite reaction zone increases with time, this means that the concentration and chemical potential gradients across the zone will diminish. This causes slowing of diffusion rates and thus reaction rates, and

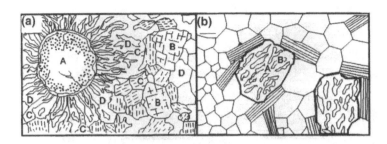
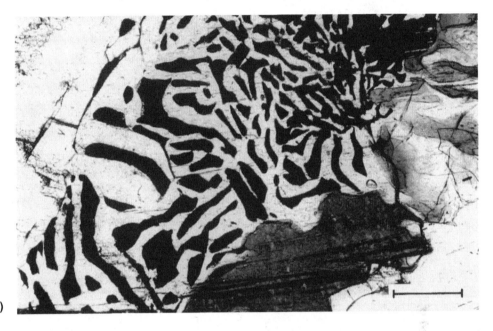

FIG. 6.10 (a) A schematic illustration of symplectite developed in the zone between two reacting phases. In this illustration, A and B are at disequilibrium and are separated by a symplectic intergrowth of C and D. (b) A schematic illustration of a symplectic intergrowth of phases A and B as total replacement (pseudomorph) after some earlier phase. (c) A natural example, showing symplectic intergrowth of orthopyroxene (light) and magnetite (black). Meta-norite, south-west Norway. Scale = 0.5 mm (PPL).

ultimately the reaction terminates when the slowest diffusing element(s) are unable to keep pace with the reaction. The situation is not helped if the symplectite is forming during cooling, because the decline in temperature will also cause a slowing of diffusion rates. It is perhaps for these reasons that many symplectite-forming reactions fail to go to completion and are commonly preserved in granulite facies rocks.

A specific type of symplectite comprising vermicular (worm-like) quartz intergrown with plagioclase is known as MYRMEKITE (Fig. 6.13). The formation of myrmekite both in metamorphic and in igneous rocks has long been a subject of debate. The development of the various schools of thought on its origin are reviewed by Smith (1974) and Phillips (1974, 1980). The present consensus of opinion regarding myrmekite in metamorphic rocks generally relates its development to the breakdown of K-feldspar. This is founded on the common observation that myrmekitic intergrowths of

Mineral inclusions, intergrowths and coronas

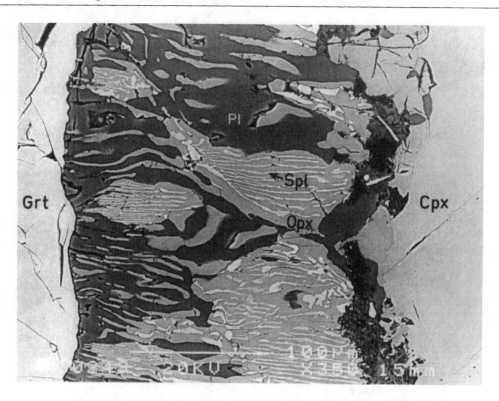

FIG. 6.11 An SEM back-scattered image of symplectic intergrowths in a granulite facies mafic protomylonite (after Brodie, 1995, courtesy of Blackwell Science). The instability of Grt (the principal reactant) in the presence of Cpx has produced a symplectic intergrowth of Pl (dark) + Opx (medium grey). Fine (1–3 μm) 'strings' of Spl (light grey) are intergrown with the Opx, but it is difficult to determine whether this is part of the original symplectite or else produced during subsequent exsolution from the Opx (Brodie, 1995). Minor Fe–Ti oxides (white) are also present. The symplectites of this example developed in dilatant zones, and their alignment defines the original extension direction at the time of reaction. Scale bar = 100 μm; field of view c. 340 μm.

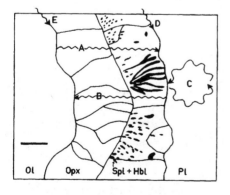

FIG. 6.12 Complex Pl–Ol symplectites (based on Fig. 5.15 of Kretz et al., 1989). Scale bar = 0.1 mm.

plagioclase and quartz are especially common in retrogressed high-grade gneisses lacking K-feldspar, and commonly in association with the breakdown of sillimanite or kyanite to white mica. The complete reaction can be written simplistically as

$$Kfs + Al_2SiO_5 + H_2O \rightleftharpoons \underbrace{Pl + Qtz + Ms}_{myrmekite}. \quad (6.3)$$

This model has its roots in the proposals of Becke (1908), but has been expanded in detail by Phillips and co-workers. Phillips (1980) describes and discusses the different

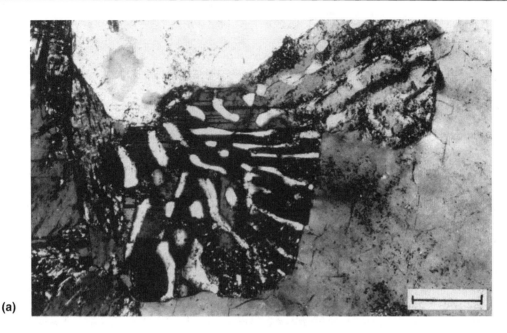

FIG. 6.13 (a) Myrmekite in augen gneiss from Tongue, Scotland. This microstructure comprises vermicular quartz (white) intergrown with plagioclase. It is common in high-grade metamorphic rocks, and forms in association with the retrogression of K-feldspar. Scale = 0.1 mm (XPL).

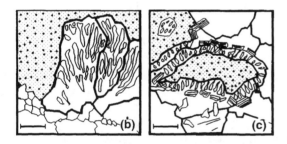

FIG 6.13 (b) A schematic illustration of bulbous myrmekite as a total replacement for K-feldspar. Scale = 0.1 mm. (c) A schematic illustration of a relict K-feldspar crystal mantled by myrmekite. Scale = 0.1 mm.

morphologies of myrmekite, of which those types most common to metamorphic rocks are illustrated in Figs 6.13(b) & (c).

The type termed 'bulbous myrmekite' (Figs 6.13(a) & (b)), in which the myrmekite appears to invade adjacent alkali feldspar, is very common in high-grade (especially quartzofeldspathic) gneisses. The other commonly encountered type consists of relict alkali feldspar surrounded by myrmekite and muscovite (Fig. 6.13(c)). Simpson & Wintsch (1989) demonstrate that in many cases such myrmekite forms in direct response to stress-induced K-feldspar replacement, and that it preferentially develops on the most strained margins of the feldspar (see also Section 6.2 regarding the development of flame perthite on strained feldspars). In Al_2SiO_5-free rocks (e.g. meta-granitoids), myrmekite is also common, and can be explained in terms of a late-stage ion-exchange reaction between K-feldspar and fluid (Droop, pers. comm., 1997), namely,

$$KAlSi_3O_8 + Na^+ \rightarrow NaAlSi_3O_8 + K^+,$$
$$2KAlSi_3O_8 + Ca^{2+} \rightarrow CaAl_2Si_2O_8 + 4SiO_2 + 2K^+,$$
Kfs   fluid   Pl   Qtz   fluid.
(6.4)

A recent study by Brodie (1995) documents how vermicular symplectites of orthopyroxene, plagioclase and spinel have formed at the margins of garnets synchronous with deformation under granulite facies conditions in metabasites within a shear zone. Such symplectites develop in rocks in areas of relatively low strain, such as at shear zone margins but, interestingly, the symplectites are oriented parallel to the main foliation and regional stretching direction (Fig. 6.11). Brodie (1995) concluded that the distribution of symplectites within the rock indicate that they only developed in dilatant areas, especially garnet grain boundaries and cracks in garnet. Such observations from the Ivrea–Verbano zone, Italy, are used as evidence for deep crustal extension localised along high-strain zones.

6.5 Coronas (of high-grade rocks)

Corona structures (Plates 2(a) & (b)) are especially common in high-grade gneisses, granulites and eclogites. They consist of a core of one mineral phase completely (or almost completely) enclosed by a 'corona' of another phase (or phases). The terms 'collar', 'atoll' and 'moat' have also been used in the literature to describe such features. However, the use of the term 'atoll' is not recommended, because it could cause confusion, as it has also been used with reference to 'atoll porphyroblasts' (e.g. atoll garnets), a specific microstructure in which the core of a chemically zoned porphyroblast has reacted to form other minerals (e.g. micas).

An example of a high-grade corona structure involves a garnet core with a plagioclase collar (Plate 2(a)) and newly developed orthopyroxene separating plagioclase from quartz. This is a familiar decompression texture of many granulites, and results from the reaction Grt + Qtz = Opx + Pl. Many other types of high-grade coronas have been recorded, the minerals and reactions being largely the same as those of symplectites. Coronas may be mono- or bi-mineralic and in some cases are multiple. In multishell coronas it is common to find that certain 'shells' are represented by symplectic intergrowths.

As with symplectic intergrowths, the development of corona structures in high-grade metamorphic rocks results from the instability of peak assemblages in response to declining P–T conditions during uplift. In all cases the 'core' phase reaches a state of disequilibrium with one or more of the surrounding phases. The ensuing reaction produces a corona of some new phase or phases, thus forming a barrier isolating those phases at disequilibrium. In common with symplectites, the process involved is one of localised diffusive transfer of material between the reactants, with high temperatures allowing reasonable reaction rates. As with symplectites, the broadening of the corona decreases chemical potential gradients and thus slows diffusion. Likewise, diffusion rates slow as temperature declines, and ultimately reaction progress is terminated. The study of olivine–plagioclase coronas by Johnson & Carlson (1990), utilising both transmitted light microscopy and SEM back-scattered electron (atomic number contrast) imaging, provides a particularly well documented example of coronas and the processes involved in their formation. The contribution by Rubie (1990) emphasises that corona-forming reactions occur at interfaces between specific mineral phases, and gives evidence for complex multi-component diffusion of elements derived from adjacent domains. Additionally, Rubie (1990) provides valuable discussion on possible approaches towards evaluating the kinetics of corona-forming reactions in eclogites.

Coronas and symplectites clearly provide useful information about reactions, but with the information now available to construct detailed petrogenetic grids it is also possible to relate such reaction microstructures in terms of the P–T path that the rock has followed. This is discussed more fully in Section 12.3.4.

A structure comparable to that of the coronas described above in relation to granulites may also be encountered in certain types of migmatite. The so-called 'flecky' gneisses of Loberg (1963), Russell (1969) and Ashworth

FIG. 6.14 A drawing of "flecky gneiss' (based on Fig. 50a of Loberg, 1963). Mafic cores (black) have an assemblage And + Bt + Qtz + Pl, and are surrounded by a quartzofeldspathic leucosome corona, or selvage (white), that separates the core from the mesosome matrix assemblage of Sil + Bt + Qtz + Feld. Scale bar = 1 cm.

(1985) result from high-temperature reaction processes, again involving diffusion of elements down local chemical potential gradients, to produce a coarse and irregular corona structure (Fig. 6.14). In the Loberg (1963) example from Västervik, Sweden, mafic cores of And + Bt + Qtz + Pl are surrounded by quartzofeldspathic (leucosome) coronas or haloes. The mesosome forming the main body of the rock comprises Sil + Bt + Qtz + Feld. The mafic clots and coronas are considered to have developed during the polymorphic transformation sillimanite to andalusite, and to have involved Fe + Mg diffusion inwards to the cores and K diffusion outwards into the leucosome (Fisher, 1970). By considering reasonable diffusion rates, Fisher (1977) estimated that for core radii of about 5 mm, the segregations would form in about 0.066 Ma.

References

Ackermand, D. & Karl, F. (1972) Experimental studies on the formation of inclusions in plagioclases from metatonalites, Hohe Tauern, Austria (lower temperature stability limit of the paragenesis anorthite plus potash feldspar). *Contributions to Mineralogy and Petrology*, 35, 11–21.

Alling, A.L. (1938) Plutonic perthites. *Journal of Geology*, 46, 142.

Anderson, T.B. (1984) Inclusion patterns in zoned garnets from Magerøy, north Norway. *Mineralogical Magazine*, 48, 21–26.

Angel, F. (1930) Über Plagioklasefüllungen und ihre genetische Bedeutung. *Mitt. Naturw. Ver. Steinmark*, 67, 36–52.

Anovitz, L.M. & Essene, E.J. (1987) Phase equilibria in the system $CaCO_3$–$MgCO_3$–$FeCO_3$. *Journal of Petrology*, 28, 389–414.

Ashworth, J.R. (1985) Introduction, in *Migmatites* (ed. J.R. Ashworth). Blackie, Glasgow, 1–35.

Atherton, M.P. & Brenchley, P.J. (1972) A preliminary study of the structure, stratigraphy and metamorphism of some contact rocks of the western Andes, near the Quebrada Venado Muerto, Peru. *Geological Journal*, 8, 161–178.

Becke, F. (1908) Über myrmekit. *Mineralogische und Petrographische Mitteilungen*, 27, 377–390.

Bowman, J.R. (1978) Contact metamorphism, skarn formation and origin of C–O–H skarn fluids in the Black Butte aureole, Elkhorn, Montana. Ph.D. thesis, University of Michigan.

Brodie, K.H. (1995) The development of orientated symplectites during deformation. *Journal of Metamorphic Geology*, 13, 499–508.

Burton, K.W. (1986) Garnet-quartz intergrowths in graphitic pelites: the role of the fluid phase. *Mineralogical Magazine*, 50, 611–620.

Christian, J.W. (1965) *Theory of transformations in metals and alloys*. Pergamon Press, Oxford.

De Waard, D. (1967) The occurrence of garnet in the granulite-facies terrane of the Adirondack Highlands and elsewhere, an amplification and a reply. *Journal of Petrology*, 8, 210–232.

Essene, E.J. (1982) Geologic thermometry and barometry, in *Characterization of metamorphism through mineral equilibria* (ed. J.M. Ferry). Mineralogical Society of America, Reviews in Mineralogy No.10, 153–206.

Ferguson, C.C., Harvey, P.K. & Lloyd, G.E. (1980) On the mechanical interaction between a growing porphyroblast and its surrounding matrix. *Contributions to Mineralogy and Petrology*, 75, 339–352.

Fisher, G.W. (1970) The application of ionic equilibria to metamorphic differentiation: an example. *Contributions to Mineralogy and Petrology*, 29, 91–103.

Fisher, G.W. (1977) Non-equilibrium thermodynamics in metamorphism, in *Thermodynamics in Geology* (ed. D.G. Fraser). D. Reidel, Boston, Ch. 19, 381–403.

Frondel, C. (1934) Selective incrustation of crystal forms. *American Mineralogist*, 19, 316.

Harvey, P.K., Lloyd, G.E. & Shaw, K.G. (1977). Arcuate cleavage zones adjacent to garnet porphyroblasts in a hornfelsed metagreywacke. *Tectonophysics*, 39, 473–476.

Hubbard, F.H. (1965) Antiperthite and mantled feldspar textures in charnockite (enderbite) from S.W. Nigeria. *American Mineralogist*, 50, 2040–2051.

Johnson, C.D. & Carlson, W.D. (1990) The origin of olivine–plagioclase coronas in metagabbros from the Adirondack Mountains, New York. *Journal of Metamorphic Geology*, 8, 697–717.

Kerrick, D.M. (1990) *The Al_2SiO_5 polymorphs*. Reviews in Mineralogy No. 22, Mineralogical Society of America, Washington, DC, 406 pp.

Kerrick, D.M. & Woodsworth, G.J. (1989) Aluminium silicates in the Mount Raleigh pendant, British Columbia. *Journal of Metamorphic Geology*, 7, 547–563.

Kretz, R. (1988) SEM study of dolomite microcrystals in Grenville marble. *American Mineralogist*, 73, 619–631.

Kretz, R., Jones, P. & Hartree, R. (1989) Grenville metagabbro complexes of the Otter Lake area, Quebec. *Canadian Journal of Earth Sciences*, 26, 215–230.

Loberg, B. (1963) The formation of a flecky gneiss and similar phenomena in relation to the migmatite and vein gneiss problem. *Geologiska Föreningens Stockholm Förhandlingar*, 85, 3–109.

Nesbitt, B.E. (1979) Regional metamorphism of the Ducktown, Tennessee massive sulfides and adjoining portions of the Blue Ridge Province. Ph.D. thesis, University of Michigan.

Nockolds, S.R., Knox, R.W.O'B. & Chinner, G.A. (1978) *Petrology for students*. Cambridge University Press, Cambridge, 435 pp.

Passchier, C.W. (1985) Water-deficient mylonite zones – an example from the Pyrenees. *Lithos*, 18, 115–127.

Phillips, E.R. (1974) Myrmekite – one hundred years later. *Lithos*, 7, 181–194.

Phillips, E.R. (1980) On polygenetic myrmekite. *Geological Magazine*, 117, 29–36.

Powell, D. (1966) On the preferred crystallographic orientation of garnet in some metamorphic rocks. *Mineralogical Magazine*, 35, 1094–1109.

Pryer, L.L. & Robin, P.-Y.F. (1995) Retrograde metamorphic reactions in deforming granites and the origin of flame perthite. *Journal of Metamorphic Geology*, 13, 645–658.

Rice, A.H.N. & Mitchell, J.I. (1991) Porphyroblast textural sector-zoning and matrix displacement. *Mineralogical Magazine*, 55, 379–396.

Rosenfeld, J.L. (1969) Stress effects around quartz inclusions in almandine and the piezothermometry of coexisting aluminium silicates. *American Journal of Science*, 267, 317–351.

Rubie, D.C. (1990) Role of kinetics in the formation and preservation of eclogites, in *Eclogite Facies Rocks* (ed. D.A. Carswell). Blackie, Glasgow, 111–140.

Russell, R.V. (1969) Porphyroblastic differentiation in fleck gneiss from Västervik, Sweden. *Geologiska Föreningen Stockholm Förhandlingar*, 91, 217–282.

Shelley, D. (1993) *Igneous and metamorphic rocks under the microscope*. Chapman & Hall, London, 445 pp.

Simpson, C. & Wintsch, R.P. (1989) Evidence for deformation-induced K-feldspar replacement by myrmekite. *Journal of Metamorphic Geology*, 7, 261–275.

Smith, J.V. (1974) *Feldspar minerals: Volume 2, Chemical and textural properties*. Springer-Verlag, Berlin, 690 pp.

Vernon, R.H. (1976) *Metamorphic processes*. George Allen & Unwin, London, 247 pp.

Vernon, R.H. (1987) Oriented growth of sillimanite in andalusite, Placitas – Juan Tabo area, New Mexico, USA. *Canadian Journal of Earth Science*, 24, 580–590.

Yardley, B.W.D., MacKenzie, W.S. & Guildford, C. (1990) *Atlas of metamorphic rocks and their textures*. Longman Science & Technology, and John Wiley, New York.

Chapter seven

Replacement and overgrowth

7.1 Retrograde metamorphism

7.1.1 Environments of retrograde metamorphism

Retrogression or retrograde metamorphism is a process involving the breakdown of higher P–T assemblages in association with declining P–T conditions. Since the majority of retrograde reactions require hydration or carbonation (Table 7.1), the presence of a fluid phase is essential for these reactions to proceed. Many high-grade metamorphic rocks seen at the Earth's surface exhibit remarkably fresh assemblages, despite the phases in the assemblage being well outside their stability fields. This indicates either that fluids did not enter the rock to promote the reactions expected with declining P–T conditions, or that uplift rates were far greater than reaction rates.

Rocks of regional metamorphic and subduction-related environments commonly display late-stage retrogression associated with their uplift history. This suggests that fluids, having been driven off by the various devolatilisation reactions associated with burial and prograde metamorphism, re-enter rocks during uplift, to induce retrogression. Although regional-scale retrograde metamorphism can be observed (e.g. amphibolitised granulites), it is often found that retrograde metamorphism is concentrated in specific zones. Active zones of deformation such as fault and thrust zones are particularly favoured (e.g. Beach, 1980). This is because metamorphic fluids are channelled into these areas as a result of P_f and chemical potential gradients, in addition to the fact that deformation processes aid metamorphic reactions and often enhance bulk permeability (see Section 8.4 for details). Fluids migrate to areas of low P_f, and in regimes of extensional and strike-slip tectonics it is not surprising that fluids concentrate into dilatant fault zones. In addition to the brecciation and cataclasis, it is widely observed that such fault zones are intensely silicified, dolomitized, sericitised, chloritised or show some other features of chemical or mineralogical change indicative of high fluid flux. Many such zones are important sites of economic mineral deposits (e.g. shear zone hosted gold deposits such as Golden Mile, Kalgoorlie, Australia; Boulter et al., 1987). In the case of thrust zones, the emplacement of a 'hot' slab over cooler rocks may lead to convective circulation of fluids in the footwall block. These fluids will enhance reactions in the thrust zone and lead to retrogression at the base of the hangingwall block.

Retrogression is also important in areas of igneous intrusion and hydrothermal activity. The heat associated with such areas promotes metamorphic reactions by providing the energy

Replacement and overgrowth

TABLE 7.1 Common retrograde reactions in the main compositional groups of metamorphic rocks.

Ultramafic rocks

Olivine	→ serpentine	H_2O-rich fluids
	→ magnesite	CO_2-rich fluids
Enstatite	→ anthophyllite	
Opx and/or olivine	→ talc ± serpentine	

Metabasites

Ca-plagioclase	→ Na-plagioclase + Ep/Zo/Czo	Very common amphibolite facies → greenschist facies retrogression (H_2O-rich fluids).
	→ zeolites	Common in very low grade burial metamorphism and ocean-floor metamorphism
	→ sericite/muscovite	In metabasites this usually requires significant K^+ introduction.
	→ calcite	CO_2-rich fluids.
	→ scapolite	Hydrothermal metamorphism CO_2-rich fluids.
Clinopyroxene	→ hornblende/actinolite	
Hypersthene	→ hornblende/actinolite	
Hornblende	→ actinolite	
	→ chlorite	
	→ biotite	Usually associated with significant K^+ introduction
Blue (Na-) amphibole (glaucophane/crossite)	→ green (Ca-) amphibole (actinolite)	
Garnet	→ chlorite	
Ilmenite or rutile	→ sphene	

Granitoid rocks

K-feldspar	→ sericite/muscovite/pyrophyllite
	→ clay minerals (e.g. kaolinite)
Plagioclase	→ sericite (epidote group minerals)
Biotite	→ chlorite

Calc-silicate rocks

Forsterite	→ serpentine
Anorthite	→ epidote minerals (± sericite)
	→ carbonate minerals
Diopside	→ tremolite—actinolite
Tremolite	→ talc

Metapelites

Garnet	→ chlorite and/or biotite
Staurolite	→ sericite
	→ sericite + chlorite
Andalusite, kyanite, sillimanite	→ sericite/white mica
Cordierite	→ pinite (fine mix of sericite + chlorite)
Chloritoid	→ chlorite (± sericite)
Biotite	→ chlorite
Ilmenite	→ sphene

necessary for reactions to proceed. Intrusion-related fluids carry various ions in solution, and by their interaction with interstitial fluids and minerals in the surrounding country rocks can induce both prograde and retrograde reactions. The principal factors controlling the type of alteration that occurs are the composition and mineralogy of the host rock, the composition of the fluid and the temperature of the fluid. The extent of infiltration is controlled by the nature of the host-rock permeability and by P_f. Upward-moving hydrothermal fluids from a cooling magma (especially granite or granodiorite) invade the surrounding country rocks to cause 'metasomatic' alteration. METASOMATISM is a process of alteration or chemical modification involving enrichment of the rock in certain ions derived from some external source. This influx of ions via a fluid phase induces various metamorphic reactions in the country rock, although much of the original microstructure of the rock may still be preserved. Potassic alteration (e.g. sericitisation) is a particularly common type of alteration in such situations, as is propyllitic alteration, involving widespread chloritisation, and argillic alteration, giving rise to extensive development of clay minerals. Extensive tourmalinisation is a common feature of metasomatism in the vicinity of granitoid intrusions (e.g. Hercynian granites of south-west England). Later sections of this chapter deal with specific types of retrogression and replacement, but for further details on the various types of hydrothermal alteration, the reader should refer to Thompson & Thompson (1996).

7.1.2 Textural features of retrogression

CORONAS and REACTION RIMS are a clear sign of disequilibrium between certain phases in the assemblage. In pelites and metabasites of the amphibolite facies and lower grade, coronas (or reaction rims) of hydrous mineral phases are common (Plate 4(b)). They are best recognised around porphyroblasts that are in disequilibrium with matrix conditions, either due to shearing at lower-grade conditions or in association with uplift and cooling. Continued disequilibrium and reaction may give rise to a partial or complete pseudomorphing of the unstable phase (Plates 4(c)–(e)). A PSEUDOMORPH is defined as a crystal that has been completely altered or replaced by another mineral or aggregate of minerals, and yet still retains its original shape. It may be either (a) a single-phase single-crystal pseudomorph, (b) a single-phase multicrystal pseudomorph or (c) a multiphase multicrystal pseudomorph. Single-phase multicrystal pseudomorphs are frequently observed, especially cases involving the replacement of a porphyroblast by an aggregate of some hydrous phase as part of a rehydration reaction; for example, garnet to chlorite (Plate 4(c)) or olivine to serpentine (Plate 4(d)). Equally common are multiphase multicrystal pseudomorphs, particularly those producing bi-mineralic hydrous assemblages; for example, hornblende $\rightleftharpoons$ actinolite + chlorite, and staurolite $\rightleftharpoons$ chlorite + sericite and garnet $\rightleftharpoons$ chlorite + biotite (Plate 4(e)). The fact that fine-grained multicrystal aggregates are common indicates high nucleation rate relative to growth rate. This causes numerous nucleation sites and a tendency towards site saturation. The degree of pseudomorphing of a particular phase can vary considerably across the domain of a thin section (Plate 4(d)) and on the outcrop scale may be localised in certain zones. This reflects the variable degree of fluid infiltration and illustrates the important role that fluids play in promoting reactions. In Plate 4(d), H_2O is required for olivine to alter to serpentine. Those areas infiltrated by fluid show complete pseudomorphing of olivine, and yet in adjacent areas in which the fluid has only gained minimal access the olivines are almost entirely unaltered. The preservation of undistorted 'soft assemblage' pseudomorphs in highly sheared rocks (Plate 4(c)) implies that

Replacement and overgrowth

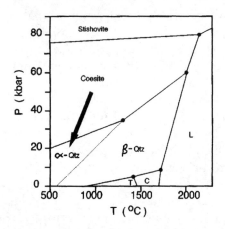

FIG. 7.1 A phase diagram for the SiO_2 system, showing the P–T stability fields (based on Žoltai & Stout, 1984) for the various silica polymorphs, and emphasising the polymorphic transformation coesite → α-quartz (arrowed) associated with decompression during uplift of certain ultra-high-pressure rocks. T, Tridymite; C, cristobalite; L, liquid.

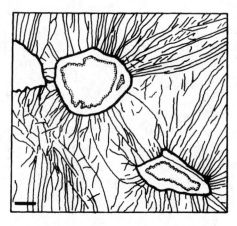

FIG. 7.2 A drawing of a coesite inclusion in garnet in the process of transformation to α-quartz (based on Fig. 3d of Chopin, 1984). The positive volume change associated with this transformation causes characteristic radial fracturing in the garnet. The solid line defines the SiO_2 inclusion, comprising a remnant high-relief coesite core (inside stippled area), surrounded by α-quartz (unornamented). Scale bar = 100 μm.

pseudomorphing occurred after deformation. Polymorphic transformations such as aragonite to calcite, and coesite to α-quartz (e.g. Chopin, 1984; Okay, 1995) occur when rocks formed at very high pressure undergo decompression during uplift (Figs 7.1 & 7.2).

Reaction rims (Plate 4(b)) and partial or complete pseudomorphs (Plates 4(c)–(e)) are obvious disequilibrium textures. Other alteration features include 'core replacement' and 'zone replacement'. Zonal alteration is especially common in igneous plagioclase crystals (Plates 4(f) & (g)). It indicates that certain zones of the plagioclase are out of chemical equilibrium with the matrix fluid, and consequently are more prone to alteration. In such cases the fluid gains access along microfractures or cleavage planes of the crystal. Feldspars, amphiboles and micas are particularly prone to such alteration. Plate 4(h) is a good example of TOPOTAXY (or TOPOTACTIC TRANSFORMATION), in which the mineral being replaced is of similar crystallographic structure to the mineral replacing it. In this case biotite is being replaced by chlorite, both of which are phyllosilicates. This type of transformation can represent either a prograde or a retrograde feature. It is retrograde if biotite forming part of a high-temperature assemblage is being replaced, but prograde if, for example, metamorphic chlorite is in the process of replacing detrital biotite (Section 7.2). Numerous XRD, SEM and TEM studies have been undertaken on such interlayered phyllosilicate phases and the transformation processes involved (e.g. Veblen & Ferry, 1983; Jiang & Peacor, 1991). The use of high-resolution TEM in such research on polysomatic intergrowths is reviewed by Allen (1992).

When studying porphyroblasts it may not always be easy to decide whether fine-grained mineral enclosures represent alteration products or inclusions incorporated during growth. However, there are certain points to take into account, and features to look for that should make the decision-making easier. First, with

Retrograde metamorphism

fine-grained alteration products it is often difficult to clearly discern grain boundaries by standard microscopy, whereas inclusions, although commonly fine-grained, usually have well-defined boundaries. Second, alteration products will usually be of minerals chemically similar to the mineral they are retrogressing (e.g. sericite after K-feldspar), whereas included mineral phases will often be quite different (e.g. rutile in garnet).

7.1.3 Specific types of retrogression and replacement

Serpentinisation

Serpentine is the most common alteration product of olivine, and serpentinisation of ultramafic rocks (Fig. 7.3) and forsterite marbles (Plate 4(d)) has received considerable attention in the literature (e.g. Peacock (1987). Serpentinisation occurs in environments such as ocean-floor metamorphism, shear zones developed during orogenesis or ophiolite obduction, and in contact metamorphic aureoles with extensive aqueous fluid infiltration. The addition of H_2O is essential for serpentinisation to proceed, and in general it takes place below 500°C, and commonly at less than 350°C. If mass remains constant during serpentinisation of a peridotite a substantial volume increase (35–45%) will occur. The basic reactions can be written as

$$2Mg_2SiO_4 + 3H_2O = Mg_3Si_2O_5(OH)_4 + Mg(OH)_2 \quad (7.1)$$
forsterite serpentine brucite

and

$$3MgSiO_3 + 2H_2O = Mg_3Si_2O_5(OH)_4 + SiO_2. \quad (7.2)$$
enstatite serpentine silica

If the original olivine or pyroxene contains a component of iron, magnetite will be an additional product of the reaction. Experiments by Wegner & Ernst (1983), which included a study of the hydration of forsterite to give serpentine plus brucite (7.1), showed that the reaction rate was appreciably faster when $P(H_2O)$ was increased from 1 to 3 kbar.

FIG. 7.3 Serpentinised ultramafic rock, showing a distinctive serpentine 'mesh'. Dawros, Connemara, Ireland. Scale bar = 125 μm (XPL).

Replacement and overgrowth

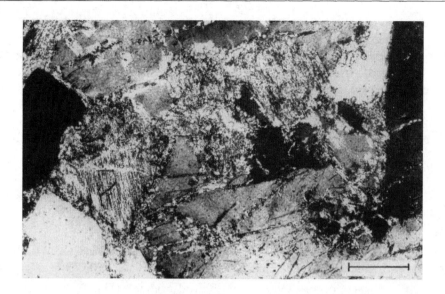

FIG. 7.4 Pyroxenes replaced by amphibole (actinolitic hornblende) as part of the process of uralitisation in meta-igneous rocks within a Precambrian basement window. Troms, Norway. Scale bar = 125 μm (XPL). The surrounding minerals are plagioclase and minor quartz.

Uralitisation

Uralitisation is the term given to the replacement of primary igneous pyroxenes by amphiboles (typically tremolite, actinolite or hornblende). This replacement commonly occurs at the margins of plutons when late-stage intrusion-related aqueous fluids interact with already crystallised parts of the intrusion. A similar replacement process can also take place during orogenic metamorphism to produce a rock termed epidiorite. This rock type preserves a coarse crystal aggregate with no preferred orientation (i.e. igneous appearance), but has original igneous pyroxenes replaced by actinolite or hornblende (Fig. 7.4). For such a rock to form, low strain rates are required to preserve the igneous appearance, and aqueous fluids are required to produce the amphiboles. If strain rates are higher a cleavage or schistosity would form, the original igneous microstructural features would be lost and the rock would be termed a greenschist.

Chloritisation

Extensive replacement of original assemblages by chlorite (i.e. chloritisation) involves major influx of aqueous fluids. Such retrogression is commonly associated with hydrothermal alteration of mafic rocks and greenschist facies shear zones (Fig. 7.5). Aqueous fluids are always involved, but the precise nature of the reaction(s) will be a function of bulk rock composition and the chemistry of the infiltrating fluids. If biotite is one of the reactants the K^+ liberated will enter the fluid phase and may be used in the formation of sericite elsewhere in the rock.

Sericitisation

The alteration of K-feldspar, plagioclase, Al_2SiO_5 polymorphs, staurolite and cordierite to fine-grained aggregates of white mica (termed 'sericite') is a common feature of retrogressed quartzofeldspathic rocks and metapelites. This sericitic alteration is especially common in hydrothermal environments and

Retrograde metamorphism

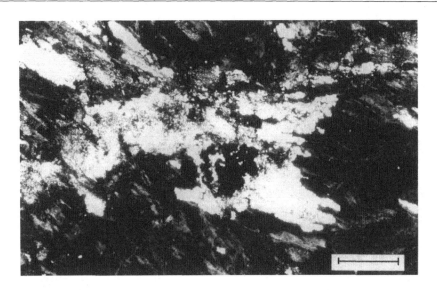

FIG. 7.5 Chloritisation of mafic rock in a greenschist facies shear zone, Syama, Mali. Dark areas = chlorite; light grey to white speckled areas = calcite. Scale = 125 μm (XPL).

greenschist facies shear zones (Fig. 7.6). The basic form of the K-feldspar → sericite reaction can be written as

$$3\,KAlSi_3O_8 + 2H^+ = KAl_2[AlSi_3O_{10}](OH)_2 + 6SiO_2 + 2K^+.$$
K-feldspar   sericite   quartz
(7.3)

Whenever the activity of H^+ relative to K^+ increases, the reaction will be driven to the right and sericite will start to form at the expense of K-feldspar. In cases of extensive hydrogen-ion metasomatism, H^+ infiltrates the system and Ca and Na are flushed out, to leave a rock with an extensively modified chemistry and a mineral assemblage largely devoid of plagioclase and enriched in sericite (± clay minerals). If the fluid is enriched in K^+ (potassium metasomatism), mafic igneous rocks and metabasites may be completely transformed to mica-dominated assemblages, with little or no plagioclase, amphibole or pyroxene. If temperature conditions are below that of biotite stability (approximately 400°C), the assemblage Chl + Serc will dominate.

Saussuritisation

Saussuritisation involves the liberation of Ca from calcic plagioclase to leave a feldspar of more albitic composition 'dusted' with a fine-grained aggregate of epidote group minerals (± Serc ± Cal). This type of alteration is common during greenschist facies retrogression and situations of low-pressure hydrothermal alteration. Although this type of alteration is often patchy, in zoned plagioclase crystals of igneous rocks it is common to see the saussuritic alteration concentrated in the more anorthitic core of individual crystals (Plate 4(f)), or in An-rich zones (Plate 4(g)). To develop such specific zonal alteration, the fluid causing the alteration probably diffused through the crystal lattice, probably along cleavage or microcracks, and gave rise to selective alteration of those zones at disequilibrium with the fluid.

Sodic and sodic–calcic (Na–Ca) metasomatism

Na(–Ca) metasomatism has been recorded from environments of sea-floor metamorphism (e.g.

Replacement and overgrowth

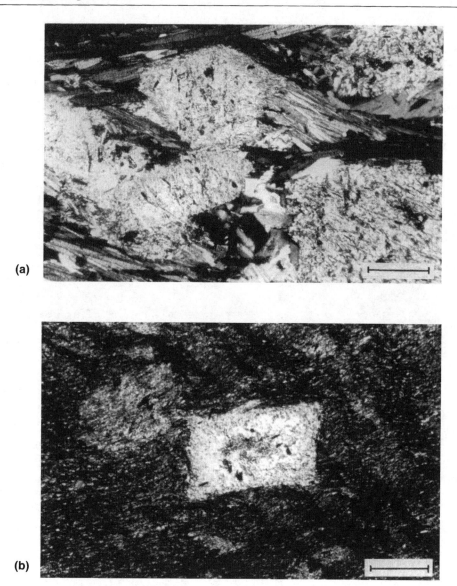

FIG. 7.6 (a) The sericitisation of staurolite schist, giving rise to total pseudomorphs after staurolite. [Stt–Grt schist, Glen Lethnot, Scotland. Scale bar = 0.5 mm (XPL).] (b) The sericitic alteration of andalusite (var. chiastolite), in the contact metamorphic aureole of the Skiddaw Granite, England. Scale bar = 0.5 mm (XPL).

Seyfried et al., 1988), in association with porphyry copper deposits around acid- intermediate arc intrusives (e.g. Carten, 1986), and within regions of extensive fluid infiltration associated with Proterozoic granitoids (e.g. De Jong & Williams, 1995; Oliver, 1995). The metasomatism is characterised by exchange of Na for Ca or K, and to a lesser extent Ca for Fe and Mg. The most characteristic mineralogical change that occurs is formation of new albite, and replacement of pre-existing feldspar by albite (i.e. ALBITISATION). Other possible

Overgrowth textures

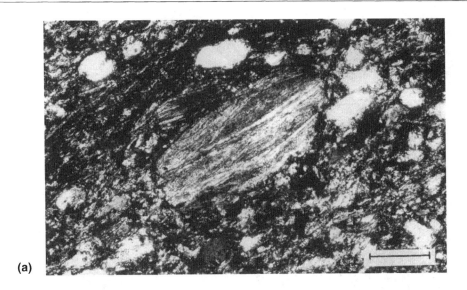

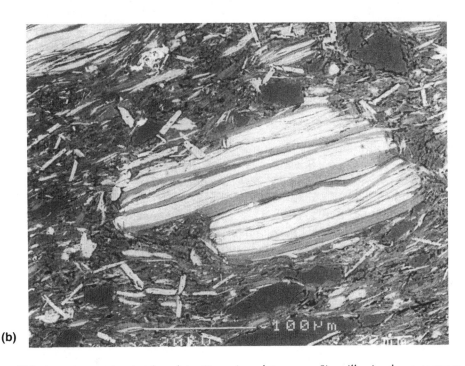

FIG. 7.7 Chlorite–mica stacks developed in Devonian slates near Siouville, in the outermost part of the Flamanville Granite contact metamorphic aureole, Normandy, France. (a) Standard optical microscopy (XPL), Scale = 125 μm. The chlorite–mica stack (centre) comprises intergrown chlorite (dark) and muscovite (light) as a probable replacement for detrital biotite. The matrix of the rock comprises rounded grains of quartz (white/light grey) in a fine groundmass of phyllosilicates. (b) An SEM (back-scattered electron) image showing enhanced detail of the chlorite–mica stack (light grey/white = Chl; medium grey = muscovite). The image also shows greater detail of the matrix, comprising 10–20 μm length chlorite laths (light grey), and 20–100 μm quartz grains (dark grey) in an ultra-fine (<5 μm) groundmass of white mica and quartz. Scale = 100 μm.

Replacement and overgrowth

(c)

FIG. 7.7 (*contd*) (c) A detail of the central part of the chlorite–mica stack shown in (b), showing individual layers to vary from 1 μm to 15 μm in thickness. Scale = 10 μm.

changes, depending on the nature of the fluid and the *P–T* conditions, include conversion of feldspars to zeolites (ZEOLITISATION), replacement by scapolite (SCAPOLITISATION) or conversion to epidote (EPIDOTISATION). In all cases the origin of the fluid is considered to be either magmatic, derived from reactions in meta-evaporites, of direct sea-water origin, or of some combination of these sources.

7.2 Overgrowth textures during prograde metamorphism

Although the evidence of retrograde reactions is commonly preserved, the evidence of prograde reactions is far less frequently encountered. Even so, there are several notable examples in which prograde reaction rims and intergrowth textures have been recognised. One such example is the prograde replacement of detrital biotites by metamorphic chlorite (e.g. Jiang & Peacor, 1994) and the development of other CHLORITE–MICA STACKS (Fig. 7.7) in late diagenesis to low greenschist facies metamorphism (e.g. Warr *et al.*, 1993; Merriman *et al.*, 1995). Other prograde reaction textures include the overprint of blue glaucophanic amphibole rimming green calcic amphibole in metabasic rocks, and the mantling of kyanite by coarse muscovite laths (or Mc + Sil) in pelites (Fig. 7.8). The latter is related to the prograde breakdown of kyanite to sillimanite in the upper amphibolite facies. The sillimanite is more commonly observed to nucleate at quartz grain boundaries (Fig. 5.3), or intergrown with biotite, rather than as a direct overgrowth on kyanite. Reaction rims of coarse muscovite have similarly been observed around andalusite as part of the prograde reaction to form kyanite. Such reaction rims should

Overgrowth textures

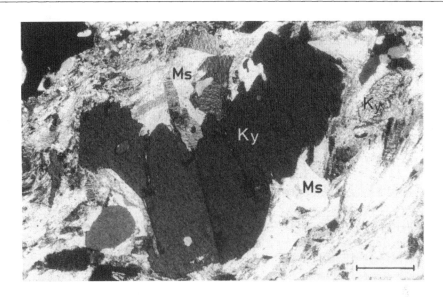

FIG. 7.8 The prograde breakdown of kyanite (at extinction) to coarse muscovite relating to the upper amphibolite facies transformation Ky → Sil. This example is from a kyanite gneiss from the Caledonides of north Norway. Scale = 0.5 mm (XPL).

FIG. 7.9 The direct overgrowth of one Al_2SiO_5 polymorph on another. In this case, from an Al_2SiO_5–Qtz vein from Snake Creek (Queensland, Australia), kyanite is overgrowing andalusite. Scale bar = 0.5 mm (XPL).

not be confused with fine-grained sericitic rims and pseudomorphs commonly associated with the retrogression of aluminium-silicate phases (Fig. 7.6). Direct overgrowth of one Al_2SiO_5 polymorph by another (Fig. 7.9) is rarely seen in metamorphic rocks, but more normally two (and sometimes three) polymorphs can exist metastably together. The reason for the coexistence of one or more of the polymorphs is largely due to the small differences in Gibbs free energy between each of them. This commonly leads to the metastable persistence of one of the phases outside its stability field.

The term EPITAXIAL OVERGROWTH (or EPITAXY) is used to describe cases of coherent overgrowth, in which a new phase preferentially nucleates on a crystallographically similar pre-existing phase. This occurs because structural similarities between substrate phase and overgrowth phase minimise the surface energy and allow easier nucleation. The example of one type of amphibole preferentially nucleating and growing on another (Plate 4(a)), is a good example of such overgrowth. It can occur as both a prograde and retrograde feature, depending on the amphiboles involved and their interrelationships.

A final example of a prograde overgrowth or replacement feature is the case of sillimanite (fibrolite) + biotite as a prograde pseudomorphic replacement of garnet (e.g. Yardley, 1977; Diella *et al.*, 1992). Such matted aggregates or 'knots' of fibrolitic sillimanite are often referred to as *FASERKIESEL*. This replacement of garnet by biotite and fibrolitic sillimanite produces pseudomorphs appreciably enriched in Al and depleted in Fe, indicating extensive local-scale diffusive mass transfer, in a complex 'Carmichael-type' ionic reaction (Section 1.3.3).

References

Allen, C.M. (1992) Mineral definition by HRTEM: problems and opportunities, in *Minerals and reactions at the atomic scale: transmission electron microscopy* (ed. P.R. Buseck). Mineralogical Society of America, Reviews in Mineralogy No. 27, Mineralogical Society of America, Washington, DC, Ch. 8, 289–333.

Beach, A. (1980) Retrogressive metamorphic processes in shear zones with special reference to the Lewisian Complex. *Journal of Structural Geology*, 2, 257–263.

Boulter, C.A., Fotios, M.G. & Phillips, G.N. (1987) The Golden Mile, Kalgoorlie; a giant gold deposit localized in a ductile shear zone by structurally induced infiltration of an auriferous metamorphic fluid. *Economic Geology*, 82, 1661–1678.

Carten, R.B. (1986) Sodium–calcium metasomatism: chemical, temporal, and spatial relationships at the Yerrington, Nevada, porphyry copper deposit. *Economic Geology*, 81, 1495–1519.

Chopin, C. (1984) Coesite and pure pyrope in high-grade blueschists of the Western Alps: a first record and some consequences. *Contributions to Mineralogy and Petrology*, 86, 107–118.

De Jong, G. & Williams, P.J. (1995) Giant metasomatic system formed during exhumation of mid-crustal Proterozoic rocks in the vicinity of the Cloncurry Fault, northwest Queensland. *Australian Journal of Earth Sciences*, 42, 281–290.

Diella, V., Spalla, M.I. & Tunesi, A. (1992) Contrasting thermochemical evolutions in the Southalpine metamorphic basement of the Orobic Alps (Central Alps, Italy). *Journal of Metamorphic Geology*, 10, 203–219.

Jiang, W.-T. & Peacor, D.R. (1991). Transmission electron microscopic study of the kaolinization of muscovite. *Clays and Clay Minerals*, 39, 1–13.

Jiang, W.-T. & Peacor, D.R. (1994). Formation of corrensite, chlorite and chlorite–mica stacks by replacement of biotite in low-grade pelitic rocks. *Journal of Metamorphic Geology*, 12, 867–884.

Merriman, R.J., Roberts, B., Peacor, D.R. & Hirons, S.R. (1995) Strain-related differences in crystal growth of white mica and chlorite: a TEM and XRD study of the development of metapelitic microfabrics in the Southern Uplands thrust terrane, Scotland. *Journal of Metamorphic Geology*, 13, 559–576.

Okay, A.I. (1995) Paragonite eclogites from Dabie Shan, China: re-equilibration during exhumation? *Journal of Metamorphic Geology*, 13, 449–460.

Oliver, N.H.S. (1995) Hydrothermal history of the Mary Kathleen Fold Belt, Mt. Isa Block, Queensland. *Australian Journal of Earth Sciences*, 42, 267–279.

Peacock, S.M. (1987) Serpentinization and infiltration metasomatism in the Trinity peridotite, Klamath province, northern California: implications for

References

subduction zones. *Contributions to Mineralogy and Petrology*, **95**, 55–70.

Seyfried, W.E., Jr., Bernt, M.E. & Seewald, J.S. (1988) Hydrothermal alteration processes at mid-ocean ridges: constraints from diabase alteration experiments, hot-spring fluids, and composition of the oceanic crust. *Canadian Mineralogist*, **26**, 787–804.

Thompson, A.J.B. & Thompson, J.F.H. (1996) *Atlas of alteration (a field and petrographic guide to hydrothermal alteration minerals)*. Geological Association of Canada (Mineral Deposits Division), St. Johns, Newfoundland, Canada, 119 pp.

Veblen, D.R. & Ferry, J.M. (1983) A TEM study of the biotite–chlorite reaction and comparison with petrologic observations. *American Mineralogist*, **68**, 1160–1168.

Warr, L.N., Primmer, T.J., & Robinson, D. (1993) Variscan very low-grade metamorphism in southwest England: a diastathermal and thrust-related origin. *Journal of Metamorphic Geology*, **9**, 751–764.

Wegner, W.W. & Ernst, W.G. (1983) Experimentally determined hydration and dehydration reaction rates in the system $MgO-SiO_2-H_2O$. *American Journal of Science*, **283-A**, 151–180.

Yardley, B.W.D. (1977) The nature and significance of the mechanism of sillimanite growth in the Connemara Schists, Ireland. *Contributions to Mineralogy and Petrology*, **65**, 53–58.

Zoltai, T. & Stout, J.H. (1984) *Mineralogy: concepts and principles*. Burgess, Minneapolis, Minnesota.

Part C

Interrelationships between deformation and metamorphism

Chapter eight

Deformed rocks and strain-related microstructures

Various effects of deformation, recovery and recrystallisation can be seen in thin sections of deformed metamorphic rocks. The following sections give an introduction to such features, how they form and how they can be recognised. For further details, the texts and edited volumes by Poirier (1985), Barber & Meredith (1990), Knipe & Rutter (1990), Boland & FitzGerald (1993), Passchier & Trouw (1996) and Snoke et al. (in press), plus the review by Green (1992) on TEM analysis of deformation in geological materials, are all invaluable sources of reference.

8.1 Deformation mechanisms

When a rock undergoes deformation due to some superimposed stress, the mineral constituents of that rock may deform in either a brittle manner (Fig. 8.1) by fracturing (= CATACLASTIC FLOW), or in a ductile fashion by crystal–plastic processes (Fig. 8.2). The way in which a particular mineral deforms is influenced by many factors, but especially by temperature and strain rate. As a general rule, lower temperatures and higher strain rates favour brittle deformation of minerals, while higher temperatures and lower strain rates promote ductile deformation. That is not to say that the change in deformation style occurs simultaneously in all minerals. The different crystallographic properties of individual phases mean that under a given set of conditions some phases may undergo plastic deformation while others deform in a brittle manner. Both hornblende and feldspar deform in a brittle manner at low to moderate temperatures, whereas at high temperatures crystal–plastic processes operate. However, in the case of quartz, deformation by crystal–plastic processes is characteristic over a much broader range of conditions. This means that in granitoid rocks sheared at low temperatures, quartz will exhibit features of ductile deformation, while feldspars show brittle fragmentation (Fig. 8.3).

8.2 Inter- and intracrystalline deformation processes and microstructures

8.2.1 Defects

Plastic deformation of crystals is facilitated by lattice defects, of which there are three main

Deformed rocks and strain-related microstructures

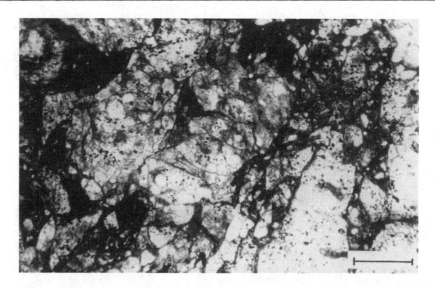

FIG. 8.1 The early stages of brecciation in sub-greenschist oolitic ironstone, Gwna Group, Anglesey, Wales. Scale bar = 1 mm (PPL).

FIG. 8.2 Mylonite (an example of crystal–plastic (ductile) processes). Mylonitised granite, Baltic Shield, Sweden. The coin is 22 mm in diameter.

classes; namely, POINT DEFECTS, DISLOCATIONS and GRAIN BOUNDARIES. Point defects (Fig. 8.4) can be subdivided into those that represent vacant sites in the crystal lattice (VACANCIES), and those representing extra atoms or molecules in the lattice (INTERSTITIALS). Point defects may migrate through the crystal lattice by diffusive processes involving exchange with neighbouring ions in a manner obeying Fick's laws of diffusion (Section 1.3.5).

Inter- and intracrystalline processes

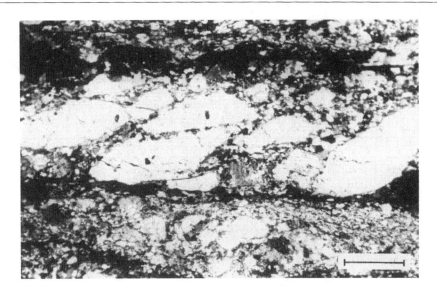

FIG. 8.3 Protomylonite. Low-temperature deformation of this granitoid rock has produced grain-size reduction of quartz by ductile shearing, while giving rise to brittle fragmentation of feldspars (centre) in the same assemblage. Troms, Norway. Scale bar = 1 mm (XPL).

DISLOCATIONS are thermodynamically unstable linear defects along which some slip has occurred (see below for details). In chemically homogeneous material (e.g. pure ice, salt, and so on), GRAIN BOUNDARIES are effectively two-dimensional defects separating grains the lattices of which are differently oriented. In polycrystalline rock, the grain-boundary region is somewhat more complex, but in essence can be viewed as a complex two-dimensional defect (see below).

8.2.2 Dislocations

Dislocations are contained within the crystal structure, and concentrated at grain boundaries. Such defects distort the lattice of the crystal and introduce an internal strain. There are two main types of dislocations. These are EDGE DISLOCATIONS (Fig. 8.5(a)), where the crystal has an additional half lattice plane, and SCREW DISLOCATIONS (Fig. 8.5(b)), where part of the crystal is displaced by a lattice unit, giving a twisted lattice at the line of

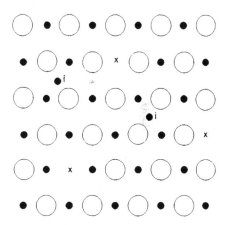

FIG. 8.4 Point defects in a crystal lattice. x, vacancies; i, interstitials.

dislocation, but elsewhere the lattice planes line up.

The greater the number of dislocations, the greater is the internal energy (stored elastic energy), and consequently if such dislocations can be eliminated a certain amount of energy will be released. Each dislocation is characterised by

119

Deformed rocks and strain-related microstructures

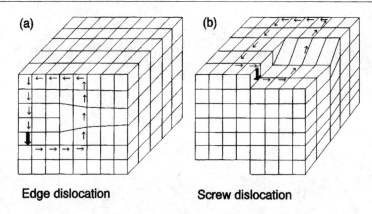

FIG. 8.5 A schematic illustration of different types of dislocation within the crystal lattice: (a) edge dislocations; (b) screw dislocations. Thick arrows define the Burgers vector for each dislocation; small arrows define the dislocation loop or Burgers circuit for each example (see the text for further details).

a slip vector or BURGERS VECTOR, defining the direction and amount of lattice displacement. In the schematic illustration (Fig. 8.5(b)), a 'square' circuit is shown around the dislocation, defined by an equal number of atoms on each side (small arrows in Fig. 8.5(b)). This loop or 'Burgers circuit' is not closed because of the step caused by the dislocation. The connecting line across this step (solid arrow in Fig. 8.5(b)) defines the Burgers vector. Dislocations may split into two or more partial dislocations, which show misfit in relation to the crystal lattice adjacent to the slip plane. This surface defining the zone of mismatch between the partial dislocations and the adjacent ordered crystal lattice is termed a STACKING FAULT. Such microstructural features are especially common in minerals with a large unit cell distance (e.g. orthopyroxene). They preferentially develop whenever the combined energy contribution due to misfit of the partials plus the energy of the stacking fault is less than the strain energy due to the single dislocation (Green, 1992).

Dislocations play a vital role in ductile deformation of rocks. The various minerals present in a deforming rock each have their own SLIP SYSTEMS, representing preferential slip (glide) planes in the crystal lattice, with the slip direction defined by the Burgers vector. In quartz, for example, basal slip is often the dominant mode of deformation. Depending on factors such as grain orientation, temperature and strain rate, a given crystal may have more than one slip system active at any time. If dislocations can propagate fairly freely during ductile deformation, then STRAIN SOFTENING processes will operate. In mechanical terms strain softening is expressed as 'a reduction in stress at constant strain rate, or increase in strain rate at constant stress' (White et al., 1980). GEOMETRIC SOFTENING is one such softening process, and involves grain size reduction via intracrystalline slip and grain reorientation. Not only are the new grains smaller, but they are also strain-free. Since quartz largely deforms by basal slip, this leads to pronounced crystallographic alignment of quartz in strongly deformed quartzofeldspathic mylonites. The basal slip planes become reoriented to lie close to parallel with the shear plane. This alignment can be recognised in thin section by use of the sensitive tint plate (see Sections 8.2.7 & 10.8 and Plates 7(a) & (b) for details).

The intersection of different slip systems will lead to entanglement of migrating dislocations.

Inter- and intracrystalline processes

Such DISLOCATION TANGLES make further deformation of the crystal increasingly difficult, and greatly contribute to overall STRAIN (WORK) HARDENING during deformation of rocks at low temperatures or else relatively fast strain rates. Although the migration of dislocations along a slip plane may be impeded by a tangle, or some other obstruction such as an inclusion, the migration of vacancies to the dislocation plane may allow the dislocation to move at right-angles to the dislocation plane and thus enable it to by-pass the obstruction by a process known as CLIMB. The combination of dislocation glide and climb is referred

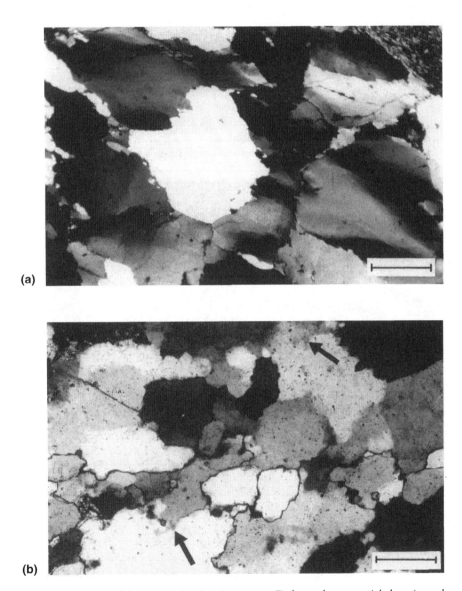

FIG. 8.6 (a) Undulose (or undulatory) extinction in quartz. Deformed quartz-rich lens in a shear zone, 'Pyrite Belt', Spain. Scale = 1 mm (XPL). (b) Sub-grain development (some examples arrowed), in a deformed quartz-rich lens. Pyrite Belt, Spain. Scale = 0.1 mm (XPL).

Deformed rocks and strain-related microstructures

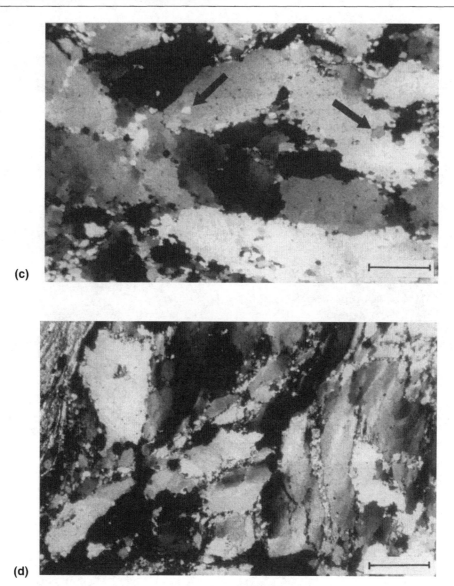

FIG. 8.6 (*contd*) (c) New grains (some examples arrowed), in quartzitic mylonite. Pyrite Belt, Spain. Scale = 0.5 mm (XPL). (d) Mortar ('core-and-mantle') structure developed in mylonite. Pyrite Belt, Spain. Scale = 0.5 mm (XPL).

to as DISLOCATION CREEP, a process of increasing importance at higher temperatures, and one which gives much greater mobility to dislocations (see Section 8.2.3 for details).

In most metamorphic rocks developed in association with deformation (e.g. mylonites, schists and gneisses) UNDULOSE EXTINCTION (= undulatory extinction) is a feature of some or many grains. This arises due to distortion or bending of the crystal lattice, giving a high concentration of dislocations and other defects. It is frequently seen in many pre- and

Inter- and intracrystalline processes

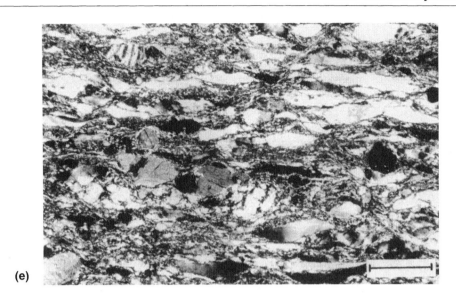

(e)

FIG. 8.6 (*contd*) (e) Ribbon quartz in quartz mylonite. Moine Thrust, N.W. Scotland. Scale = 0.5 mm (XPL).

syntectonic minerals, especially quartz (Fig. 8.6(a)), feldspar, olivine, kyanite and mica. In thin section, UNDULOSE EXTINCTION is recognised by a zone of extinction sweeping across the crystal as the stage is rotated. In quartz, undulose extinction is generally elongate sub-parallel to [0001] (Carter *et al.*, 1964) as a result of heterogeneous slip on this plane during distortion of the lattice. Increased bending leads to the development of discrete DEFORMATION BANDS (kink bands) which have sharply defined high-angle boundaries compared to zones of undulose extinction. They have significant crystallographic mismatch relative to the main crystal, and may terminate either at grain boundaries or inside grains. A detailed study by Mawer & Fitzgerald (1993) on kink band boundaries in quartz of a quartz ribbon mylonite showed that rather than being single high-angle boundaries apparently representing simple rotations of {0001} (as seen with light microscope), 1–2 μm wide strips existed at kink boundaries (TEM study), with an intermediate orientation relative to the two limbs of the kink. From the geometry observed, the conclusion drawn was that at least two slip systems must have operated in the formation of the kink bands.

8.2.3 Creep mechanisms

CREEP processes are extremely important during crystal–plastic deformation of rocks and many other materials. Creep experiments on rocks and ceramics are mostly undertaken at strain rates ($\dot{\varepsilon}$) of 10^{-9} to 10^{-4} s^{-1}, but in natural rock, where features of creep are undoubtedly recorded, the strain rates are much slower, and typically in the range 10^{-15} to 10^{-12} s^{-1}. During deformation experiments it is observed that many materials have a period over which strain rate is constant, known as STEADY-STATE CREEP. At constant applied load, the material being tested often behaves in a manner approximating to constant strain rate. This implies creep with no overall strain hardening, and is achieved by any strain hardening that occurs being counterbalanced by processes of recovery and recrystallisation (Barber, 1990). This type of behaviour, referred to as POWER-LAW CREEP, where

123

$\dot{\varepsilon} \propto \sigma^n$ is characteristic in ceramics and rocks at temperatures above $0.4T_m$ (Barber, 1990), where T_m is the melting temperature (in Kelvin) of the material concerned (note that T/T_m is referred to as the *homologous temperature*). At stresses of geological interest (σ = 1–100 MPa), experimental studies have shown that at moderately high temperatures the behaviour of almost all rocks and minerals is dominated by power-law creep. For example, the experimental research by Ranalli (1982) for olivine assemblages demonstrated a clear power-law relationship, with stress exponent (n) of 3, for $\sigma \leq 200$ MPa.

Having introduced key behavioural aspects of creep, let us consider the main processes involved. Two main creep processes have been identified, namely, DISLOCATION CREEP and DIFFUSION CREEP. In rock deformation it is of course likely that more than one creep mechanism may operate at any given time, but depending on particular conditions one process will usually assume dominance.

DISLOCATION CREEP characterises deformation of rocks at low to moderate bulk shear stress and over a wide range of temperatures and strain rates. It is a slow process involving propagation of dislocations through crystals by a combination of glide and climb. Dislocations may glide rapidly until an obstacle is met which temporarily or finally halts any further propagation. Minor obstacles may be overcome by thermal agitation, while for larger obstacles diffusion-controlled climb may be

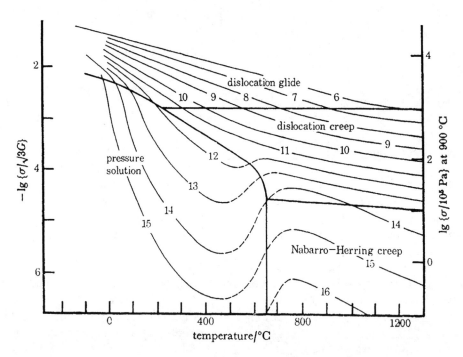

FIG. 8.7 A deformation mechanism map for quartz modified by the addition of a pressure solution field. The region of dashed strain rate contours represents the inhibition of pressure solution through a decrease in pore-water concentration. Grain size diameter, $d = 100$ μm; $V = 22$ cm^3; σ is the differential stress ($\sigma_{11} - \sigma_{33}$); contours of $-$log strain rate (after Fig. 9 of Rutter, 1976). Note that for a larger grain size (e.g. 1 mm), the coble creep/pressure solution field shows appreciable contraction to lower temperature and lower differential stress (see Fig. 7 of White, 1976; courtesy of The Royal Society).

Inter- and intracrystalline processes

required before they can be negotiated or eliminated (Poirier, 1985). Dislocation creep is an important process in low- to medium- grade (e.g. greenschist facies) shear zones, but also at higher metamorphic grades. At lower-temperature, high crustal levels, rock deformation is dominated by brittle failure. The change from brittle failure to crystal–plastic creep processes is referred to as the *brittle– ductile transition*. It is not a sharply defined changeover, because the changeover from brittle to plastic behaviour varies from one mineral to the next, and thus from one rock to the next. For quartz-rich rocks the changeover from dominantly brittle behaviour to dominantly plastic behaviour is often approximated to basal greenschist facies conditions (e.g. Sibson, 1977, 1990).

DIFFUSION CREEP is an important process at various strain rates, over a wide range of geological temperatures and shear stresses. Under such conditions it provides the driving force for grain-boundary sliding. It is especially important in rocks of small grain size, whereas in coarse-grained rocks dislocation creep dominates over a wider range of conditions (White, 1976). Diffusion creep can be subdivided into NABARRO–HERRING CREEP and COBLE CREEP. NABARRO–HERRING creep is the dominant process at high temperatures and low shear stress (Fig. 8.7). By a combination of grain-boundary sliding and diffusive transport of matter through the crystal lattice and along grain boundaries, metamorphic rocks deforming at high-temperature conditions (e.g. granulite facies) can experience major microstructural transformations involving grain shape changes and rearrangement without intergranular cracks opening up. COBLE CREEP is fluid-absent grain-boundary diffusion. It typifies lower-temperature conditions and a wide range of shear stress conditions. Because of the lower-temperature conditions, grain-boundary diffusion predominates over lattice diffusion; indeed, lattice diffusion becomes extremely inefficient below 500°C. Grain-boundary diffusion in the presence of a fluid is termed PRESSURE SOLUTION, and it is dominant during diagenesis and low-grade metamorphism (Fig 8.7). The process involves stress-induced solution transfer of material down chemical potential gradients along grain boundaries. In this case, the superimposed stress causes material to be taken into solution at high-solubility sites and transported to low-solubility sites where it is precipitated. The distance over which transport occurs can vary considerably, such as very local transport from high- to low-stress boundaries of individual grains, to transfer over greater distances, depositing material in veins (Chapter 11) or even taking material out of the local system.

Solution transfer processes are important in the development of crenulation cleavages (Section 4.4). In Fig. 4.6(b), a discrete crenulation cleavage is defined by thin dark pressure solution seams of insoluble carbonaceous material. STYLOLITES (Fig. 8.8) are irregular, serrated or jagged pressure solution surfaces, which are particularly common in massive carbonate and quartzite units from diagenetic to greenschist facies conditions. They generally develop perpendicular to principal compressive stress and are commonly defined by thin (typically 0.5–3.0 mm) seams of dark insoluble material. Stylolites are often considered to develop during compaction in the early stages of diagenesis (Park & Schot, 1968). However, tectonic stylolites are also recognised. These cross-cut bedding, and may intersect and offset tectonic features such as veins (Ramsay & Huber, 1983).

8.2.4 Grain boundaries

Grain boundaries are the regions of contact between adjacent crystals. To be classed as true grain boundaries, the crystallographic misorientation of adjacent grains must be greater than 10°. Low-angle (<10° boundaries) are termed sub-grains, and characterise recovery

125

Deformed rocks and strain-related microstructures

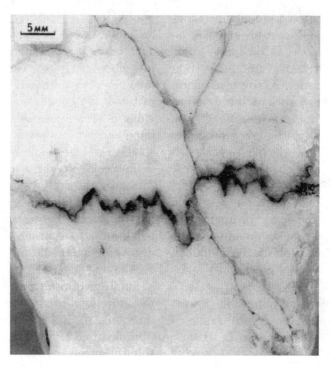

FIG. 8.8 Stylolite development in vein quartz. Ashanti, Ghana. Scale bar = 5 mm.

processes (see below). Grain-boundary conditions in polymineralic rocks are considerably more complex than those of chemically homogeneous materials. Even in largely monomineralic rocks such as quartzites and marbles, there are likely to be some impurities. The role of grain boundaries as sites for nucleation and growth of new minerals has already been discussed in Section 5.1.

In metallurgy, the structure of grain boundaries is relatively well studied, but the study of boundaries between rock-forming minerals has only received detailed attention since the early 1980s. White & White (1981) described the grain-boundary regions of deforming rocks as disordered regions comprising tubules at grain triple-junctions, that can form an interconnected network through the grain aggregate, isolated microscopic voids (μm-scale) along grain boundaries, and a thin (nm-scale) film of distorted crystal structure (possibly with fluid present) connecting the various voids and tubules (Fig. 8.9). In addition, mineral inclusions are commonly observed along grain boundaries. Work by Watson & Brenan (1987) and others has shown that, depending on the wetting characteristics of the fluid concerned, the grain-boundary fluid may form a totally connected network or else be isolated at grain triple-junctions or as tubes and inclusions along the grain boundary. Empty or fluid-filled tubes and 'ellipsoidal' inclusions are commonly recorded along grain boundaries in various minerals (e.g. Spiers *et al.* (1990) for *rocksalt*; and Craw & Norris (1993) for *vein quartz),* suggesting that fluids utilise grain-boundary regions and have a role in grain-boundary processes.

The lattice misorientation introduced by grain boundaries can be considered as the misorientation introduced by a planar array of dislocations. During deformation and metamorphism, the configuration of grain bound-

Inter- and intracrystalline processes

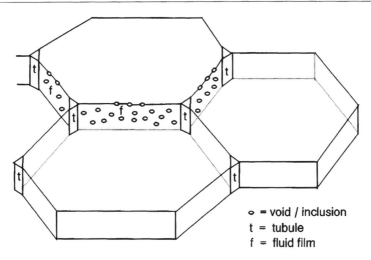

FIG. 8.9 A schematic illustration of the microstructure of grain-boundary regions (modified from Fig. 4 in White & White, 1981). Fluid may reside as μm-scale inclusions along grain boundaries (possibly connected by nm-scale fluid film), and as larger tubules (t) at grain triple-junctions.

aries becomes modified. The two most important processes in operation are those of GRAIN-BOUNDARY SLIDING and GRAIN-BOUNDARY MIGRATION.

GRAIN-BOUNDARY SLIDING is movement within the plane of the grain boundary. It is a process that typifies regimes of creep by diffusional flow, and is generally accompanied by grain-boundary diffusion of material via a fluid film (see above). It occurs under applied shear stress and is most prevalent in the deformation of fine-grained aggregates at elevated temperatures and low stresses. It can be considered in terms of the physical movement of individual grains past each other, and is largely achieved by the climb and glide of grain-boundary dislocations, often with an accompanying degree of boundary migration. Analogue experiments using octachloropropane (e.g. Ree, 1994) provide some evidence for the nature of grain-boundary sliding and void development, but perhaps the best evidence for grain-boundary sliding in natural materials comes from experimental work such as that of Walker et al. (1990), on deformation of synthetic calcite polycrystal aggregates. On the basis of their experiments, Walker et al. (1990) concluded that in situations of low shear stress, grain-boundary sliding is the dominant deformation mechanism in calcite polycrystal aggregates. It is achieved largely by dislocation processes, but with a certain degree of diffusive mass transfer.

GRAIN-BOUNDARY MIGRATION, involving movement normal to the plane of the grain boundary, is stress-induced but aided greatly by elevated temperatures. The principal factors controlling the *rate* of grain-boundary migration are temperature, lattice orientation and minor phases or impurities within the aggregate of grains. Temperature is important because at higher temperatures diffusive processes become more important and thus atomic rearrangement in the grain-boundary region is easier. Lattice orientation is also found to have an important influence on grain-boundary migration, with those grains with the same or very similar orientation having the least mobile boundaries. The role of boundary migration during grain coarsening, which minimises the total free energy by decreasing the surface energy contribution, was discussed in Section 5.3, and will not be reiterated here.

The influence of minor phases in the aggregate is clear from Fig. 5.17(b), where micas have 'pinned' quartz boundaries and thus inhibited migration during recrystallisation.

8.2.5 Recovery

RECOVERY includes an important set of processes that decrease the stored elastic energy of the system. A deformed crystal has increased internal energy relative to its undeformed state due to dislocations contained within the lattice. The internal energy increase is directly proportional to the dislocation density (= combined length of dislocations per unit volume) of the crystal. The lowering of energy by elimination or ordering of dislocations by propagation into existing grain boundaries and voids coupled with migration and climb of randomly arranged dislocations into stable arrays ('walls') at a high angle to active glide planes is an important part of the recovery process. Such ordering leads to the development of strain free SUB-GRAINS in larger crystals (Fig. 8.6(b)).

SUB-GRAINS are defined as areas with misorientations of a few degrees relative to the parent grain, and separated from the parent grain by dislocation walls. Because of the small difference in optical orientation between sub-grain and parent grain, such boundaries are not sharply defined. Whole grains can be converted to a mosaic of sub-grains, but especially common is the concentration of sub-grains at grain margins. Their size varies from those sub-grains clearly visible by standard microscopy (Fig. 8.6(b)) to those minute sub-grains only visible by SEM/TEM studies. They develop during primary creep after relatively little strain, and are a clear sign that RECOVERY processes have operated. Experimental studies by Pontikis & Poirier (1975) and Ross et al. (1980) have demonstrated an empirical relationship between sub-grain size and superimposed stress, the sub-grain size decreasing as the applied stress increases. This in turn has been used to estimate the stress responsible for rock deformation in natural examples, since it appears that with stress decrease the sub-grain size remains stable and is not modified. This being the case, sub-grain size should be representative of maximum stress experienced by the mineral, and thus offers considerable potential as a palaeopiezometer.

Because fluid inclusions represent imperfections within crystals, they increase the internal energy of such crystals. TEM studies (e.g. Reeder (1992) for *carbonates;* and Bakker & Jansen (1994) for *quartz),* have shown that fluid inclusions often have a close association with dislocations (Fig. 11.16), and since dislocations may be eliminated by propagation towards grain boundaries, the elimination of fluid inclusions in a similar manner has often been suggested as part of the recovery process, but especially during recrystallisation. An important aspect of the experimental study of Bakker & Jansen (1994) was the recognition that during recovery minute quantities of fluid leaked from the micron-scale fluid inclusions, without rupturing the original inclusion. TEM studies showed that the leakage was facilitated by dislocations, which display numerous nm-scale 'bubbles' of leaked fluid along their length. Such features are of course far too small to be recognised with standard optical microscopy. Bakker & Jansen (1994) argue that bubble nucleation on dislocations makes an important contribution to recovery, because each bubble eliminates part of the dislocation, and thus the elastically strained atoms around that part of the dislocation. In so doing, the internal energy of the crystal is diminished. The experimental work of Gerretsen *et al. (1993)* also demonstrated that dislocation generation accompanying re-equilibration of fluid inclusions plays an important role in the deformation of 'wet' synthetic quartz.

DEFORMATION LAMELLAE (Fig. 8.10) are narrow crystallographically oriented

Inter- and intracrystalline processes

FIG. 8.10 Deformation lamellae in quartz from deformed metaconglomerate. Cherbourg region, Normandy, France. Scale bar = 125 μm (XPL).

planar features of <10 μm width that are commonly seen in deformed quartz (e.g. Carter et al., 1964; Drury, 1993). They are visible with the light microscope, and until the mid-1970s, were considered to represent dislocation slip bands. On this basis, they were used to evaluate the dominant slip systems operating at particular strain rate and P–T conditions. TEM studies by McLaren et al. (1967) and others revealed that many of the observed defect substructures could be directly equated with the fine lamellae seen with an ordinary petrological microscope. Experimentally deformed quartz lamellae are typically defined by dislocation slip bands, Brazil twins and zones of glass. However, the review by Drury (1993) points out that TEM studies of naturally produced lamellae usually show them to be defined by elongated sub-grains, sub-grain walls, and zones with variable densities of dislocations and sub-micron fluid inclusions ('bubbles'). In view of this, Drury (1993) concluded that natural deformation lamellae cannot be used to determine which slip systems may have operated, but that the presence of sub-basal deformation lamellae in quartz can be interpreted in terms of dynamic recovery of dislocations initially present in slip bands.

8.2.6 Recrystallisation

RECRYSTALLISATION is the natural progression from recovery processes, and minimises the energy of the system still further, by stress-induced grain-boundary migration. This amalgamates smaller grains into larger ones, so reducing the surface energy of the system, and eliminates dislocations within crystals to reduce the internal energy. These changes create a more stable arrangement of grains and grain boundaries, and thus decreases the energy of the system as a whole. Elevated temperatures and the presence of a grain boundary fluid greatly aid such transformations.

DYNAMIC RECRYSTALLISATION is the term used for recrystallisation synchronous with deformation. An important component of

Deformed rocks and strain-related microstructures

dynamic recrystallisation is sub-grain rotation. It was described above how, as part of the recovery process, dislocations migrate to form stable 'walls' which define boundaries of sub-grains. These sub-grains have lattice misorientations of a few degrees relative to the parent grain, but if climb-accommodated dislocation creep continues, the further addition of dislocations at the sub-grain boundaries leads to gradual rotation of the sub-grain and increasing mismatch between sub-grain and parent grain. Once a high-angle contact (>10°) has been established, the term 'sub-grain' is no longer appropriate, and the term NEW GRAIN is used. New grains are a characteristic feature of dynamic recrystallisation and, although similar to sub-grains, they can be distinguished by virtue of their sharp contacts with adjacent grains (Fig. 8.6(c)). In cross-polarised light, this is clearly seen by abrupt changes in interference colours between adjacent grains, compared to the slight and gradual changes associated with sub-grains. In sheared quartzites and quartzofeldspathic rocks CORE-AND-MANTLE STRUCTURE (= MORTAR STRUCTURE) is commonly observed. This consists of large strained porphyroclasts of quartz or feldspar surrounded by a fine-grained aggregate of recrystallised new grains (Fig. 8.6(d)). Similar features of dynamic recrystallisation are also observed in certain sheared amphibolites, eclogites and marbles. In such cases, the porphyroclasts and fine-grained aggregates are of hornblende, clinopyroxene and calcite respectively.

Another mechanism by which NEW GRAINS are produced during dynamic recrystallisation involves grain-boundary migration to isolate lobes of irregular or serrated grain boundaries. Irregularly SUTURED or SERRATED boundaries initiate by boundary migration, causing BULGING of one grain into its neighbour in order to eliminate areas with high dislocation density. In such cases the grain with the more stable, low dislocation density margin always bulges into the grain with less stable high dislocation density boundary, so reducing the internal energy of the system. This bulging of grain boundaries can ultimately give

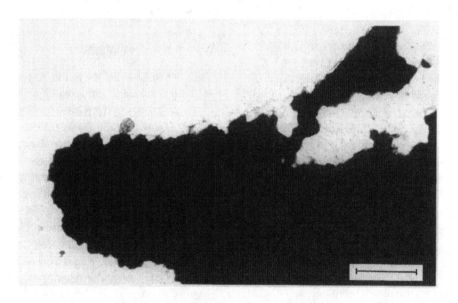

FIG. 8.11 Bulging of grain boundaries in recrystallised vein quartz from Snake Creek, Queensland, Australia. Scale bar = 0.5 mm (XPL).

▲ Plate 1a

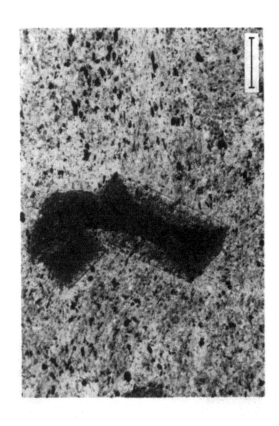

▲ Plate 1b

▼ Plate 1c

▼ Plate 1d

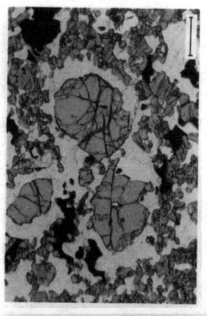

◀ Plate 2a

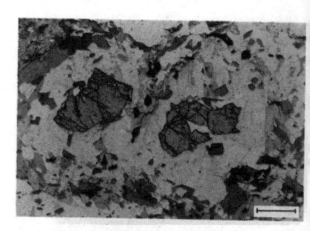

▲ Plate 2b

▼ Plate 2c

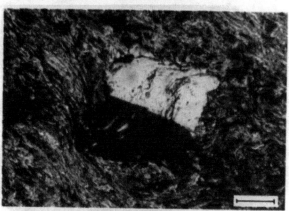

Plate 2d ▶

◀ Plate 2e

▼ Plate 2f

◀ Plate 3a

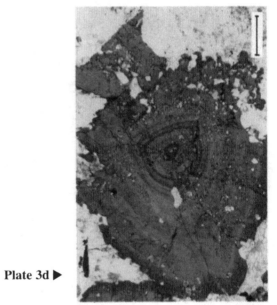

▲ Plate 3b

▼ Plate 3c

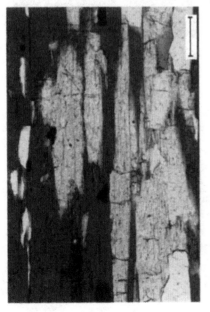

Plate 3d ▶

◀ Plate 4a

▼ Plate 4b

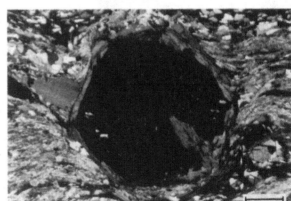

◀ Plate 4c

▲ Plate 4d

▼ Plate 4e

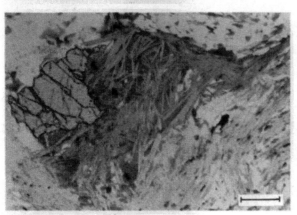

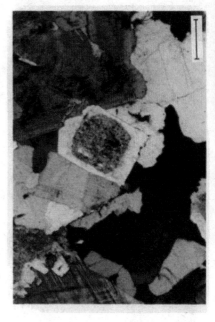

Plate 4f ▶

◀ Plate 4g

▼ Plate 4h

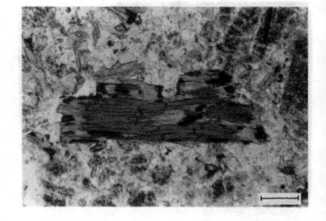

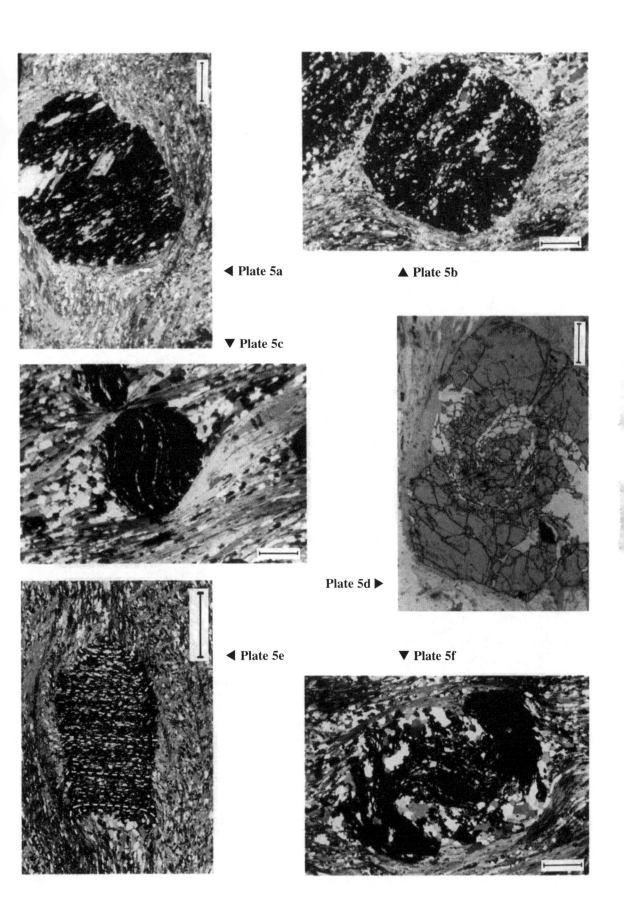

◀ Plate 5a ▲ Plate 5b

▼ Plate 5c

Plate 5d ▶

◀ Plate 5e ▼ Plate 5f

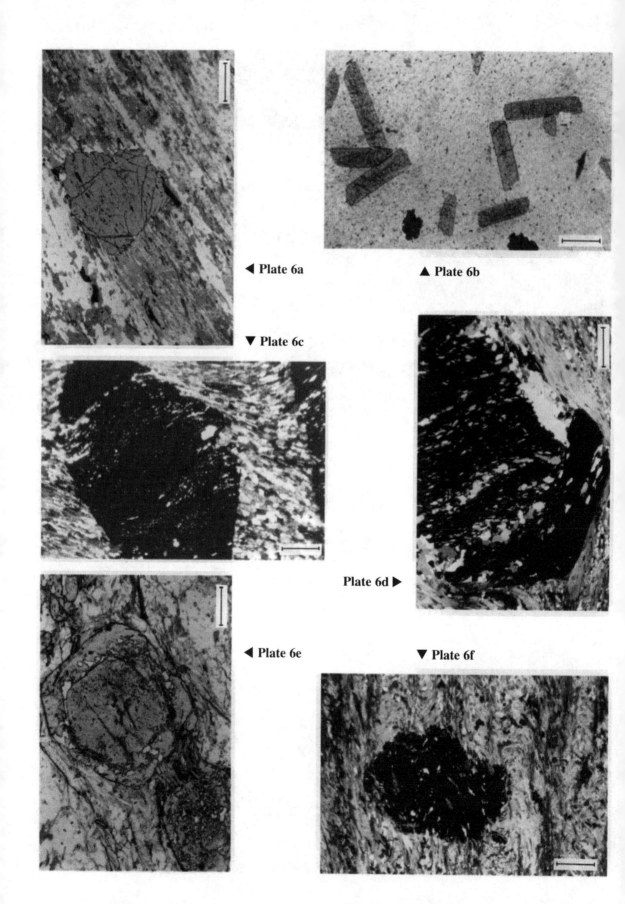

◀ Plate 6a
▲ Plate 6b
▼ Plate 6c
Plate 6d ▶
◀ Plate 6e
▼ Plate 6f

▲ Plate 7a   ▼ Plate 7c   ▲ Plate 7b   ▼ Plate 7d

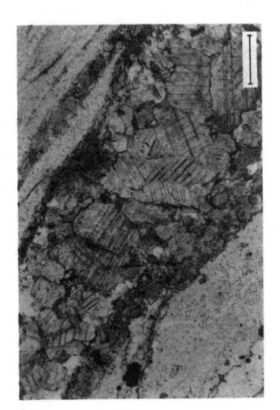
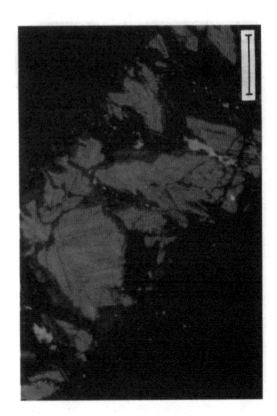

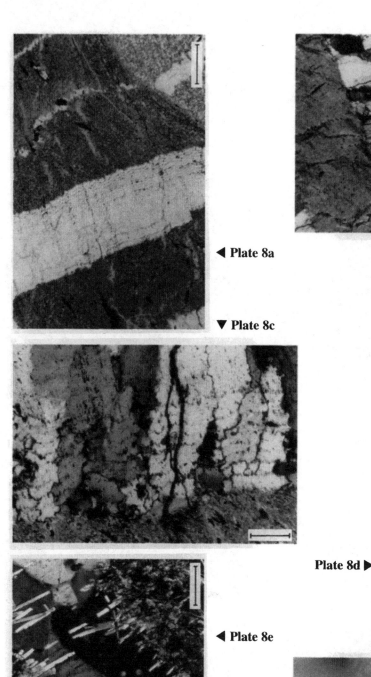

◀ Plate 8a

▼ Plate 8c

▲ Plate 8b

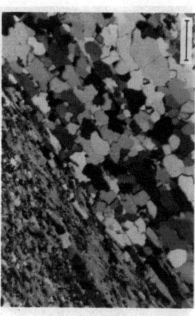

Plate 8d ▶

◀ Plate 8e

▼ Plate 8f

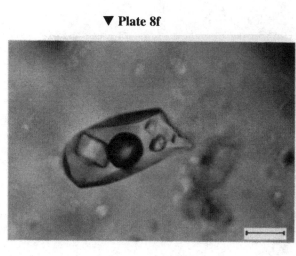

rise to new grains, especially if the bulge develops a narrow neck (Fig. 8.11). The neck becomes the likely site for sub-grain wall development, and if rotation occurs a new grain with a high-angle boundary forms. For further details on dynamic crystallisation, the reader is referred to the paper by Drury & Urai (1990).

STATIC RECRYSTALLISATION occurs post-deformation and takes place in response to elevated temperature conditions promoting further re-equilibration and energy reduction. The principal way in which this is achieved is by lowering the surface energy of the system by reducing total grain-boundary area. This involves grain-boundary migration to smooth out irregular grain boundaries within the aggregate, and to eliminate small grains by amalgamation. Elimination of dislocations and other lattice defects is an integral part of the process, and if the aggregate is completely recrystallised, crystals showing evidence of strain (e.g. undulose extinction) should be absent. A polygonal-equigranular aggregate (e.g. Fig. 5.14(b)) is a prime example of the microstructural arrangement characteristic of static recrystallisation.

8.2.7 Crystallographic-preferred orientations

Crystallographic (lattice)-preferred orientations (CPOs) are a common feature of highly deformed rocks, and in many cases the CPO has a close relationship with the grain shape orientation. For example, biotite grains in schists show a strong alignment of their {001} planes to define the schistosity, and since the {001} planes contain the long axes, the grain shape orientation is often aligned, especially in L–S tectonites (Section 4.3; Fig. 4.4). However, in other cases, most notably highly recrystallised mylonitic rocks, the original grain-shape fabric may have been largely lost during recrystallisation processes, but a residual CPO may still remain.

In the case of granular aggregates of more or less equant grains (e.g. quartz, calcite and olivine), dislocation creep (Section 8.2.3) is an important process in changing crystal shape, as well as orientation relative to neighbouring grains. Deformation twinning will also play a role in the development of both grain-shape fabrics and CPOs. If an aggregate of grains with random crystallographic orientation exists prior to deformation, when subjected to a certain amount of deformation (whether coaxial or non-coaxial), there will be a gradual change to some form of CPO. Different minerals have different slip and deformation-twin systems. For some minerals relatively few slip systems operate, whereas for others numerous slip systems may operate, and thus the resultant CPO after deformation will be less straightforward to interpret. It can be appreciated that the nature of CPOs in polyphase aggregates is more complex because of the interaction of crystals with different slip systems. The precise conditions at the time of deformation will influence which slip systems are active, and as a general rule increasing differential stress will increase the number of active slip systems. The more intense and prolonged the deformation is, the greater the tendency is for a more pronounced CPO. However, this may not be entirely true, because the degree of dynamic recrystallisation may serve either to weaken or improve the CPO. Jessell & Lister (1990) examined the influence of temperature on quartz fabrics.

It is outside the scope of this text to discuss the full range of CPOs that have been identified in relation to different mineral aggregates and in different deformation regimes. However, for further insight, useful overviews are given by Law (1990) and Passchier & Trouw (1996). Passchier & Trouw (1996) also provide a useful description of the procedure for use of the universal stage (U-stage) in order to measure quartz or calcite c-axes, the two minerals for which CPOs have been most extensively studied. Although CPOs have a number of potential uses for interpreting the structural evolution of a particular rock, the complexity of deformation often makes it difficult to unravel the story, especially if late-stage recrystallisation has been strong. One of the most useful applications (especially quartz *c*-axes) has been in the interpretation of amount and sense of shear during non-coaxial deformation (Section 10.8 and Plates 7(a) & (b)).

8.3 Fault and shear zone rocks and their microstructures

Rocks of fault zones and shear zones can be subdivided into foliated types (i.e. MYLONITES and PHYLLONITES) and non-foliated types (termed CATACLASITES), including such rocks as FAULT BRECCIAS.

At high crustal levels localisation of high strain rates gives rise to brittle faulting. With the exception of fault gouge, which is often well foliated, faulting at high crustal levels generates non-foliated fault rocks consisting of variable-size rock fragments in a finer grained matrix. Sibson (1977) draws the distinction between cohesive and incohesive fault rocks. On the basis of the proportion of matrix, he subdivides incohesive non-foliated fault rocks into FAULT BRECCIAS (>30% visible fragments) and FAULT GOUGE (<30% visible fragments). Cohesive non-foliated fault rocks are also subdivided on the basis of matrix proportion and range from 'CRUSH BRECCIAS' (0–10% matrix) through PROTOCATACLASITE (10–50% matrix), CATACLASITE (50–90% matrix) to ULTRACATACLASITE (90–100% matrix). CATACLASIS is a process involving brittle fragmentation and rotation of mineral grains. During the development of a cataclasite this is accompanied by grain-boundary sliding and diffusive mass transfer mechanisms (e.g. Lloyd & Knipe, 1992). At deeper levels crystal–plastic processes operate in ductile shear zones to generate mylonites and related foliated rocks. As discussed in an earlier section, the so-called ''brittle–ductile' transition occurs at approximately 300°C (10–15 km depth) for quartz-rich rocks, which corresponds broadly with the lower-temperature boundary of greenschist facies conditions. In broad terms, this approximates to the base of the shallow seismogenic zone of the crust (seismic–aseismic transition). It should be emphasised that the depth of transition will not be the same for all rocks, but will vary according to their rheological properties. Rutter (1986) makes the important point that broad use of the term 'brittle–ductile' transition can be potentially misleading, since 'ductility' is not mechanism dependent, but simply reflects substantial non-localised strain. He advocates expressions such as 'brittle–plastic' or 'cataclastic–plastic' in order to give a precise indication of the change in deformation mechanism associated with mode of failure transition in rocks.

Wise *et al.* (1984) introduced a classification of deformed rocks in terms of strain rate versus recovery rate. The rocks thus far described are all placed in the field of high strain rate and low recovery rate. Rather than use the term 'crush breccia' to describe coherent non-foliated fault rocks with <10% matrix, terms such as 'silicified fault breccia' and 'carbonate-cemented fault breccia' (depending on the nature of the cement) are preferred. This is due to the fact that fault breccia is already widely used in the literature for both coherent and incoherent rocks of this type.

A special type of cataclasite formed by rapid fault movement at high crustal levels (e.g. earthquake-related) is known as PSEUDOTACHYLITE (Fig. 8.12). This localised and relatively rare rock type generally occurs in narrow zones (mm–cm scale) and commonly displays irregular mm-scale injection veinlets off the main surface. It comprises fine fragments in a dark glassy groundmass (black in transmitted light). Sibson (1977) describes it as forming due to rapid movement inducing ther-

Fault and shear zone rocks

mal fragmentation and frictional melting (T probably > 1000°C) under dry conditions at depths greater than 1 km but less than about 10 km. For further details on pseudotachylite microstructures, see Maddock et al. (1987) and Lin (1994).

MYLONITES are rocks of considerable tectonic significance, and ever since their recognition by Lapworth in 1885 have been the focus of attention for many structural and metamorphic geologists. The rock 'mylonite' is best defined as a cohesive, foliated and usually lineated rock produced by tectonic grain-size reduction via crystal–plastic processes in narrow zones of intense deformation. It contains abundant relict crystals, 'porphyroclasts', (10–50%

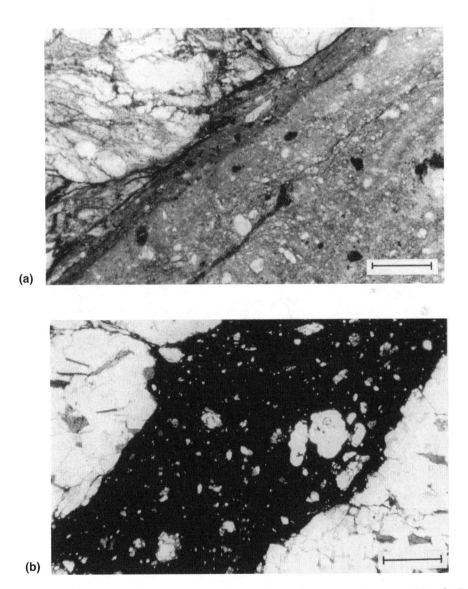

FIG. 8.12 (a) Pseudotachylite. Hetai mine, Guangdong, China. Scale bar = 0.5 mm (PPL). (b) Pseudotachylite cutting Qtz–Pl–Bt gneiss. Butt of Lewis, Scotland. Scale bar = 1 mm (PPL).

Deformed rocks and strain-related microstructures

of the rock) which characteristically are of similar composition to the matrix minerals. It is not a term restricted to a particular compositional range of rocks. Depending on the observed mineralogy, it is thus possible to have 'granitoid mylonites', 'carbonate mylonites', 'amphibolitic mylonites', and so on. Highly sheared rocks dominated by phyllosilicate minerals are termed PHYLLONITES. These are characterised by S–C fabrics formed

FIG. 8.13 (a) Protomylonite: single and polycrystalline quartz porphyroclasts surrounded by a fine-grained matrix of quartz and sericite. Arran, Scotland. Scale = 1 mm (XPL). Note the serrated grain boundaries in the centre of the photograph, the undulatory extinction exhibited by several porphyroclasts, and the deformation bands displayed by the crystal in the top right of the photograph. (b) Mylonite: porphyroclasts of quartz and feldspar in an ultra-fine-grained matrix of quartz and sericite. Ghana. Scale = 1 mm (XPL).

Fault and shear zone rocks

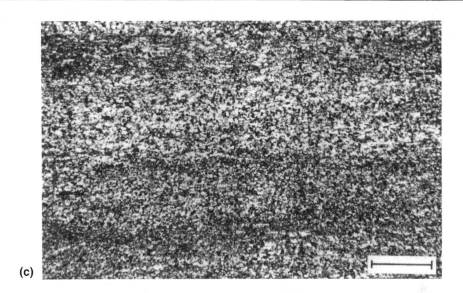

FIG. 8.13 (contd) (c) Ultramylonite. Abisko, Sweden. Scale = 1 mm (XPL). This quartzitic ultramylonite shows a characteristic lack of porphyroclasts and ultra-fine grain size common to many ultramylonites. The quartz shows very good crystallographic alignment (see Plates 7(a) & (b)).

synchronous with shearing. The intersection of these fabrics produces a characteristic 'button schist' or 'oyster-shell' texture (see Section 10.3 for further details).

Mylonites are associated with thrusts and shear zones, generally operating at deeper crustal levels than those responsible for the development of cataclasites. At such depths deformation is more ductile, and crystal–plastic processes predominate. Mylonites are generally associated with relatively high strain rates coupled with appreciable recovery rate. Estimates of the rates of microstructural changes in mylonites have been made by Prior et al. (1990), with particular reference to the Alpine Fault Zone, New Zealand.

Like cataclasites, MYLONITES (sensu lato) can be classified in terms of matrix : porphyroclast ratios. PROTOMYLONITE (Fig. 8.13(a)) consists of abundant (50–90%) clasts, while at the other end of the spectrum highly sheared ULTRAMYLONITE (Fig. 8.13(c)) has <10% clasts, and these are usually small, in a very fine-grained matrix. As a general rule, the recrystallised grain size decreases as differential stress increases. MYLONITE (sensu stricto) (Fig. 8.13(b)) lies between these two end-members and has 10–50% clasts. Those mylonites involving extensive recrystallisation and mineral growth synchronous with shearing are often referred to as BLASTOMYLONITES.

In strongly deformed rocks such as mylonites, cataclasites and gneisses, it is common to observe large relict crystals in a finer-grained matrix. This is known as PORPHYROCLASTIC MICROSTRUCTURE (Figs 8.13(a) & (b)), and the relict crystals as PORPHYROCLASTS. It is important to appreciate the difference between porphyroclastic structure, which consists of large relict crystals, from porphyroblastic structure (Fig. 5.4) which consists of large newly grown crystals.

Porphyroclasts are generally of the same minerals as those present in the matrix. However, the proportions of different phases present as porphyroclasts may differ with respect to the matrix according to the mechanical behaviour of individual minerals under the prevailing conditions. Minerals such as feldspar, kyanite, hornblende and garnet, that deform in a brittle manner over a wide range of conditions, commonly form porphyroclasts, while phyllosilicates and epidote minerals are often more important as matrix constituents. In less sheared rocks the porphyroclasts are generally larger and more angular, becoming smaller and more rounded as deformation intensifies. Depending on the extent of recrystallisation, porphyroclastic rocks such as mylonites commonly display a mix of microstructural features indicative of deformation (e.g. undulose extinction and deformation bands), recovery (e.g. sub-grains and deformation lamellae) and recrystallisation (e.g. new grains and core-and-mantle structure). The nature and attitude of porphyroclasts and their 'tails' can be useful in determining shear sense (Section 10.6).

8.3.1 Deformation of quartzitic and quartzofeldspathic rocks

Experimental work such as that of Tullis et al. (1973) and Hirth & Tullis (1992), and the detailed overview given by Tullis (1990), documents the progressive changes in mechanisms and microstructures of quartzitic and quartzofeldspathic aggregates during deformation under various temperature, strain and strain rate conditions. These transformations in deformation mechanisms and resultant microstructures, while focusing on quartz-rich rocks, are generally applicable to a wide range of materials.

At low temperatures, low strains, but fast strain rates, deformed quartz aggregates display patchy undulose extinction and limited sub-grain development and recrystallisation at grain boundaries. Under these conditions in deforming quartz aggregates, dislocation production is too fast for diffusion-controlled dislocation climb to be an effective recovery mechanism. Instead, the principal mode of recovery is grain boundary migration recrystallisation (Hirth & Tullis, 1992). At increased temperature or decreased strain rate, undulose extinction is pronounced, and sub-grains are a prominent feature at grain boundaries. Under these conditions, dislocation climb has become a more effective recovery mechanism, and dynamic recrystallisation takes place by sub-grain rotation.

In experiments on quartz aggregates at temperatures of approximately 700°C and fast strain rates (10^{-6} s^{-1}), recovery and recrystallisation are seen, facilitated by dislocation creep processes. Some deformation lamellae occur, but the widespread development of small recrystallised new grains at boundaries of original grains becomes a prominent feature. Unrecrystallised grains show undulose extinction instead of sharp deformation bands, have sutured boundaries and become progressively flattened as strain increases. Tullis (1990) points out that at the pressures and strain rates of most experiments dislocation creep processes initiate at much higher temperatures than is the case for natural pressures and low strain rates (e.g. 10^{-14} s^{-1}). At high temperatures (> 800°C), complete, or near complete, recrystallisation is observed. Very few original grains remain, most having been transformed to an aggregate of small recrystallised grains and sub-grains. Vernon (1976) describes how, during experiment, syntectonic recrystallisation of highly deformed quartzite (axial compression > 50%) produces finely recrystallised RIBBON QUARTZ, which is identical to microstructures of many natural quartz–mylonites (Fig. 8.6(e)).

The observations and conclusions from studies focusing on the progressive microstructural changes seen in natural quartz mylonites as a product of ductile shearing and mylonitisation (e.g. White et al., 1980; Knipe, 1990) have much in common with the findings from experimental work. In the early deformation of quartzofeldspathic mylonites, intracrystalline slip occurs and quartz develops undulose extinction. Increased ductile deformation leads to the development of sub-grains, deformation bands and deformation lamellae in quartz, while large feldspar grains show brittle fragmentation and may show deformation twins (Section 5.6.3). Continued recovery and dynamic recrystallisation reduces internal dislocation density by development of sub-grains and the formation of new grains. Serration and new grain development at porphyroclast margins are widespread. If the temperature, or the amount of strain, is high during deformation, elongation of grains will occur to give ribbon quartz texture.

8.3.2 Deformation of mafic rocks

Over the past two decades there have been a number of important studies concerning the deformation processes and microstructures of deformed mafic rocks (Fig. 8.14) over a range of P–T conditions, strains and strain rates (e.g. Brodie & Rutter, 1985; Skrotzki, 1990; Rutter & Brodie, 1992; Lafrance & Vernon, 1993; Stünitz, 1993). The next few paragraphs summarise the main observations and conclusions from these studies.

The dominant minerals of mafic rocks vary according to P–T conditions, and consequently the nature of rock deformation and resultant microstructures varies according to the specific assemblage and the conditions under which deformation occurs. Plagioclase is a major constituent of mafic rocks under all conditions except the eclogite facies (Cpx + Grt assemblages), while hornblende is a key phase at mid- to lower crustal conditions. Pyroxenes assume dominance in the lower crust (granulite facies), whereas towards the upper crust, greenschist facies assemblages such as Chl–Act–Ep–Ab–Qtz are widespread. In mafic and ultramafic rocks of the upper mantle, olivine–pyroxene assemblages dominate, whereas primary igneous rocks such as gabbro have Pl–Cpx assemblages.

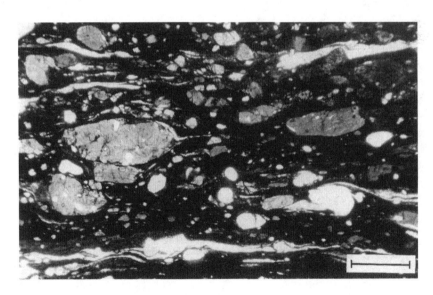

FIG. 8.14 Mafic mylonite: white porphyroclasts are of plagioclase, while other porphyroclasts are of hornblende. Lewis, Scotland. Scale bar = 0.5 mm (XPL).

While an understanding of deformation processes prevalent under peak conditions is important, it is also essential to recognise the microstructures and processes involved in the deformation of high P–T mafic assemblages at different conditions of P, T, stress and strain. At low-temperature conditions, minerals such as pyroxene, hornblende, plagioclase and garnet deform by cataclastic processes, involving microfracturing of individual grains. However, at elevated temperatures and/or strain rates, empirical observation and experimental work has demonstrated that plagioclase is considerably more deformable (softer) than hornblende, pyroxene or garnet. This means that in highly sheared bi-mineralic or polymineralic mafic assemblages, plagioclase often shows evidence of considerable intracrystalline plastic deformation (facilitated by dislocation glide), whereas amphiboles and pyroxenes in the same assemblage display much less evidence for intracrystalline plasticity (e.g. Brodie, 1981; Rutter & Brodie, 1992). In terms of the rock microstructure, this means that while the matrix and porphyroclast phases will be more or less identical, there will be a tendency for a dominance of resistant phases such as hornblende, pyroxene, and/or garnet as porphyroclasts, while softer phases such as plagioclase will be more significantly represented in the flowing matrix (e.g. Rutter & Brodie, 1992).

In addition to mechanical processes, a number of mineralogical (chemical) changes also take place as high-temperature mafic minerals and Ca-plagioclase react to form more stable lower-temperature phases. These retrograde reactions generally take place at greenschist facies conditions in the presence of an aqueous fluid. Indeed, the fluid presence is essential for the formation of hydrous phases. The interrelationships between deformation and metamorphism are discussed more fully in Sections 8.4 & 8.5.

The study by Stünitz (1993) of metagabbros from the Sesia Zone, western Alps, found that as deformation intensifies there is quite contrasting behaviour between the mafic minerals and plagioclase. Irrespective of the intensity of deformation, the mafic minerals deform by fracturing. Synchronous with the deformation, the pyroxene and hornblende crystals break down to actinolite, but also exhibit microfracturing. Fractured plagioclase porphyroclasts are also observed, but the fine-grained products of plagioclase recrystallisation and retrogression (albite and zoisite) are not fractured. These fine-grained aggregates are interpreted as playing a crucial role in initiating a change from deformation, largely by fracturing to bulk deformation predominantly by viscous flow. Stünitz (1993) noted this change in mechanism in the moderately deformed metagabbros with S–C fabrics, but recorded it as most pronounced in the more intensely mylonitic rocks. The shift to viscous flow deformation in the fine-grained albite–zoisite aggregates also led to focusing of subsequent deformation into zones enriched in such minerals (i.e. formerly plagioclase-rich zones), and suggests that plagioclase retrogression, coupled with development of chlorite-rich assemblages, is a crucial part of the 'softening' process in the deformation of mafic rocks at greenschist facies conditions.

Lafrance & Vernon (1993) examined gabbroic mylonites and ultramylonites deformed under low- to mid-amphibolite facies conditions. They recorded extensive recrystallisation of plagioclase by grain-boundary migration, but noted that plagioclase-rich layers have a strong crystallographic preferred orientation. Polygonal pyroxenes at the margins of larger pyroxene porphyroclasts, and the general lack of sub-grains also indicates pronounced high-temperature recrystallisation involving crystal–plastic processes, but amphiboles within the gabbroic mylonites show brittle fragmentation, with a preference for intragranular fracturing along the {110} cleavage. Many of the amphibolite facies metabasic mylonites studied by Lafrance & Vernon (1993) show pronounced (mm-scale) differentiation into layers rich in plagioclase and layers rich in mafic minerals. Although dislocation creep is dominant in amphibolite facies mylonites, fracturing of amphiboles and solution-transfer processes also play an important role. Brodie & Rutter (1985) and Skrotzki (1990) have studied high-temperature ($T > 650°C$) amphibolitic mylonites in shear zones from the Ivrea Zone, north-west Italy. The TEM work of Skrotzki (1990) shows that hornblendes of the shear zone comprise recrystallised grains, subgrains, free dislocations and abundant stacking faults. These microstructures indicate both recovery and recrystallisation and are consistent with dislocation creep being the dominant deformation process. The lack of any significant difference in microstructures between porphyroclasts and matrix suggests widespread dynamic recrystallisation.

8.3.3 Deformation of carbonate rocks

The study of deformed carbonate rocks (both natural and experimental) has been a subject of interest for many decades, but has been the focus of considerable research in recent years (e.g. Dietrich & Song, 1984; Schmid et al., 1987; Wenk et al., 1987; Burkhard, 1990; Walker et al., 1990; van der Pluijm, 1991; Rutter et al., 1994; Rutter, 1995; Busch & van der Pluijm, 1995). With so much literature on the subject, it is difficult to give an adequate review of all the research findings that have been made, but it is hoped that the description below provides a useful summary of some of the observations regarding deformation mechanisms and microstructures over a range of conditions.

Experimental data

A considerable amount of experimental work has been undertaken on Carrara marble and Solnhofen limestone, or else on synthetic pure calcite rock (e.g. Schmid et al., 1987; Walker et al., 1990; Rutter et al., 1994; Rutter, 1995), in order to establish the behaviour of carbonate rocks and the dominant deformation mechanism(s) over a broad range of conditions relevant to geological situations. As well as showing variations according to different conditions of P, T, superimposed stress and strain rates, the behaviour of carbonate rocks is shown to be strongly influenced by grain size.

In fine- and ultrafine-grained carbonate mylonites, the temperature increase from 300°C to 700°C under experimental conditions marks a progressive change from grain-size insensitive crystal–plastic flow processes to grain-size sensitive superplastic flow (e.g. Rutter et al., 1994). The experiments of Rutter et al. (1994) found that although strong low-temperature fabrics in fine-grained aggregates become weakened during high-temperature recrystallisation and crystal–plastic or superplastic flow, they do survive to some extent. In their simple shear experiments on Solnhofen limestone (grain size ≈ 4 μm) and Carrara marble (grain size ≈ 200 μm), Schmid et al. (1987) identified four distinct microfabric regimes. The first of these, the *twinning regime*, was observed from room temperature to 400°C in Solnhofen limestone, and at temperatures ≤ 600°C in Carrara marble at shear stresses > 80 MPa. In these simple shear experiments, most grains displayed a single set of *e*-twins, whereas in coaxial testing conjugate sets are often developed (Schmid et al., 1987). At higher temperatures and/or lower strain rates, Schmid et al. (1987) found that twinning was absent and that a regime of *intracrystalline slip* prevailed, and that original serrate grain boundaries remain as such, with no evidence of grain-boundary migration. Carrara marble commonly displayed core-and-mantle structure, with sub-grains in grain-boundary regions. The marked difference in temperature defining the transition from twinning to intracrystalline slip (≈ 400°C in Solnhofen limestone and ≈ 700°C in Carrara marble, at laboratory strain rates) is directly related to the grain size of the starting material. Thus, coarser-grained calcite rocks display twinning to higher temperatures. In the ultrafine-grained Solnhofen limestone, Schmid et al. (1987) observed a gradual transition into a *grain-boundary sliding* regime at high temperatures (700–900°C), but did not record this in the Carrara marble. In this regime, the observed microstructural features include (a) straight grain boundaries in place of originally serrate boundaries, and (b) weak or completely absent grain-shape fabric. Both of these grain-boundary equilibration features will have involved grain-boundary migration, and there is evidence of grain growth (Schmid et al., 1987). Direct microstructural evidence for grain-boundary sliding was not recorded by Schmid et al. (1987). However, it was inferred as the dominant mechanism in Solnhofen limestone at high temperatures during simple shear, because of microstructural features virtually identical to those seen in the coaxial experiments of Schmid et al. (1977), where rheological and microstructural considerations led the authors to conclude that under the specific high-temperature conditions grain-boundary sliding would be the dominant deformation mechanism. During experiments on Carrara marble at 800–900°C, significant grain growth occurred as a result of dynamic recrystallisation. This was facilitated by *grain-boundary migration*, and represents the fourth regime identified by Schmid et al. (1987).

Rutter (1995) undertook a series of experiments on Carrara marble, with a mean grain size of 130±29 μm, a confining pressure of 200 MPa, a strain rate of 10^{-4} s^{-1} and temperatures in the range 500–1000°C. At temperatures of ≈600°C, deformation twinning was well developed in calcite, and twin boundary migration has been identified as a key process involved during recrystallisation, without modifying grain size. Where stresses are too low for twinning to develop, grain-boundary migration recrystallisation occurs, and leads to overall grain coarsening throughout the rock (Rutter, 1995). This recrystallisation involves the development of nuclei at grain-boundary bulges, and subsequent sub-grain rotation. Such behaviour typified experiments at temperatures in the range 700–900°C, and thus compares well with the findings of Schmid et al. (1987).

With regard to coarser grain sizes, Walker et al. (1990) suggest that at high temperatures, grain-size sensitive flow may extend to coarse calcite aggregates (grain size > 1 mm), such that equigranular calcite aggregates of amphibolite facies marbles may be no different from calcite aggregates formed by static grain growth.

Natural examples

The work of Burkhard (1990) on deformed micritic limestones from the Helvetic nappes, Switzerland, provides a good example of the types of deformation mechanisms that operate in carbonate rocks over the range 150–350°C. Despite large bulk strain and low temperature (< 300°C), micritic limestones (grain size 3–6 μm) generally lack any crystallographic preferred orientation (CPO), and show no signs of recrystallisation. The principal deformation process inferred to be operating at these conditions is grain-boundary sliding, assisted by diffusive-transfer processes. A single low-temperature (<180°C) mylonitic fault rock proved an exception to the rule and displayed substantial grain-size reduction (grain size 1–3 μm): this was interpreted as recrystallised sub-grains and thus indicative of substantial intracrystalline slip or creep. In the low greenschist facies ('epizone') samples (T > 300°C), Burkhard (1990) noted increase in grain size (6–10 μm) as the most notable feature. Together with the weak grain-shape fabric and the variably developed CPO, this suggests a significant degree of dynamic recrystallisation. Schmid *et al.* (1987), when considering the implications of their experimental work in relation to naturally deformed calcite rocks, reasonably inferred that for low greenschist facies conditions or slightly lower, intracrystalline twinning was the dominant mechanism. However, since calcite readily anneals, the evidence of twinning is often lost due to twin-boundary migration. Therefore, a lack of observed twins in deformed calcite rocks does not necessarily mean that twinning has not occurred.

Studies by van der Pluijm (1991) and Busch & van der Pluijm (1995) on the Bancroft shear zone, Ontario, give a valuable insight into the sequence of processes that operate and microstructural changes that occur from protolith to ultramylonite in a 15–20 m wide, upper greenschist facies (T approximately 450–500°C) shear zone in marble. Rapid changes in intensity of shearing are observed, including the transition from protomylonite to ultramylonite over the scale of a thin section. The protolith has experienced low strain, facilitated by grain-boundary migration and the production of deformation twins. The protomylonite has a strong CPO and a well developed porphyroclastic microstructure with abundant core-and-mantle structures as well as undulose extinction and deformation bands. These microstructural features suggest that dislocation creep has been an important mechanism, and that rotation recrystallisation played a key role during dynamic recrystallisation. The coarse mylonites comprise a dynamically recrystallised aggregate of grains, of similar size to the protomylonite matrix grains; some of the larger grains have a subgrain microstructure. The coarse mylonites have a weak shape fabric and an almost random crystallographic arrangement. This randomness is used by Busch & van der Pluijm (1995) as evidence for limited dislocation creep. Cathodoluminescence reveals that most grains have a secondary calcite overgrowth, indicating that solution-transfer processes were important during mylonitisation. Fluid-assisted grain-boundary sliding is considered as the dominant process. Moving further into the shear zone, and higher strains, the coarse mylonite passes with abrupt transition (over 1–2 cm) into fine-grained S–C mylonite and then ultramylonite. The S–C mylonites have a well developed oblique shape fabric, and a pronounced CPO with *c*-axes perpendicular to the shear plane. Ultramylonites are fine-grained (20–30 µm), homogenous and with a shape fabric oblique to the shear plane. The development of shape fabrics and CPOs is attributed to dislocation creep processes, and the grain-size reduction shown by the S–C mylonites and ultramylonites is attributed to rotation recrystallisation. A key point from the study of Busch & van der Pluijm (1995) was that over very short distances within a shear zone there is evidence that various deformation mechanisms have operated, although at any given time in the evolution of the shear zone only one particular mechanism would be dominant.

In coarse-grained amphibolite facies marbles, deformation twinning and a lack of any preferred crystallographic orientation is typical. This suggests extensive recrystallisation involving grain-boundary migration and intracrystalline (twinning-dominated) slip as the dominant mechanisms.

8.3.4 Distinguishing between schists and mylonites

Although protomylonites, with their distinctive porphyroclastic texture, and ribbon mylonites, with highly elongate quartz grains are readily identified, the distinction between certain mylonites and schists may in some instances be less easy. Both rock types experience strong ductile deformation, and both possess a strong planar fabric (usually also with a linear component). In strict terms, to call a rock a mylonite it must be demonstrated that there has been grain-size reduction. In the field, mylonitic fabrics may usually be distinguished from schistose fabrics by virtue of their porphyroclastic texture, and by the fact that mylonites grade laterally into undeformed or less deformed rocks of similar composition. In narrow shear zones a recognisable change in orientation of the principle fabric may be visible in outcrop, but in other cases this may not be clear. Although mylonites develop in both narrow zones and broad crustal shears, schists and schistose fabrics are always regionally extensive.

In thin section, mylonites typically contain angular or rounded porphyroclasts (e.g. quartz and feldspar) in a fine-grained or variable-sized matrix. Porphyroclasts show features of strain, recovery and recrystallisation, and may be polycrystalline. By contrast, schists normally contain rounded or euhedral unstrained porphyroblasts in a medium- to coarse-grained matrix. To distinguish fine-grained schists without porphyroblasts from quartz-rich ultramylonites, it is necessary to examine the degree of crystallographic alignment of quartz. Initial microscopic examination of the rock in Fig. 8.13(c) may suggest a fine-grained psammite or semi-pelitic schist. However, by insertion of a sensitive tint plate, a pronounced crystallographic alignment of quartz is seen, demonstrating that it is in fact an ultramylonite, albeit slightly recrystallised (see Plates 7(a) & (b), which show the same sample). Use of the sensitive tint plate as a quick check on the degree of crystallographic alignment of quartz is highly recommended as a first approach to evaluating strain and deformation processes, even if detailed Universal-stage work is not going to be undertaken. For further details on the various applications of crystallographic fabric data in the study of strain paths and deformation processes in rocks, the review by Law (1990) provides a useful introduction.

8.4 The influence of deformation on metamorphic processes

During metamorphic reactions, the rate at which heat is produced or consumed varies according to the reaction kinetics and the enthalpy of the reaction (that energy evolved when substances react). Prograde reactions are endothermic or, in other words, they absorb heat from the surroundings. In the contact aureole around intrusions it is fluids, and energy in the form of heat, that drive metamorphic reactions. However, in regional metamorphic environments where deformation accompanies metamorphism, and more particularly in fault and shear zones where high strains and strain rates are concentrated into narrow zones, the influence of active deformation and mechanical energy on metamorphic processes can be significant. The role of deformation during metamorphic transformations can be particularly significant in relation to reaction kinetics (e.g. nucleation rate, growth rate and overall transformation rate). Rutter & Brodie (1995) identify three main ways in which deformation may enhance the rate of metamorphic transformations, namely: (1) grain-size reduction leading to increased surface area and thus more surface free energy to promote reaction; (2) production of strained grains with high dislocation densities, which have enhanced solubility relative to unstrained grains of the same phase; and (3) increased temperature due to shear heating, where energy in the form of heat is produced by the release of strain energy, and as a result enhances metamorphic reactions.

Grain-size reduction may also have the effect of enhancing permeability, and fluid movement through areas such as shear zones, but where the dihedral angle between grains is >60° (e.g. carbonates), porosity will not be connected, so grain-size reduction alone is no guarantee of increased permeability (Rutter & Brodie, 1995). Shear zones are often documented as areas of increased fluid flow, so a transient enhancement of permeability synchronous with shearing seems likely in many cases. There are a number of ways in which this could occur, including grain-boundary sliding, which generates grain-scale dilatancy, or cataclastic deformation, which gives dilation on a range of scales.

There are many ways in which deformation may influence the sites of reactions. First, it will influence the spatial distribution of those sites at which dissolution is most favoured. Second, it will establish chemical potential gradients on a variety of scales and, third, it will often increase bulk permeability and thus aid diffusive mass transfer within the rock.

It is now well established that grain-boundary processes play a crucial role in both deformation and metamorphism. The concentration of loose bonds and dislocations provides ideal sites for fluid–mineral interaction, leading to reaction and nucleation of new phases. The migrating grain boundary interfaces (which are highly disordered regions) are particularly favourable sites for reactions. The interconnection of grain boundaries provides the necessary pathways for diffusive mass transfer of material via the fluid phase. Driven by chemical potential gradients, this results in the redistribution of chemical components and promotes phase transformations within the rock.

The deformation of rocks produces strained crystals, giving rise to an increased defect density and thus enhancing intracrystalline diffusion. The increase in dislocation density at grain margins raises the surface energy and provides more available free bonds. This increase in surface energy causes a lowering of the activation energy for nucleation, and means that strained crystals will offer more favourable sites for nucleation. Experimental work (Davis & Adams, 1965) has shown that strain resulting from high shear stresses increases dislocation densities, and can increase reaction rates by several orders of magnitude. The various strain energies (elastic strain energy, defect energy and surface energy) and heat generated during active deformation become sources of energy for chemical work. This transformation of mechanical energy into chemical energy occurs by the reaction of deformed crystals with the grain-boundary fluid. This fluid forms a vital link between deformational and chemical processes operating in a rock. It is crucial for the operation of deformation processes such as pressure solution, as well as having an essential role in most metamorphic reactions.

When considering the controls on porphyroblast nucleation in Chapter 5, the heterogeneous nature of rock materials was discussed. Because rocks and rock sequences exhibit such strong heterogeneity, it means that their deformation in response to applied stress will similarly be heterogeneous. This gives rise to strain partitioning, with some areas experiencing only low strain while others become highly strained. This partitioning occurs from the macroscale right down to the microscale, with significant strain variations developing around fold hinges, boudins, porphyroblasts and porphyroclasts. The development of strain and strain rate gradients produces dislocation density and thus chemical potential gradients. Fluctuations in chemical potential gradients will influence all reactions, and it is unlikely that chemical equilibrium will be achieved in actively deforming rocks, especially where stresses and strains are large.

Through its effect on the local activities of particular aqueous species, deformation contributes to determining which minerals will be dissolved and replaced. Sites of dissolution preferentially develop in local areas of stress or strain concentration. Where there is high differential stress (e.g. crenulation limbs), solubility increases and leads to pressure solution. Because of this, authors such as Bell & Hayward (1991) have suggested that porphyroblasts (e.g. garnet) will not nucleate on actively shearing crenulation limbs or other situations of active shear.

There are a number of ways in which deformation can increase solubility. The first is by increasing the concentration of dislocations in deformed minerals, and thus increasing the lattice energy. Densities of dislocations in 'tangles' and sub-grain walls can be large enough to increase solubilities by >10%. This means that strained crystals of a mineral such as K-feldspar are more soluble than unstrained crystals, and will thus more readily retrogress (Wintsch, 1985). Passchier (1985) documents a case in which the origin of flame perthites is attributed to the degree of deformation of K-feldspar grains, and Simpson & Wintsch (1989) have demonstrated a link between myrmekite (in metamorphic rocks) and stress-induced K-feldspar replacement.

While dissolution occurs in areas of high stress and strain, the precipitation of dissolved material occurs in low-strain regions and areas of extension. Pressure shadows around rigid objects such as porphyroblasts (Fig. 5.4) and precipitation in extension fractures to give veins provide the most obvious sites, but crenulation hinges are also important 'sinks' for material being actively dissolved from strongly sheared limbs (Fig. 4.6). Fibrolite is often observed to be concentrated in zones of high shear strain, both on the micro- and the macroscale. This has led various authors (e.g. Vernon, 1987; Wintsch & Andrews, 1988; Kerrick, 1990) to conclude that the development of fibrolite aggregates is deformation-induced. The explanations developed by each author differ, but the basic link between fibrolite development and zones of high shear has now been recognised in many high-grade schists and gneisses.

A second way in which deformation enhances solubility and general transformation kinetics is by the general process of grain-size reduction. This is characteristic of mylonitisation, and enhances solubility by increasing the surface area of grain boundaries available for reaction. This gives an overall increase in surface energy, and may also increase bulk permeability, and thus enable greater fluid access. This in turn promotes reactions by way of the ions carried in solution and the catalytic affect of the fluid.

A portion of the work of deformation will also be dissipated as heat. This too will contribute to increasing silicate solubility in the aqueous fluid, and if stress and strain rates are high enough, frictional heating can locally raise temperatures by as much as 1000°C. This induces frictional melting and pseudotachylite (Fig. 8.12) formation, and has been related to cases of rapid brittle faulting at high crustal levels (e.g. earthquakes). The contribution of shear-heating in ductile regimes is less well established. Theoretical calculations suggest that a temperature increase of up to 150°C may be possible. However, these are likely to be overestimates, since the calculations assume that all mechanical work is converted to heat, and generally overlook the fact that synchronous prograde reactions are endothermic, and thus consume most of the heat generated, and that circulating fluids will be effective at transferring heat.

8.5 The influence of metamorphism on deformation processes

The foregoing discussion has concentrated on the influence that deformation has on metamorphic processes, but there are also various metamorphic processes that influence the rate and type of deformation. Brodie & Rutter (1985) produced a five-fold classification of mechanistic interactions between metamorphic trans-formations and deformability, recently re-evaluated (Rutter & Brodie, 1995) with reference to experimental and natural examples. The effects of metamorphism on deformability are categorised as follows: (1) facilitation of cataclasis due to elevated pore-fluid pressure in dehydration reactions or in response to melting; (2) enhanced plasticity resulting from transformation-induced volume changes; (3) development of fine-grained reaction products, facilitating grain size sensitive flow processes; (4) changes in plastic deformability of silicate minerals due to recrystallization and increased activity of pore fluid; and (5) promotion of diffusion creep via by the enhanced potential gradient of a reaction along the diffusion path (Rutter & Brodie, 1995).

'Reaction enhanced ductility (plasticity)' (White & Knipe, 1978; Rubie, 1990) is a particularly important process that operates in actively deforming rocks where metamorphic reactions are simultaneously in progress. Metamorphic reactions give rise to increased ductility in a number of ways. First, they will produce small grains, and allow grain size sensitive flow processes such as grain-boundary sliding to operate (i.e. a change in deformation mechanism is induced). Second, reactions will aid 'strain softening' by producing soft, strain-free grains. Third, retrograde reactions often convert 'hard' phases such as feldspar to 'soft' phases such as quartz, sericite and calcite, a process known as 'reaction softening' (White *et al.*, 1980). These reaction-related processes often make an important contribution to the overall softening and enhanced ductility in mylonites (e.g. White *et al.* (1980) and Williams & Dixon (1982) for *granitoid mylonites*; and Brodie & Rutter (1985) and Stünitz (1993) for *mafic mylonites*), but if prograde reactions take place this typically gives rise to overall hardening. Dynamic recrystallisation during metamorphism aids deformation processes by continually providing new strain-free grains.

By liberating or consuming fluid, metamorphic reactions have a major influence on fluid pressure (P_f) at the time of deformation. Where fluid production is fast with respect to diffusivity/permeability, the rapid fluid release may induce a local increase in P_f (Chapter 11). If the production of fluid is sufficiently rapid, hydraulic fracturing may occur. Fractures will propagate until such a time as P_f subsides, but may develop further if P_f rises once more to exceed rock strength. Therefore, it is apparent that increasing P_f due to devolatilisation reactions may aid cataclastic flow and increase deformability. As with grain-size reduction, this will significantly modify the rheological behaviour. The release of fluid during dehydration reactions can also enhance grain-boundary diffusion and may promote diffusion-accommodated grain-boundary sliding (Rubie, 1990).

Certain phase transformations, especially in carbonate rocks, involve significant negative volume changes (up to 30%). These may have a significant, although probably short-lived, effect on bulk rock porosity and permeability, which in turn influences rock deformation. For example, stress concentrations and additional void formation resulting from volume changes may induce rock failure by brittle fracturing. Other effects may also occur, but our present knowledge of the contribution of volume changes during phase transformations is rather limited, and certainly requires further investigation.

References

Bakker, R.J. & Jansen, J.B. (1994) A mechanism for preferential H$_2$O leakage from fluid inclusions in quartz, based on TEM observations. *Contributions to Mineralogy and Petrology*, **116**, 7–20.

Barber, D.J. (1990) Régimes of plastic deformation – processes and microstructures: an overview, in *Deformation processes in minerals, ceramics and rocks* (eds D.J. Barber & P.G. Meredith). Mineralogical Society of Great Britain and Ireland, Monograph Series No. 1, Ch. 6, 138–178.

Barber, D.J. & Meredith, P.G. (1990) *Deformation processes in minerals, ceramics and rocks*. Mineralogical Society of Great Britain and Ireland, Monograph Series No. 1, 423 pp.

Bell, T.H. & Hayward, N. (1991) Episodic metamorphic reactions during orogenesis: the control of deformation partitioning on reaction sites and duration. *Journal of Metamorphic Geology*, **9**, 619–640.

Boland, J.N. & FitzGerald, J.D. (eds) (1993) *Defects and processes in the solid state: geoscience applications*. Developments in Petrology No. 14 ('The McLaren Volume'). Elsevier, Amsterdam, 470 pp.

Brodie, K.H. (1981) Variation in amphibole and plagioclase composition with deformation. *Tectonophysics*, 78, 385–402.

Brodie, K.H. & Rutter, E.H. (1985) On the relationship between deformation and metamorphism, with special reference to the behaviour of basic rocks, in *Metamorphic reactions* (eds A.B. Thompson & D.C. Rubie). *Advances in Physical Geochemistry*, 4, 138–179.

Burkhard, M. (1990) Ductile deformation mechanisms in micritic limestones naturally deformed at low temperatures (150–350°C), in *Deformation mechanisms, rheology and tectonics* (eds R.J. Knipe & E.H. Rutter). Geological Society Special Publication No. 54, 241–257.

Busch, J.P. & van der Pluijm, B.A. (1995) Calcite textures, microstructures and rheological properties of marble mylonites in the Bancroft shear zone, Ontario, Canada. *Journal of Structural Geology*, 17, 677–688.

Carter, N.L., Christie, J.M. & Griggs, D.T. (1964) Experimental deformation and recrystallization of quartz. *Journal of Geology*, 72, 687–733.

Craw, D. & Norris, R.J. (1993) Grain boundary migration of water and carbon dioxide during uplift of the garnet-zone Alpine Schist, New Zealand. *Journal of Metamorphic Geology*, 11, 371–378.

Davis, B.L. & Adams, L.H. (1965) Kinetics of the calcite–aragonite transformation. *Journal of Geophysical Research*, 70, 433–441.

Dietrich, D. & Song, H. (1984) Calcite fabrics in a natural shear zone environment, the Helvetic nappes of western Switzerland. *Journal of Structural Geology*, 6, 19–32.

Drury, M.R. (1993) Deformation lamellae in metals and minerals, in *Defects and processes in the solid state: geoscience applications* (eds J.N. Boland & J.D. FitzGerald). Developments in Petrology No. 14 ('The McLaren Volume'). Elsevier, Amsterdam, 195–212.

Drury, M.R. & Urai, J.L. (1990) Deformation-related recrystallisation processes. *Tectonophysics*, 172, 235–253.

Gerretsen, J., McLaren, A.C. & Paterson, M.S. (1993) Evolution of inclusions in wet synthetic quartz as a function of temperature and pressure; implications for water weakening, in *Defects and processes in the solid state: geoscience applications* (eds J.N. Boland & J.D. FitzGerald). Developments in Petrology No. 14 ('The McLaren Volume'). Elsevier, Amsterdam, 27–47.

Green, H.W. (1992) Analysis of deformation in geological materials, in *Minerals and reactions at the atomic scale: transmission electron microscopy* (ed. P.R. Buseck). Mineralogical Society of America, Reviews in Mineralogy, No. 27, Ch. 11, 425–454.

Hirth, G. & Tullis, J. (1992) Dislocation creep regimes in quartz aggregates. *Journal of Structural Geology*, 14, 145–159.

Jessell, M.W. & Lister, G.S. (1990) A simulation of the temperature dependence of quartz fabrics, in *Deformation mechanisms, rheology and tectonics* (eds R.J. Knipe & E.H. Rutter). Geological Society Special Publication No. 54, 353–362.

Kerrick, D.M. (1990) *The Al_2SiO_5 polymorphs*. Mineralogical Society of America, Reviews in Mineralogy No. 22, 406 pp.

Knipe, R.J. (1990) Microstructural analysis and tectonic evolution in thrust systems: examples from the Assynt region of the Moine Thrust Zone, Scotland, in *Deformation processes in minerals ceramics and rocks.* (eds D.J. Barber & P.G. Meredith). Mineralogical Society of Great Britain and Ireland, Monograph Series No. 1, 228–261.

Knipe, R.J. & Rutter, E.H. (1990) *Deformation mechanisms, rheology and tectonics*. Geological Society Special Publication No. 54, The Geological Society, London, 535 pp.

Lafrance, B. & Vernon, R.H. (1993) Mass transfer and microfracturing in gabbroic mylonites of the Guadalupe Igneous Complex, California, in *Defects and processes in the solid state: geoscience applications* (eds J.N. Boland & J.D. FitzGerald). Developments in Petrology No. 14 ('The McLaren Volume'). Elsevier, Amsterdam, 151–167.

Law, R.D. (1990) Crystallographic fabrics: a selective review of their applications to research in structural geology, in *Deformation mechanisms, rheology and tectonics* (eds R.J. Knipe & E.H. Rutter). Geological Society Special Publication No. 54, The Geological Society, London, 335–352.

Lin, A. (1994) Glassy pseudotachylite veins from the Fuyun fault zone, northwest China. *Journal of Structural Geology*, 16, 71–84.

Lloyd, G.E. & Knipe, R.J. (1992) Deformation mechanisms accommodating faulting of quartzite under upper crustal conditions. *Journal of Structural Geology*, 14, 127–143.

McLaren, A.C., Retchford, J.A., Griggs, D.T. & Christie, J.M. (1967) Transmission electron microscope study of Brazil twins and dislocations experimentally produced in natural quartz. *Physica Status Solidi*, 19, 631–644.

Maddock, R.H., Grocott, J. & van Nes, M. (1987) Vesicles, amygdales and similar structures in fault-generated pseudotachylites. *Lithos*, 20, 419–432.

Mawer, C.K. & FitzGerald, J.D. (1993) Microstructure of kink band boundaries in naturally deformed Chewings Range Quartzite, in *Defects and processes in the solid state: geoscience applications* (eds J.N. Boland & J.D. FitzGerald). Developments in

References

Petrology No. 14 ('The McLaren Volume'). Elsevier, Amsterdam, 49–67.

Park, W.C. & Schot, E.H. (1968) Stylolites: their nature and origin. *Journal of Sedimentary Petrology*, 38, 175–191.

Passchier, C.W. (1985) Water-deficient mylonite zones – an example from the Pyrenees. *Lithos*, 18, 115–127.

Passchier, C.W. & Trouw, R.A.J. (1996) *Microtectonics*. Springer-Verlag, Berlin, 289 pp.

Poirier, J.-P. (1985) *Creep of crystals*. Cambridge University Press, Cambridge, 260 pp.

Pontikis, V. & Poirier, J.-P. (1975) Phenomenological and structural analysis of recovery-controlled creep, with special reference to the creep of single-crystal silver chloride. *Philosophical Magazine*, 32, 577–592.

Prior, D.J., Knipe, R.J. & Handy, M.R. (1990) Estimates of the rates of microstructural changes in mylonites, in *Deformation mechanisms, rheology and tectonics* (eds R.J. Knipe & E.H. Rutter). Geological Society Special Publication No. 54, The Geological Society, London, 309–319.

Ramsay, J.G. & Huber, M.I. (1983) *The techniques of modern structural geology: Volume 1: Strain analysis*. Academic Press, London.

Ranalli, G. (1982) Deformation maps in grain size-stress space as a tool to investigate mantle rheology. *Physics of the Earth and Planetary Interiors*, 29, 42–50.

Ree, J.-H. (1994) Grain boundary sliding and development of grain boundary openings in experimentally deformed octachloropropane. *Journal of Structural Geology*, 16, 403–418.

Reeder, R.J. (1992) Carbonates: growth and alteration microstructures, in *Minerals and reactions at the atomic scale: transmission electron microscopy*. (ed. P.R. Buseck). Mineralogical Society of America, Reviews in Mineralogy No. 27, Ch. 10, 381–424.

Ross, J.V., Ave Lallement, H.G. & Carter, N.L. (1980) Stress dependence of recrystallized grain and subgrain size in olivine. *Tectonophysics*, 70, 39–61.

Rubie, D.C. (1990) Mechanisms of reaction-enhanced deformability in minerals and rocks, in *Deformation processes in minerals, ceramics and rocks* (eds D.J. Barber & P.G. Meredith). Mineralogical Society of Great Britain and Ireland, Monograph Series No. 1, Ch. 10, 262–295.

Rutter, E.H. (1976) The kinetics of rock deformation by pressure solution. *Philosophical Transactions of the Royal Society of London*, 283A, 203–219.

Rutter, E.H. (1986) On the nomenclature of mode of failure transitions in rocks. *Tectonophysics*, 122, 381–387.

Rutter, E.H. (1995) Experimental study of the influence of stress, temperature, and strain on the dynamic recrystallization of Carrara marble. *Journal of Geophysical Research*, 100(B12), 24 651–24 663.

Rutter, E.H. & Brodie, K.H. (1992) Rheology of the lower crust, in *Geology of the lower continental crust* (eds D. Fountain, R. Arculus, R. & R. Kay). Elsevier, Amsterdam, 201–268.

Rutter, E.H. & Brodie, K.H. (1995) Mechanistic interactions between deformation and metamorphism. *Geological Journal*, 30, 227–240.

Rutter, E.H., Casey, M. & Burlini, L. (1994) Preferred crystallographic orientation development during plastic and superplastic flow of calcite rocks. *Journal of Structural Geology*, 16, 1431–1446.

Schmid, S.M., Boland, J.N. & Paterson, M.S. (1977) Superplastic flow in fine grained limestone. *Tectonophysics*, 43, 257–291.

Schmid, S.M., Panozzo, R. & Bauer, S. (1987) Simple shear experiments on calcite rocks: rheology and microfabric. *Journal of Structural Geology*, 9, 747–778.

Sibson, R.H. (1977) Fault rocks and fault mechanisms. *Journal of the Geological Society*, 133, 191–213.

Sibson, R.H. (1990) Faulting and fluid flow, in *Fluids in tectonically active regimes of the continental crust* (ed. B.E. Nesbitt). Mineralogical Association of Canada, Short Course No. 18, Ch. 4, 93–132.

Simpson, C. & Wintsch, R.P. (1989) Evidence for deformation-induced K-feldspar replacement by myrmekite. *Journal of Metamorphic Geology*, 7, 261–275.

Skrotzki, W. (1990) Microstructure in hornblende of mylonitic amphibolite, in *Deformation mechanisms, rheology and tectonics* (eds R.J. Knipe & E.H. Rutter). Geological Society Special Publication No. 54, 321–325.

Snoke, A.W., Tullis, J. & Todd, V.R. (eds) (in press) *Fault-related rocks: a photographic atlas*. Princeton University Press, Princeton, New Jersey.

Spiers, C.J., Schutjens, P.M.T.M., Brzesowsky, R.H., Peach, C.J., Liezenberg, J.L. & Zwart, H.J. (1990) Experimental determination of constitutive parameters governing creep of rocksalt by pressure solution, in *Deformation mechanisms, rheology and tectonics*. (eds R.J. Knipe & E.H. Rutter). Geological Society Special Publication No. 54, 215–227.

Stünitz, H. (1993) Transition from fracturing to viscous flow in a naturally deformed metagabbro, in *Defects and processes in the solid state: geoscience applications* (eds J.N. Boland & J.D. FitzGerald). Developments in Petrology No. 14 ('The McLaren Volume'). Elsevier, Amsterdam, 121–150.

Tullis, J.A. (1990) Experimental studies of deformation mechanisms and microstructures in quartzo-feldspathic rocks, in *Deformation processes in minerals, ceramics and rocks.* (eds D.J. Barber & P.G. Meredith). Mineralogical Society of Great Britain and Ireland Monograph Series No. 1, 190–227.

Tullis, J.A., Christie, J.M. & Griggs, D.T. (1973) Microstructures and preferred orientations of deformed quartzites. *Bulletin of the Geological Society of America,* 84, 297–314.

van der Pluijm, B.A. (1991) Marble mylonite in the Bancroft shear zone, Ontario, Canada: microstructures and deformation mechanisms. *Journal of Structural Geology,* 13, 1125–1135.

Vernon, R.H. (1976) *Metamorphic processes.* George Allen & Unwin, London, 247 pp.

Vernon, R.H. (1987) Growth and concentration of fibrous sillimanite related to heterogeneous deformation of K-feldspar–sillimanite metapelites. *Journal of Metamorphic Geology,* 5, 51–68.

Walker, A.N., Rutter, E.H. & Brodie, K.H. (1990) Experimental study of grain-size sensitive flow of synthetic, hot-pressed calcite rocks, in *Deformation mechanisms, rheology and tectonics* (eds R.J. Knipe & E.H. Rutter). Geological Society Special Publication No. 54, 259–284.

Watson, E.B. & Brenan, J.M. (1987) Fluids in the lithosphere, 1. Experimentally-determined wetting characteristics of CO_2–H_2O fluids and their implications for fluid transport, host-rock physical properties, and fluid inclusion formation. *Earth and Planetary Science Letters,* 85, 497–515.

Wenk, H.-R., Takeshita, T., Bechler, E., Erskine, B.G. & Matthies, S. (1987) Pure shear and simple shear calcite textures: comparison of experimental, theoretical and natural data. *Journal of Structural Geology,* 9, 731–745.

White, J.C. & White, S.H. (1981) On the structure of grain boundaries in tectonites. *Tectonophysics,* 78, 613–628.

White, S. (1976) The effects of strain on the microstructures, fabrics, and deformation mechanisms of quartzites. *Philosophical Transactions of the Royal Society of London,* 283A, 69–86.

White, S.H. & Knipe, R.J. (1978) Transformation- and reaction-enhanced ductility in rocks. *Journal of the Geological Society of London,* 135, 513–516.

White, S.H., Burrows, S.E., Carreras, J., Shaw, N.D. & Humphreys, F.J. (1980) On mylonites in ductile shear zones. *Journal of Structural Geology,* 2, 175–187.

Williams, G.D. & Dixon, J. (1982) Reaction and geometric softening in granitoid mylonites. *Textures and Microstructures,* 4, 223–239.

Wintsch, R.P. (1985) The possible effects of deformation on chemical processes in metamorphic fault zones, in *Metamorphic reactions* (eds A.B. Thompson & D.C. Rubie). *Advances in Physical Geochemistry,* 4, 251–268.

Wintsch, R.P. & Andrews, M.S. (1988) Deformation induced growth of sillimanite: 'stress' minerals revisited. *Journal of Geology,* 96, 143–161.

Wise, D.U., Dunn, D.E., Engelder, J.T., Geiser, P.A., Hatcher, R.D., Kish, S.A., Odom, A.L. & Schamel, S. (1984) Fault-related rocks: suggestions for terminology. *Geology,* 12, 391–394.

Chapter nine

Porphyroblast–foliation relationships

When studying a suite of metamorphic rocks, especially those formed during orogenic metamorphism, the common occurrence of porphyroblasts and one or more tectonic fabrics provides valuable information concerning deformation–metamorphism interrelationships. When integrated with pressure and temperature estimates based on mineral assemblages and geothermobarometry, coupled with any age dates that may exist to date particular events in the region studied, such microstructural features can enable the construction of a pressure–temperature–time (P–T–t) path for individual rocks. This in turn can give important insights into aspects of crustal evolution, or metamorphic evolution in the vicinity of major igneous intrusions. Although the interpretations from porphyroblast–foliation relationships have excellent potential, the interpretation of porphyroblast inclusion trails and their relationships with external matrix fabrics in the rock have always been contentious topics. Different interpretations of the same textural and microstructural features have led to fundamental differences of opinion concerning the processes responsible for particular features observed and consequently the overall interpretations can vary greatly between different researchers. In Sections 5.1, 5.2 & 5.3, aspects of porphyroblast nucleation and growth were considered, and in Section 6.1 general aspects concerning mineral inclusions were discussed. This chapter concentrates on the various aspects of porphyroblast–foliation relationships, providing an introduction to the characteristic microstructural features observed, and providing an overview of the various points to take into consideration when making interpretations. The chapter highlights particular problem areas, where opinion is divided concerning the interpretation.

9.1 Thin-section 'cut effects'

Many metamorphic rocks exhibit planar and/or linear structures defined by the preferred alignment of minerals, as a result of one or more deformation events. Since porphyroblasts commonly overgrow these structures, the thin-section cut relative to these planar or linear elements has a major influence on the types of inclusion patterns observed in porphyroblasts. In order to interpret these inclusion trails correctly, the main cut effects must be fully appreciated. Let us consider these with reference to the classical case of S-shaped inclusion trails commonly seen in garnets. Traditionally interpreted as syntectonic rotation of the porphyroblast with respect to the matrix foliation (e.g. Zwart, 1962; Spry, 1963), many so-called 'rotated' porphyroblasts have now been reinterpreted in terms of change in fabric orientation relative to porphyroblast during growth, or as porphyroblast overgrowth of crenulations, the porphyroblast not having rotated at all (e.g. Bell & Rubenach, 1983; Bell, 1985). There is considerable debate on this matter, which will be covered in Section 9.2.2, but at this stage discussion is confined to the different porphyroblast–foliation relationships that can be obtained for a single prophyroblast or group of porphyroblasts according to the orientation of the thin-section cut.

Powell & Treagus (1970) made a study of the 'cut effect' on inclusion patterns by comparison of model-generated patterns with those observed in natural syntectonic garnet porphyroblasts from Norway and Scotland. In cases of equidimensional porphyroblasts such as garnet, the inclusion pattern observed can be likened to a cross-section through a sphere, and those sections cut perpendicular to the 'rotation' axis (X–Z sections) will produce an S-shaped fabric. A similar, though not identical, fabric is said to occur in sections cut oblique to this orientation, but for sections cut parallel to, and containing, the rotation axis, ⊃C, 11 and ⊄D patterns are observed (Fig. 9.1). More recently, Johnson (1993a) made an in-depth analysis of variations in spiral-shaped inclusion trail geometries of natural garnets as a function of orientation of thin-section cut. He records various inclusion trail geometries, including closed loop ⊄D patterns. However, Johnson (1993a) records that spiral axes are generally oblique to the main external foliation, and that closed loops result from non-cylindrical spiral-shaped inclusion trails.

Another thin-section 'cut effect', examined by Powell & Treagus (1970), Johnson (1993a) and others, is the case of parallel cuts passing through different parts of the porphyroblast. When thin sections are made it is understandable that in some cases the cut will have passed through the centre or close to the centre of the

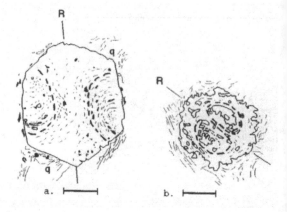

FIG. 9.1 ⊃C and ⊄D inclusion fabrics in garnet for sections cut parallel to, and containing, the rotation axis (R) (modified after MacQueen & Powell, 1977, Figs 5(a) & (b)). Scale = 0.5 mm.

porphyroblast, whereas in neighbouring porphyroblasts the cut may have been at or close to the edge. As a result, there may be appreciable microstructural differences from one porphyroblast to the next. Cuts through the centre will record the full history, whereas cuts close to the edge will only record the final stages of porphyroblast development. This has been verified by cutting serial sections through individual porphyroblasts. It has been shown that in the case of porphyroblasts showing S-fabrics, a more pronounced 'S' is seen in cuts through the centre compared to cuts near to the rim (Fig. 9.2(a)). In porphyroblasts with a more complex history of growth – let us say, two or more textural zones – there will be more significant differences between 'centre' and 'rim' cuts (Fig. 9.2(b)). Because of the problems of interpretation caused by this type of 'cut effect', it is necessary to identify carefully, and to base interpretations on those porphyroblasts cut centrally. Unless you have good control on the cut of a specific porphyroblast in a given thin section, it may be advisable to place greater emphasis on the porphyroblasts with the largest cross-sectional area. This approach is based on the assumption that porphyroblasts with the

Porphyroblast growth in relation to foliation

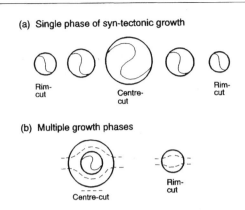

FIG. 9.2 Schematic illustrations of centre-rim cut effects in porphyroblasts: (a) single period of growth; (b) multiphase growth.

smallest radius (cross-sectional area) represent cuts closest to the edge of the crystal. However, in doing this, a certain amount of caution must be exercised, because not all porphyroblasts nucleate simultaneously, and those nucleated late will often be smaller. In consequence, they will naturally have an included fabric that is only representative of the later parts of the tectonometamorphic history. Therefore, the simpler fabrics of smaller porphyroblasts need not always be put down to a cut effect. In order to obtain a full three-dimensional picture of porphyroblast-inclusion trails, many researchers consider it necessary routinely to cut a set of serial sections through a single porphyroblast, and to make a range of thin-section cuts in different orientations (e.g. Johnson, 1993a; Johnson & Moore, 1996). It has been commonplace to cut thin sections perpendicular to the main fabric (schistosity/cleavage) and parallel to the principal mineral elongation direction (P-sections), but for a better 3-D understanding of the inclusion geometry it is advisable to cut at least one section perpendicular to both the principal fabric and mineral elongation lineation (N-section). Bell & Johnson (1989) further advocate that for complete 3-D understanding of the inclusion fabric, a section cut parallel to the principal foliation should also be made (S-section).

9.2 Porphyroblast growth in relation to foliation development

9.2.1 Recognition and interpretation of pre-foliation (pre-tectonic) crystals

Pre-foliation crystals are those that existed in the rock prior to the onset of deformation, and the development of a continuous foliation (e.g. cleavage or schistosity). This includes detrital grains, relict igneous crystals and any metamorphic crystal that existed in the rock prior to deformation (e.g. 'chlorite–mica stacks'; Section 7.2, Fig. 7.7). With the exception of cases in which there has been extensive matrix recrystallisation, which makes relationships between crystal and matrix difficult to determine, 'pre-foliation' crystals are usually enveloped by the main foliation of the rock. Since pre-foliation crystals precede most if not all deformation, they will commonly exhibit signs of strain. Pre-foliation porphyroblast growth seems to be a relatively uncommon phenomenon, or at least is rarely identified. However, Vernon et al. (1993) have described examples of pre-foliation porphyroblast development in various cases of low-P/high-T orogenic metamorphism.

One or more of the following features may be seen in pre-foliation crystals, but should in no way be considered exclusive to pre-foliation crystals:

(i) *Undulose extinction.* This occurs due to distortion of the crystal lattice (Section 8.2.2). It is especially common in quartz, but is also seen in various other minerals (e.g. feldspar, biotite and kyanite). Undulose extinction in quartz porphyroclasts within a protomylonite is shown in Fig. 9.3(a).

(ii) *Fractured crystals and boudinage.* Minerals

Porphyroblast—foliation relationships

such as feldspar and hornblende behave in a brittle manner at low temperatures and/or high strain rates (for details, see Sections 8.1, 8.3.1 & 8.3.2). A highly fractured microcline crystal in what is now a protomylonite, but which originated as a granite, is shown in Fig. 9.3(b).

(iii) *Strain (pressure) shadows*. Although also a feature of crystals formed synchronous with foliation development, many pre-foliation crystals exhibit strain shadows. These are low-

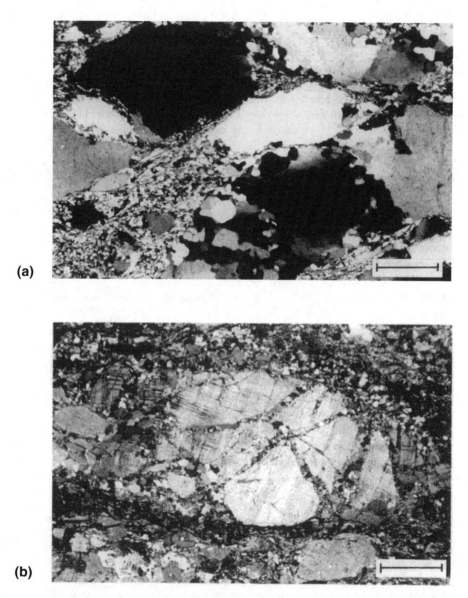

FIG. 9.3 Some characteristic features associated with pre-tectonic crystals (these may also be seen for syntectonic crystals). (a) Undulose extinction (in quartz). Scale = 0.5 mm (XPL). (b) Fragmented crystals (microcline in sheared granite). Scale = 1 mm (XPL).

Porphyroblast growth in relation to foliation

strain areas in which new minerals crystallise preferentially. They result from strain partitioning around rigid porphyroblasts or clasts (see also Section 10.7). The example in Fig. 9.3(c) shows quartz strain shadows developed around early framboidal pyrite. Note the fact that a later generation of euhedral pyrite crystals in the centre of the photograph has developed after final deformation and shows no strain shadows.

(iv) *Kinking*. This is a common feature of detrital and early-formed minerals, especially

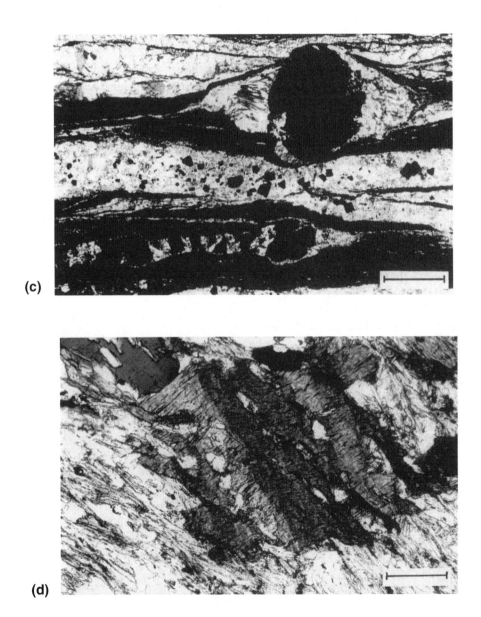

FIG. 9.3 (*contd*) (c) Pressure shadows (around framboidal pyrite). Scale = 0.5 mm. (d) Kinking (in early biotite porphyroblast). Scale = 0.1 mm (PPL).

phyllosilicate phases (see also Section 8.2.2). An early porphyroblast of biotite in a mica schist, which is strongly kinked due to subsequent deformation, is shown in Fig. 9.3(d).

9.2.2 Recognition and interpretation of syntectonic crystals

During orogenic metamorphism, most porphyroblasts are syntectonic, and although the majority of matrix minerals in deformed metamorphic rocks recrystallised or grew synchronous with deformation, it is often difficult to say unequivocally whether a given crystal is 'pre-tectonic' or 'syntectonic'. This is because many crystals developing synchronous with deformation may also be strained to some degree, and may show one or more of the same strain features listed above for pre-tectonic crystals. However, there is one commonly observed textural feature of syntectonic porphyroblasts which is absent from pre-tectonic crystals. The feature in question is that of aligned inclusion fabrics. The preservation of such fabrics, and the minerals defining the fabrics, are crucial to the understanding of the early deformation and metamorphic history of a particular rock.

'Straight' and 'S-shaped' inclusion trails

Not every syntectonic porphyroblast phase exhibits aligned inclusion trails, but they are widely reported in minerals such as garnet, albite, hornblende, staurolite and andalusite. Even in these minerals it is more common to observe poikiloblastic crystals without any distinct alignment of inclusions. Particularly common are examples in which the porphyroblast cores are heavily included, whereas the rims are largely devoid of inclusions (Fig. 5.22(b)). Although this may represent more than one growth stage, an alternative explanation is in terms of rapid initial growth causing numerous inclusions, followed by a slowing of growth rate and consequently fewer included crystals. The porphyroblast-forming reaction will involve one or more of the matrix minerals, and in general the reactant phases are not significantly represented in the included assemblage. The minerals most commonly present as inclusion trails are quartz, opaque minerals (e.g. ilmenite and graphite), epidote group minerals and, to a lesser extent, amphibole, mica, chloritoid and feldspar. The inclusion trails are widely interpreted as representing the matrix grain size at the time of inclusion, and because matrix grain size shows a general increase during prograde metamorphism it is usual for the enclosed fabric to be of similar or finer grain size than the external fabric (Plates 5 & 6).

For those porphyroblasts in which inclusion trails are clearly defined, various types are commonly encountered. These range from straight inclusion trails that are sharply discordant with the external fabric (Plates 5(a) & (b)), to inclusion fabrics with a shallow or pronounced 'S' form (Plates 5(c) & (d)), and in some cases spirals of much greater complexity (Fig. 9.4).

Discordance between porphyroblast inclusion trails and the external (matrix) fabric

Cases of straight porphyroblast inclusion trails (S_i) sharply discordant (Plate 5(a)) with the external fabric (S_e) have traditionally been interpreted in terms of static overgrowth of some pre-existing foliation followed by a later deformation event intensifying the matrix fabric and causing the discordant relationships. With such an interpretation the internal fabric would often be considered as S_1 and the external fabric as S_2. This interpretation has been advanced largely because of the assumption that porphyroblast growth rates are slow relative to deformation rates. Depending on the nature of the deformation partitioning in the early stages of fabric development, it can prove difficult to reconcile straight inclusion fabrics with porphyroblastesis synchronous with foliation development. The

Porphyroblast growth in relation to foliation

FIG. 9.4 Garnet with an exceptionally spiralled inclusion fabric: Dikänas Schist, Vasterbotten, Sweden. This figure is redrawn from a photograph by N.O. Ølesen, and figured by Schoneveld (1977).

reason for this is that the attitude of the matrix fabric relative to the porphyroblast should progressively change and thus generate a curved or sigmoidal inclusion trail. However, it has now been demonstrated that most porphyroblasts probably grow extremely rapidly in geological terms (i.e. in < 1 Ma, and possibly in < 0.1 Ma; Section 5.4). This means that at typical regional orogenic strain rates (e.g. 10^{-14} s^{-1}) there would often be no appreciable rotation of porphyroblast or matrix fabric during growth, and thus 'straight' inclusion trails would be preserved. However, at faster strain rates (e.g. 10^{-12} s^{-1}) of shear zones, significant relative rotation between porphyroblast and fabric could occur, and S-shaped inclusion trails should be more widely developed (Barker, 1994). Considering periods of porphyroblastesis as rapid events punctuating protracted periods of foliation development, the continuation of shearing after cessation of porphyroblastesis would explain the discordant S_i–S_e relationships that are commonly observed in porphyroblastic schists. With this interpretation, it means that the included fabric will commonly represent an early stage in the development of the external fabric, and need not be some entirely separate earlier schistosity or cleavage, as has often been proposed. With this in mind, 'straight' and 'slightly curved' inclusion trails are equally as likely to represent porphyroblastesis synchronous with foliation development as they are to represent growth between fabric-forming events. Because of the difficulty in using S_i–S_e relationships to determine with certainty whether the porphyroblast has rotated with respect to the external fabric or vice versa, Passchier et al. (1992) suggest that it is safest to discuss S_i–S_e angular relationships in terms of *relative* rotation between S_i and S_e.

The influence of crystallographic structure

When interpreting straight inclusion fabrics in minerals with a strong cleavage or cleavages (e.g. micas, amphiboles and kyanite), always be aware of the fact that crystallographically controlled inclusion trails can develop which have nothing to do with any pre-existing or developing tectonic fabric. These crystallographically controlled 'straight' inclusion trails develop because, within minerals possessing a cleavage, inclusions are often preferentially incorporated and aligned within the cleavage of the crystal in order to minimise the increase in

Porphyroblast—foliation relationships

internal energy caused by incorporating an inclusion into the lattice. In Chapter 6 it has already been discussed how certain types of crystallographic and growth-influenced inclusion patterns develop in certain minerals, but because of their distinctive appearance (Plate 1) these are unlikely to be confused with tectonic fabrics.

Interpretations for and against porphyroblast rotation

In cases of gentle or pronounced S-fabrics (Plates 5(c) & (d)), the traditional interpretation, exemplified by Zwart (1962) and Spry (1963), is to consider the garnet as a sphere physically rotating during growth in response to simple shear within the matrix. The inclusion trail (defined by included matrix material) is considered to initiate sub-parallel to the matrix fabric, but to become progressively rotated as the porphyroblast itself physically rotates relative to externally fixed reference co-ordinates (Fig. 9.5). This model has been advanced by many researchers (e.g. Rosenfeld, 1970; Schoneveld, 1977), and on the basis of this interpretation various authors have analysed S-fabrics in an attempt to assess the amount of relative rotation between porphyroblast and external matrix. Inclusion fabrics suggesting rotations of up to 90° (Plate 5(c)) are very common, those up to 180° (Plate 5(d)) are less common, and while fabrics apparently indicating rotations up to 800° (Fig. 9.4) have been recorded, these are extremely rare.

Wilson (1971) promoted a realistic alternative interpretation for gently curved and more pronounced S-shaped inclusion trails, advocating that sub-spherical porphyroblasts such as garnet do not rotate, but that S-shaped inclusion trails develop as the matrix foliation changes orientation relative to the growing porphyroblast. This could adequately explain the common spirals with less than 90° relative rotation (Fig. 9.6), but for porphyroblasts with more pronounced spiralling this model would

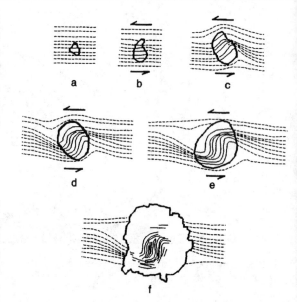

FIG. 9.5 A model for the development of 'snowball' S-fabrics in porphyroblasts (especially garnet) (after Spry, 1963).

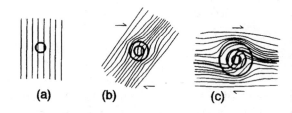

FIG. 9.6 A model for the development of S-shaped inclusion fabric in porphyroblast by transposition of the external foliation through 90° during incremental porphyroblast growth. The porphyroblast does not rotate.

be inadequate. Note that for a given sense of shear the 'S' or 'Z' developed in this model is the reverse of that produced when modelling in terms of physical rotation of porphyroblasts (compare Figs 9.5e & 9.6c; see also Section 10.5). A detailed study by Johnson (1993b) concluded that most if not all spiral-shaped inclusion trails could be explained just as well in terms of porphyroblast rotation relative to externally fixed reference co-ordinates as they could by external fabric rotation relative to fixed porphyroblasts.

Bell & Rubenach (1983), and a series of subsequent publications by Bell (1985) and coworkers, have also presented a model advocating non-rotation of syntectonic porphyroblasts. Their model is based on the fact that deformation in rocks is heterogeneous and results in strain partitioning. This occurs on various scales, such that during different stages of progressive bulk non-coaxial deformation (e.g. simple shear), different portions of the deforming mass, from the macroscale right down to the microscale, experience one of several deformation types (Fig. 9.7). The style of deformation ranges from areas of no strain to areas dominated by progressive shortening strain, and areas of progressive shearing strain. The difference in deformability between matrix and porphyroblasts gives rise to strong deformation partitioning about the rigid porphyroblasts, and produces pronounced strain gradients close to their margins. This partitioning produces an ellipsoidal island of porphyroblast width that is protected from the effects of progressive shearing (Fig. 9.7(b)). For roughly equidimensional porphyroblasts, provided that they do not deform, and provided that deformation partitioning is maintained, with perfect decoupling between porphyroblast and matrix, they should not rotate. With reference to the Mount Isa region of Australia, Bell & Rubenach (1983) successfully demonstrated how the variety of common inclusion fabrics seen in syntectonic porphyroblasts could be explained without the need to invoke rotation. They interpreted the various inclusion trails in terms of the timing of porphyroblast nucleation and growth relative to different stages in the development of an S_2 crenulation cleavage (Fig. 9.8). In view of the fact that most porphyroblasts grow very rapidly once nucleated (Section 5.4), the inclusion trails preserved represent a 'snapshot' of a particular stage in textural and mineralogical evolution of the matrix. Inclusion trails with

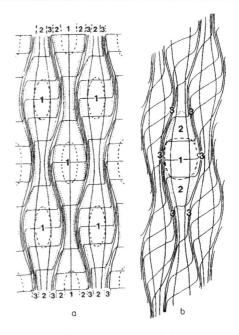

FIG. 9.7 (a) The distribution of deformation partitioning on a strain-field diagram constructed for the X–Z plane, representing a block of rock which has undergone non-coaxial progressive bulk inhomogeneous shortening. 1, No strain; 2, progressive shortening strain; 3, progressive shortening plus shearing strain (after Bell, 1985). (b) Deformation partitioning around a porphyroblast resulting from non-coaxial progressive bulk inhomogeneous shortening. The porphyroblast is outlined by the dashed line; the key to numbering is the same as in (a) (after Bell, 1985).

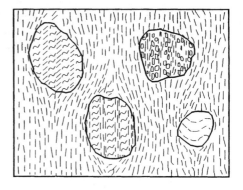

FIG. 9.8 A sketch showing porphyroblasts that have overgrown various stages of development of a D_2 crenulation cleavage (after Bell, 1985).

Porphyroblast—foliation relationships

S-shaped form would represent porphyroblast growth at a time when the rock possessed a strong crenulation fabric. Because a prominent crenulation cleavage can form with as little as 40–50% shortening, this might equally well indicate nucleation early in the deformation history or late-stage porphyroblastesis.

Although perfect decoupling between porphyroblast and matrix, and deformation partitioning about equidimensional porphyroblasts such as garnet, may theoretically suggest that they do not rotate during bulk simple shear, there is little evidence to suggest that rigid elongate porphyroblasts (e.g. hornblende) would not be rotated. Many lines of evidence suggest that during non-coaxial bulk shear inequidimensional crystals experience a couple and consequently become rotated into the shear direction, without necessarily experiencing any internal deformation. Evidence for this has been gained from shear-box modelling and from theoretical and computer modelling, as well as from evidence provided in natural examples such as porphyroclasts (Fig. 10.16) in rocks that have experienced moderate and high shear strains (e.g. Mawer, 1987; Prior, 1987; Passchier, 1987; van den Driessche & Brun, 1987; Passchier & Sokoutis, 1993). Observations confirm that at suitably high shear strains rigid inequidimensional objects can rotate through values at least up to approximately 120°. However, if porphyroblast growth is geologically very rapid, most rotation should not occur during growth, and thus spiralled inclusion fabrics should not develop (Fig.9.9, see also Barker, 1994). Many researchers would also argue that because there is probably imperfect decoupling between porphyroblast and matrix, then equidimensional porphyroblasts such as garnet would also rotate.

The non-rotation model of Bell and co-workers has been the focus of considerable debate in recent years, with many aspects of

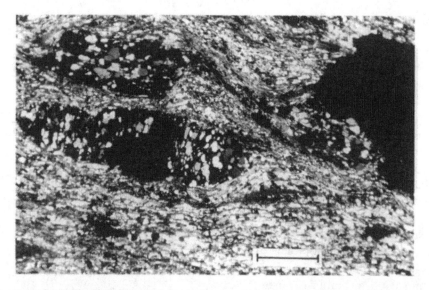

FIG. 9.9 Syntectonic hornblende porphyroblasts that have grown rapidly and in different orientations to overprint an earlier stage of fabric development, and preserve straight inclusion fabrics. Continued shearing after the short-lived phase of porphyroblastesis has rotated the two crystals into parallelism within the schistosity plane. The larger crystal, which grew with its long axis at a high angle to the initial fabric, shows sharply discordant S_i–S_e relationships. Hornblende mica–schist, Troms, Norway. Scale = 1 mm

the model and interpretations called into question. The following paragraphs examine some of the main points of criticism, but for an insight and full discussion of the main points against the model see Passchier et al. (1992), with a response by Bell et al. (1992), giving further comment in favour of the model and answering the criticisms raised by Passchier and co-workers.

Fyson (1975, 1980), Bell (1985), Steinhardt (1989) and Bell et al. (1992) have argued that the internal inclusion trails of porphyroblasts may maintain a remarkable consistency of orientation, both on the scale of a thin section and over large regions within an orogenic belt. Consistency of S_i–S_e relationships on the scale of a thin section are commonly reported, but depending on your preferred model they can either be interpreted in terms of rotation of the porphyroblasts by a similar amount relative to S_e (e.g. Oleson, 1982), or by rotation of S_e relative to 'stationary' porphyroblasts that record consistent S_i relating to some previously overgrown fabric (e.g. Fyson, 1980; Bell, 1985). Steinhardt (1989), Johnson (1992) and Bell et al. (1992) described examples in which S_i of porphyroblasts document near-constant orientation over large geographical areas despite intense later deformation. This sort of information has been used by the authors cited to indicate non-rotation of porphyroblasts relative to each other. Indeed, Bell (1985) went as far as to suggest that constantly oriented S_i indicates that the porphyroblasts have not rotated relative to fixed geographical co-ordinates and that the preserved S_i indicates the original orientation of an earlier fabric. Passchier et al. (1992) emphasised that in cases in which S_i shows constancy of orientation over a large area, Bell and others would use this as evidence to advocate non-rotation of porphyroblasts. However, when S_i shows variable orientation across an area, Bell and others have argued in terms of variable foliation orientation (due to folding) prior to porphyroblastesis. With this sort of interpretation as a possibility, it is clearly difficult to evaluate the validity of the non-rotation model. Another point that makes the consistency of orientation of S_i over a large geographical area all the more remarkable is that any post-porphyroblastesis rigid block rotation (e.g. by faulting) would cause rotation relative to a fixed geographical reference frame. This would cause rotation of S_i from porphyroblasts of one block relative to S_i in porphyroblasts of a neighbouring block.

The porphyroblast non-rotation model developed by Bell and co-workers since the mid-1980s has certainly renewed interest and stimulated considerable discussion concerning the interpretation of porphyroblast–foliation relationships. The interpretation of 'shallow-S' and certain other common porphyroblast inclusion trails in terms of porphyroblast overgrowth of specific stages of crenulation development (after Bell & Rubenach, 1983) has received popular support. However, the more radical interpretation that porphyroblasts do not rotate relative to fixed external co-ordinates during polyphase deformation, as advanced by Bell (1985) and others, has not received the same level of support. On the basis of currently available data, the majority of metamorphic geologists still favour the traditional view that porphyroblasts do rotate relative to fixed external co-ordinates.

Millipede microstructures

The term 'millipede microstructure' was introduced by Bell & Rubenach (1980) for certain syntectonic porphyroblasts which, at a given margin at which S_i passes into S_e, show S_e deflected in opposite directions (Plate 5(e)). Millipede microstructures were interpreted by Bell & Rubenach (1980) as evidence for bulk heterogeneous shortening that, at least locally, is near coaxial. Critical discussions by Passchier et al. (1992) and Johnson (1993b) indicate that such microstructures can develop in various deformation regimes. Recent papers by Johnson & Bell (1996) and Johnson & Moore (1996) discuss the usefulness of 'millipedes' and other types of oppositely concave microfolds (OCMs) seen adjacent to some porphyroblasts. The type 1 OCMs of Johnson & Bell (1996) are the cut-effect (ID-type (closed loop) inclusion trails (Section 9.1, Fig. 9.1) described by Powell & Treagus (1970) and MacQueen & Powell (1977). They are seen in thin sections cut parallel to the spiral axis of porphyroblasts with S-shaped inclusion trails. Porphyroblasts displaying such inclusion patterns are classically interpreted as forming during a single phase of growth in a regime of strongly non-coaxial deformation. Johnson (1993b) reports that such microstructures are equally well explained by the porphyroblast non-rotation model as they are by the porphyroblast rotation model. The Type 2 OCMs of Johnson & Bell (1996) are the classic 'millipede' microstructures of Bell & Rubenach (1980). As stated above, these are commonly considered indicative of bulk coaxial shortening, but Johnson & Bell (1996) and Johnson & Moore (1996) have demostrated that they may form by heterogeneous extension, and cannot be used to indicate specific deformation histories (i.e. degree of non-coaxiality of the deformation).

9.2.3 Recognition and interpretation of post-tectonic crystals

Post-tectonic crystals are most easily distinguished by their complete overgrowth of the main fabric (cleavage or schistosity) of the rock. That is to say, the fabric does not envelop the porphyroblasts, but abuts against the crystal, and in cases in which an inclusion fabric is seen the fabric passes right through the porphyroblast (Plate 6(a)). Post-tectonic crystals are characteristically associated with contact metamorphism adjacent to igneous intrusions, late thermal overprints during regional metamorphism and retrograde metamorphism associated with uplift. Pseudomorphing of one mineral by an aggregate of another mineral (e.g. garnet pseudomorphed by chlorite; see Plate 4(c) and Chapter 7) is typically a late-tectonic or post-tectonic feature associated with retrogression during uplift and waning P-T conditions. Since post-tectonic crystals have nucleated and grown after the main phases of deformation, they will not show any strain effects. This means that they will not exhibit features such as undulose extinction or kinking, and will lack pressure shadows (Plate 6(a)).

Elongate post-tectonic minerals usually show a lack of preferred orientation (Plate 6(b)), in stark contrast to the pronounced alignment of comparable pre-foliation and syntectonic crystals. Post-tectonic overgrowth of crenulations commonly gives rise to gently curved to S-shaped internal fabrics in porphyroblasts (Plates 6(c) & (f)). Superficially, these may appear similar to the syntectonic 'spiralled' inclusion fabrics described earlier. However, careful examination will reveal that the matrix does not deflect around post-tectonic porphyroblasts (whereas it does around syntectonic crystals), and that the crenulations being overgrown are present throughout the thin section. Where post-tectonic crenulation overgrowth has occurred, a range of different inclusion fabrics usually exists, as a function of different parts of the crenulation being overgrown. It is not uncommon to see several crenulations overgrown by a single porphyroblast (e.g. Plate 6(f)).

9.2.4 Complex porphyroblast inclusion trails and multiple growth stages

Late-stage metamorphism does not always produce entirely new porphyroblasts, but in many examples post-tectonic rim growth on previously formed syntectonic porphyroblasts is recognised (Plates 6(d) & (e)). Where rocks (especially those of orogenic metamorphism) have experienced a late thermal overprint, it is common to find rim growth on pre-existing porphyroblasts. In such cases a sharp break in textural zonation is often seen. The core will often show syntectonic relationships, while the rim exhibits post-tectonic features and overgrows the matrix foliation (e.g. Plate 6(d)). Rim overgrowths can be complete (Plate 6(e)) or partial (Plate 6(d)). The post-tectonic rims generally form at the margin in contact with mica-rich matrix, whereas at margins in contact with quartz-rich layers or pressure shadows the rim is absent or poorly developed. The most probable explanations for this are either in terms of locally unfavourable chemistry of the quartz-rich areas, or easier nucleation and growth in the more highly strained areas out of the pressure shadows.

More complex porphyroblast inclusion patterns are also encountered, especially in rocks that have experienced a polyphase deformation and metamorphic history during regional orogenesis (e.g. MacQueen & Powell, 1977; Bell & Johnson, 1989; Hayward, 1992). These more complex inclusion patterns commonly show distinct inflexions or breaks in inclusion trails, and are generally interpreted in terms of multistage porphyroblast growth, and overgrowth of more than one deformation event (see also Chapter 12). Bell & Johnson (1989) illustrated porphyroblasts that they interpret to have overgrown up to eight successive foliations relating to different deformation events. They suggested that given the right garnet and a favourable situation, the full history of orogenesis can be preserved in a single porphyroblast. They interpreted the complex inclusion trails that they describe from Vermont (USA), Pakistan and Nepal in terms of non-rotation of porphyroblasts, and as indicating repeated cycles of uplift and collapse during orogenesis. Many researchers have serious reservations about this, and Passchier *et al.* (1992) suggested that many of the so-called *truncation planes* reported by Bell & Johnson (1989) are really *deflection planes*. According to Passchier *et al.* (1992), such deflection planes can be explained in terms of a single phase of progressive deformation, and consequently the porphyroblasts illustrated by Bell & Johnson (1989) could be interpreted in terms of far fewer deformation events and overgrowth stages than they suggest.

From the above discussion, it is apparent that despite much research over many decades, there remains considerable debate and controversy concerning the interpretation of even the simplest types of inter-tectonic and syntectonic porphyroblast inclusion trails. For porphyroblasts with multiple growth stages and more complex inclusion trails, the problem is magnified, and it not surprising that the interpretations vary greatly from one author to the next. There are clearly many secrets locked away in inclusion trails of complex syntectonic porphyroblasts, and despite their great potential for unravelling deformation–metamorphism interrelationships during orogenesis, they remain some of the most difficult microstructures to interpret in a reliable and objective manner.

References

Barker, A.J. (1994) Interpretation of porphyroblast inclusion trails: limitations imposed by growth kinetics and strain rates. *Journal of Metamorphic Geology*, 12, 681–694.

Bell, T.H. (1985) Deformation partitioning and porphyroblast rotation in metamorphic rocks: a radical reinterpretation. *Journal of Metamorphic Geology*, 3, 109–118.

Bell, T.H. & Johnson, S.E. (1989) Porphyroblast inclusion trails: the key to orogenesis. *Journal of Metamorphic Geology*, 7, 279–310.

Bell, T.H. & Rubenach, M.J. (1980) Crenulation cleavage development – evidence for progressive, bulk inhomogeneous shortening from 'millipede' microstructures in the Robertson River Metamorphics. *Tectonophysics*, 68, T9–T15.

Bell, T.H. & Rubenach, M.J. (1983) Sequential porphyroblast growth and crenulation cleavage development during progressive deformation. *Tectonophysics*, 92, 171–194.

Bell, T.H., Johnson, S.E., Davis, B., Forde, A., Hayward, N. & Wilkins, C. (1992) Porphyroblast inclusion trail orientation data: *eppure non son girate! Journal of Metamorphic Geology*, 10, 295–307.

Fyson, W.K. (1975) Fabrics and deformation of Archaean metasedimentary rocks, Ross Lake – Gordon Lake area, Slave Province, Northwest Territories. *Canadian Journal of Earth Sciences*, 12, 765–776.

Fyson, W.K. (1980) Fold fabrics and emplacement of an Archaean granitoid pluton, Cleft Lake, Northwest Territories. *Canadian Journal of Earth Sciences*, 17, 325–332.

Hayward, N. (1992) Microstructural analysis of the classical spiral garnet porphyroblasts of south-east Vermont: evidence for non-rotation. *Journal of Metamorphic Geology*, 10, 567–587.

Johnson, S.E. (1992) Sequential porphyroblast growth during progressive deformation and low-P high-T (LPHT) metamorphism, Cooma Complex, Australia: the use of microstructural analysis in better understanding deformation and metamorphic histories. *Tectonophysics*, 214, 311–339.

Johnson, S.E. (1993a) Unravelling the spirals: a serial thin section study and three-dimensional computer-aided reconstruction of spiral-shaped inclusion trails in garnet porphyroblasts. *Journal of Metamorphic Geology*, 11, 621–634.

Johnson, S.E. (1993b) Testing models for the development of spiral-shaped inclusion trails in garnet porphyroblasts: to rotate or not to rotate, that is the question. *Journal of Metamorphic Geology*, 11, 635–659.

Johnson, S.E. & Bell, T.H. (1996) How useful are 'millipede' and other similar porphyroblast microstructures for determining synmetamorphic deformation histories? *Journal of Metamorphic Geology*, 14, 15–28.

Johnson, S.E. & Moore, R.R. (1996) De-bugging the 'millipede' porphyroblast microstructure: a serial thin-section study and 3-D computer animation. *Journal of Metamorphic Geology*, 14, 3–14.

MacQueen, J.A. & Powell, D. (1977) Relationships between deformation and garnet growth in Moine (Precambrian) rocks of western Scotland. *Bulletin of the Geological Society of America*, 88, 235–240.

Mawer, C.K. (1987) Shear criteria in the Grenville Province, Ontario, Canada. *Journal of Structural Geology*, 9, 531–539.

Oleson, N.Ø. (1982) Heterogeneous strain of a phyllite as revealed by porphyroblast–matrix relationships. *Journal of Structural Geology*, 4, 481–490.

Passchier, C.W. (1987) Stable positions of rigid objects in non-coaxial flow: a study of vorticity analysis. *Journal of Structural Geology*, 9, 679–690.

Passchier, C.W. & Sokoutis, D. (1993) Experimental modelling of mantled porphyroclasts. *Journal of Structural Geology*, 15, 895–910.

Passchier, C.W., Trouw, R.A.J., Zwart, H.J. & Vissers, R.L.M. (1992) Porphyroblast rotation: *eppur si muove? Journal of Metamorphic Geology*, 10, 283–294.

Powell, D. & Treagus, J.E. (1970) Rotational fabrics in metamorphic minerals. *Mineralogical Magazine*, 37, 801–813.

Prior, D. (1987) Syntectonic porphyroblast growth in phyllites: textures and processes. *Journal of Metamorphic Geology*, 5, 27–39.

Rosenfeld, J.L. (1970) Rotated garnets in metamorphic rocks. Geological Society of America Special Paper 129, 105 pp.

Schoneveld, Chr. (1977) A study of some typical inclusion patterns in strongly paracrystalline garnets. *Tectonophysics*, 39, 453–471.

Spry, A. (1963) The origin and significance of snowball structure in garnet. *Journal of Petrology*, 4, 211–222.

Steinhardt, C.K. (1989) Lack of porphyroblast rotation in non-coaxially deformed schists from Petrel Cove, South Australia, and its implications. *Tectonophysics*, 158, 127–140.

van den Driessche, J. & Brun, J.P. (1987) Rolling structures at large shear strain. *Journal of Structural Geology*, 9, 691–704.

Vernon, R.H., Collins, W.J. & Paterson, S.R. (1993) Pre-foliation metamorphism in low-pressure/high-temperature terrains. *Tectonophysics*, 219, 241–256.

Wilson, M.R. (1971) On syntectonic porphyroblast growth. *Tectonophysics*, 11, 239–260.

Zwart, H.J. (1962) On the determination of polymetamorphic associations and its application to the Bosost area (central Pyrenees). *Geologische Rundschau*, 52, 38–65.

Chapter ten

Shear-sense indicators

10.1 Introduction

It is common in a wide range of metamorphic rocks from a variety of metamorphic environments to see evidence for a component of non-coaxial deformation. This chapter describes some of the most commonly observed kinematic indicators in metamorphic rocks, and discusses how they may be used to assess sense of shear. In essence, there are seven main categories to consider, and these can be grouped under the following headings:

(i) sense of fold overturning and vein asymmetry;
(ii) S–C fabrics, shear bands and mica-fish;
(iii) differentiated crenulation cleavages;
(iv) spiralled inclusion fabrics;
(v) mantled porphyroclasts;
(vi) strain shadows;
(vii) grain-shape fabrics and crystallographic preferred orientations.

In attempting to determine the shear sense, and where possible the amount of shear, it is important to have an oriented sample. The thin section should be cut parallel to the X-direction of the finite strain ellipsoid (generally defined by stretching/extension lineation of the rock) and perpendicular to the X–Y plane (generally defined by cleavage/schistosity). The slice examined in thin section is therefore the X–Z plane, where X, Y and Z are the principal axes of the finite strain ellipsoid ($X \geq Y \geq Z$). Oblique cuts will certainly give erroneous assessment of the amount of shear, and could give a false impression of shear sense. Because of this they should be avoided where possible.

Another point to bear in mind is that fault and shear zones that have been reactivated at various times during their evolution will typically be dominated by textures and microstructures developed in the last stage of movement, or the most intense period of deformation. However, there may be domains within the shear zone that preserve elements of earlier stages in the evolution, such that the shear sense may not always be consistently in the same direction. For example, it is quite common in fold and thrust belts to find early compression-related thrusts that become reactivated as extensional structures during late-orogenic relaxation and collapse. As a result, the shear-sense indicators in such zones will often show mixed polarity. It is therefore crucial to be fully aware of the deformation phase to which a given shear-sense indicator relates, and not to mix shear-sense indicators from various events.

10.2 Vein asymmetry and sense of fold overturning

The attitude of veins, and sense of vergence of asymmetrical folds, are features commonly used in the field to determine sense of shear (both on a local and on a regional scale). Such structures exist on a variety of scales, and thin-section

Shear-sense indicators

studies often prove helpful in determining the precise relationships between veining and deformation.

The history of vein development in a rock is often complex, with some veins being present pre-shearing, others forming during shearing and some forming post-shearing. By careful study of cross-cutting relationships between different vein sets, and studies of relationships between veins and other structures of the rock (e.g. folds and foliations), it should be possible to determine the full history of vein development.

The superposition of a period of simple shear on variably oriented early veins will lead to folding or crenulation of some veins, while others become thinned and boudinaged. How a particular vein behaves depends on its initial attitude relative to the superposed strain ellipsoid (Fig. 10.1). Veins that lie in the shortening field of the strain ellipsoid (e.g. A) will become folded, while those in the maximum elongation (extension) direction (e.g. B) will become thinned and boudinaged. In the example illustrated, increasing shear will rotate the veins in a clockwise sense, so that vein A may ultimately move into the extensional field and undergo unfolding and boudinage.

The foregoing discussion demonstrates that with careful observation early veins can give important information relating to the sense of any superimposed shear. In using this approach, care should always be taken to establish that simple shear is primarily responsible for vein folding, remembering that folded veins are also produced by pure shear flattening of early veins oriented at a high angle to the principal axis of compression.

In the case of veins formed during simple shear, they will tend to form as *en echelon* sets in fractures developed perpendicular to the direction of maximum extension (Fig. 10.2(a)). These become rotated as shear progresses, and during rotation the vein develops a sigmoidal trace (Figs 10.2(b) & 11.2) from which sense of shear can be determined ('S'-shape = sinistral; 'Z'-shape = dextral). The sense of fold overturning or vergence can also be used to give a general indication of shear sense. Initially gentle, slightly overturned fold structures become increasingly overturned and asymmetrical as shear progresses (Fig. 10.3). Such asymmetrical folds can range from regional-scale structures down to mm-scale crenulations observable in thin section. Although commonly recorded, the use of fold asymmetry as the sole line of evidence for shear sense is not recommended. This is because similar features can be produced by pure shear flattening of features oblique to the principal compressional axes (Fig. 10.4), and in such situations any shear-sense interpretation would be entirely false.

In brittle, sub-greenschist facies regimes, Sibson (1990) illustrates how vein and fracture arrays and networks developed to accommo-

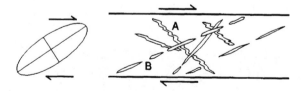

FIG. 10.1 A schematic illustration of veins deforming during superimposed simple shear. The style of deformation is a function of vein orientation relative to the superimposed strain ellipsoid, and can be used to determine the shear sense.

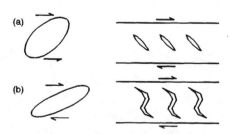

FIG. 10.2 A schematic illustration showing the development of sigmoidal tension gashes.

S—C fabrics, shear bands and mica-fish

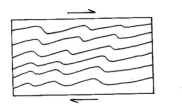
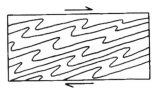

FIG. 10.3 A schematic illustration of the development of asymmetrical folds during simple shear.

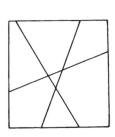
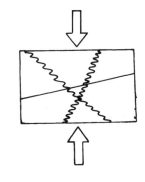

FIG. 10.4 A schematic illustration of asymmetrical folded veins formed during pure shear.

date local extension at dilational jog sites may be used to evaluate shear sense (Fig. 10.5). Although characteristically observed on the exposure-scale, many such features can also be seen on hand-specimen and thin-section scale.

10.3 S–C fabrics, shear bands and mica-fish

In rocks such as mylonites and phyllonites that have experienced high strain during intense non-coaxial ductile shear, it has been recognised that two fabrics commonly develop

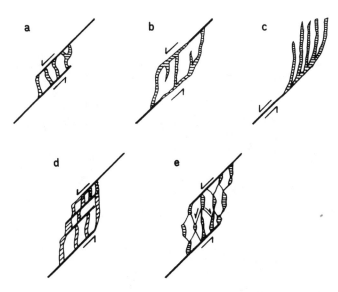

FIG. 10.5 A schematic illustration of the various vein/fracture networks that are commonly observed at dilational fault jog sites, and which can be used to evaluate shear sense. Such features are not scale-dependent, and can be observed from map-scale down to thin-section scale: (a) ladder vein; (b) cymoid loop; (c) horsetail; (d, e) alternate shear/extensional mesh models for dilational jogs (based on Fig. 4.16 of Sibson, 1990; courtesy of the Mineralogical Society of Canada).

Shear-sense indicators

simultaneously in association with this shearing. These are S–C fabrics (Fig. 10.6), the C surfaces being parallel with the shear zone margin while the S-surfaces are oblique to this (e.g. Berthé et al., 1979; Lister & Snoke, 1984; Passchier, 1991; Blenkinsop & Treloar, 1995). The field appearance of such rocks is usually very distinctive, with the intersection of the two cleavages giving rise to a texture which has commonly been described as 'fish-scale', 'button schist' or 'oyster shell' texture (Fig. 10.7). At initiation the S fabric develops at an angle (α) of about 45° to C, but as shearing progresses the angle α diminishes as S rotates towards C. Most typically α is between 15° and 45°, but in extreme cases the two surfaces may become sub-parallel. In general, as the angle α decreases, so the density of C surfaces increases.

With knowledge of the attitude of S fabrics, and the shear-zone boundary orientation (which is parallel to C fabrics), it is thus possible to ascertain the sense of shear of the zone in question (Fig. 10.6). The detail of such fabrics is often well exhibited in thin section, and when present should corroborate evidence from δ-type porphyroclasts (Section 10.6). Unfortunately, this simple picture of S–C fabrics is complicated by secondary planar fabrics commonly developed at a later stage in such rocks. These are usually antithetic to the orientation of S surfaces, and are often more pronounced and more easily recognised than such surfaces. They are termed C' fabrics (e.g. Ponce de Leon & Choukroune, 1980) or extensional crenulation

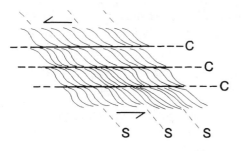

FIG. 10.6 A schematic illustration of S-C fabrics, and their relationship to the direction of shear.

FIG. 10.7 S–C and C' fabrics developed in phyllonite of the Abisko Nappe, Swedish Caledonides. The compass-clinometer (lower left), for scale, is 8 cm long.

S—C fabrics, shear bands and mica-fish

cleavages (e.g. Platt & Vissers, 1980; Platt, 1984; Dennis & Secor, 1987), and are especially well developed in phyllonites (Figs 10.7 & 10.8). In the brittle regime, synthetic Reidel shears (e.g. R_1 of Logan et al., 1979; Chester et al., 1985; Arboleya & Engelder, 1995) have the equivalent orientation to C' surfaces of ductile shear zones.

In ductile thrust zones and 'slides', C' surfaces characteristically dip towards the foreland, and may thus be used to indicate bulk transport direction. They usually form an angle of approximately 15–35° to C surfaces and the

FIG. 10.10 A schematic illustration of mica-fish as shear indicators.

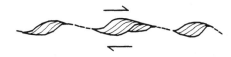

FIG. 10.11 Trails and stair-stepping between mica-fish.

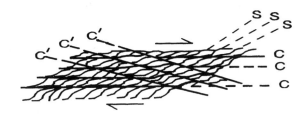

FIG. 10.8 A schematic illustration of C' fabrics in relation to S and C fabrics.

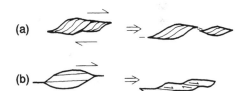

FIG. 10.12 Multiplication of mica-fish by break-up and dispersion.

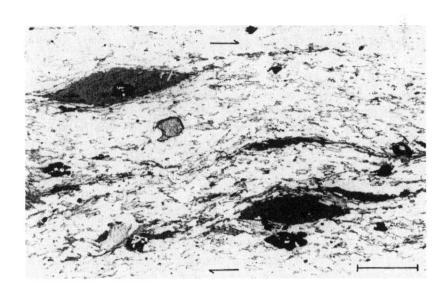

FIG. 10.9 Mica-(biotite-)fish in schist of the Høgtind nappe, Norwegian Caledonides. Scale = 1 mm (PPL).

Shear-sense indicators

shear-zone boundary. In thrusts of compressional orogenic belts they are commonly associated with late-stage orogenic collapse. The term *shear band* is used descriptively in many publications to define discrete highly sheared surfaces within mylonitic rocks. 'Shear bands' are recognised in several orientations, and while normally related to C' surfaces as described above, C surfaces have also been recognised as 'shear bands'. Since the term 'shear band' does not imply a specific orientation, it is recommended (in order to avoid confusion) that it only be used to describe the nature of C or C' surfaces, and not to have any general orientational implications.

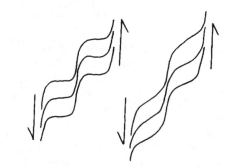

FIG. 10.13 Crenulation cleavage asymmetry as a basis for evaluation of local shear sense (based on Fig. 11 of Bell & Johnson, 1992).

Mica-fish, termed as such because of their generally 'fish-like' lozenge-shaped appearance, are useful kinematic indicators in many mylonites, phyllonites and schists. The mica crystals are large pre-existing grains (effectively 'porphyroclasts') or early porphyroblasts which are deformed by a combination of brittle and crystal–plastic processes. The asymmetry of the mica 'fish' shape (Fig. 10.9), with 001 cleavage typically either tilted sub-parallel to S or C surfaces can be used to determine sense of shear (Fig. 10.10). Lister & Snoke (1984) describe how mica-fish are usually linked by thin trails of mica fragments, and when extensively developed show a stair-stepping from one mica-fish to the next (Fig. 10.11). Mica-fish may multiply and disperse by a process typically involving slip on the basal plane, causing the break-up of early large micas: this is illustrated in Fig. 10.12.

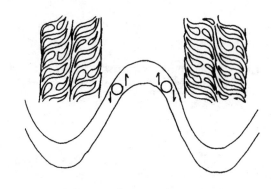

FIG. 10.14 Determination of the shear sense around a fold on the basis of minor structure asymmetry (based on Fig. 13 of Bell & Johnson, 1992).

10.4 Differentiated crenulation cleavages

Bell & Johnson (1992) describe how shear sense can be obtained from crenulation asymmetry against differentiated crenulation cleavages. Indeed, they describe how any old foliation being rotated during the development of a new foliation can be used to determine local shear sense, as long as it shows asymmetry due to deflection when passing from domains of low strain to domains of high strain (Fig. 10.13). Deflection of old foliations in the manner illustrated in Fig. 10.13 is a common feature of many metamorphic rocks of the greenschist facies, amphibolite facies and in shear zones. They are especially pronounced where a strongly differentiated crenulation cleavage is developed, and since this a common feature, are potentially one of the most useful indicators of local shear sense. Bell & Johnson (1992) illustrate how local shear sense on each limb of a fold can be usefully determined using differentiated crenulation cleavage as a shear-sense indicator, and how this local sense of shear is comparable to the bulk sense of shear in limbs between antiform and synform. It will be noticed (Fig 10.14) that this shearing is directed sub-parallel to the fold axial surface, even though the area as a whole is experiencing inhomogeneous bulk shortening perpendicular to the axial surface. In view of this, the importance of gaining a complete understanding of structures on all scales should be apparent in order to evaluate local and regional shear sense and tectonometamorphic evolution of a given terrain.

10.5 Spiralled inclusion trails

In Chapter 9 the traditional, and now controversial interpretation of S-shaped inclusion trails in syntectonic porphyroblasts as evidence for porphyroblast rotation is discussed. The development of such fabrics in this manner, and their use as shear-sense indicators is illustrated in Fig. 9.5. More recently, it has been argued that porphyroblasts, and especially equidimensional porphyroblasts such as garnet, do not rotate (e.g. Bell, 1985). Instead, the external fabric changes orientation with respect to the porphyroblast by way of progressive foliation transposition or shearing. When this occurs synchronous with porphyroblast growth it too will give rise to an S-shaped inclusion trail (Fig. 9.6), but with a completely opposite sense of rotation to that deduced by assuming that the porphyroblast rotated (cf. Fig. 9.5). Johnson (1993) reviews spiral-shaped inclusion trails of porphyroblasts in some depth, and concludes that most geometries are consistent with both the rotation and the non-rotation models. Because of this, and in view of this major difference in shear sense being a function of the interpretation of the mechanism by which the S fabric formed, any interpretation based on such fabrics should be made with extreme caution. Adding still further to the problem, it is now generally realised that porphyroblast growth is much faster than had previously been considered (i.e. commonly < 1 Ma, and even < 0.1 Ma; Barker (1994) and Section 5.4). This means that under most circumstances porphyroblasts develop very rapidly with respect to ongoing deformation, and that many S-shaped inclusion trails probably represent syntectonic crenulation overgrowth. Clearly, this sheds still further doubt on the use of S-shaped inclusion trails as kinematic indicators, and consequently their use is not recommended. More reliable are 'rolling structures' and related features characteristic of porphyroclast systems.

10.6 Mantled porphyroclasts and 'rolling structures'

Passchier & Simpson (1986) review porphyroclast systems as kinematic indicators, with more recent work undertaken by Passchier et al. (1993) and Passchier (1994). On geometric grounds, Passchier & Simpson (1986) subdivide porphyroclasts into σ– and δ-types (Fig. 10.15). σ-type porphyroclasts have wedge-shaped tails and a 'stair-stepping' symmetry (Fig. 10.15(a)). The median lines of the recrystallised tails (of the same mineral as the porphyroclast) lie on each side of a central reference plane. The side nearest the reference plane characteristically has a concave curvature, while that farthest away is planar. σ-type porphyroclasts are themselves subdivided into σ_a- and σ_b-types. The former represent isolated porphyroclasts (e.g. feldspar or hornblende) in a homogeneous foliation of uniform orientation, though locally deflected around the porphyroclast (Fig. 10.16). In some cases porphyroclasts remain rigid, and pressure shadows ('beards') of a different mineral phase (e.g. quartz) develop with geometry similar to σ_a-type tails. σ_b-type porphyroclasts are generally feldspars associated with S–C quartz feldspar mylonites (e.g. Berthé et al., 1979). They have flat surfaces along the C planes, and tend to occur in clusters. Like σ_a-type clasts, the recrystallised tails of σ_b-type clasts can be of

Shear-sense indicators

different or mixed composition in comparison to the clast itself.

δ–type porphyroclast systems (also termed 'rolling structures'), differ from σ-type systems by virtue of the fact that the median-line of the tails crosses the central reference plane (Fig. 10.15(b)). In δ-type porphyroclasts the tails are thin, and tight embayments exist between the tail and porphyroclast. Such tails only occur around equidimensional or very slightly elongate porphyroclasts, and usually extend for a considerable distance from the porphyroclast (Figs 10.15(b) & 10.17). On the basis of studies of granitoid mylonites from the Grenville Province, Mawer (1987) gave a clear illustration of how δ-type porphyroclasts can develop from σ-type porphyroclasts during progressive shearing (Fig. 10.15(c)). In cases of prolonged or episodic shearing, more complex examples are likely, including folding of thin δ-type tails (Fig. 10.18), and complex σ–δ type porphyroclast relationships (Fig. 10.19). It is important to note that Z-shaped strain shadow and σ-type porphyroclast asymmetry indicates the same sense of shear as S-shaped δ-type porphyroclast ('rolling structure') asymmetry (compare Figs 10.15(a) & (b)). Without this awareness and the ability to distinguish δ-type porphyroclasts, the sense of shear can easily be misinterpreted.

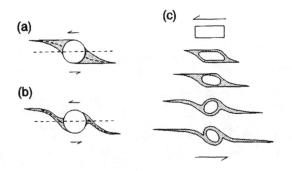

FIG. 10.15 Porphyroclast systems as kinematic indicators. (a) A σ-type porphyroclast (modified after Fig. 2a of Passchier & Simpson, 1986). (b) A δ-type porphyroclast (modified after Fig. 2e of Passchier & Simpson, 1986). (c) The development of δ-type porphyroclasts from σ-type porphyroclasts (modified from Mawer, 1987).

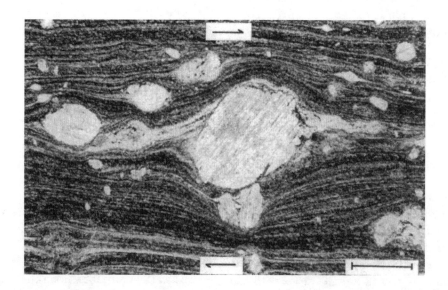

FIG. 10.16 A σ_a-type feldspar porphyroclast. Granitoid mylonite, South Armorican Shear Zone, France. Scale = 0.5 mm (PPL).

Strain shadows

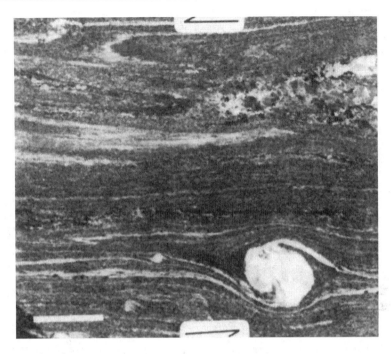

FIG. 10.17 A δ-type feldspar porphyroclast (after Fig. 5a (polished rock slab) of Mawer, 1987; courtesy of Elsevier Science). Scale = 2 cm.

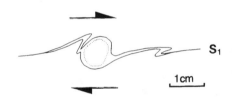

FIG. 10.18 A folded tail of δ-type porphyroclast (after Fig. 3 of van den Driessche & Brun, 1987).

10.7 Strain shadows

Ramsay & Huber (1983), in their review of pressure (strain) shadow characteristics distinguish two main types of pressure (strain) shadows; namely, 'pyrite type' and 'crinoid type'. The latter involves progressive fibre growth (of the same mineral type as the rigid object) between the object or its strain shadow and the displaced matrix surface. This type of strain shadow occurs around crinoid ossicles in deformed limestones (low-grade metacarbonates), but in general is rare in metamorphic rocks. Crystallographic continuity is retained from object to fibre.

'Pyrite-type' strain shadows are very

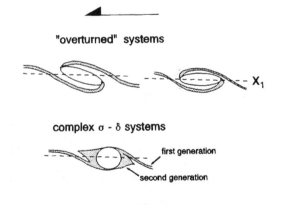

FIG. 10.19 Overturned and complex σ–δ porphyroclast systems (after Fig. 2h of Passchier & Simpson, 1986).

171

Shear-sense indicators

common in deformed greenschist facies and lower-grade metamorphic rocks. The development of such fibrous shadows (or 'fringes') involves incremental fibre growth (of mineral species different from the rigid object 'core') at the interface between the core (usually porphyroblast or porphyroclast) and the inner surface of its strain shadow. In some instances composite pressure shadows develop involving fibres of several mineral species (Figs 10.20(a) & (b)).

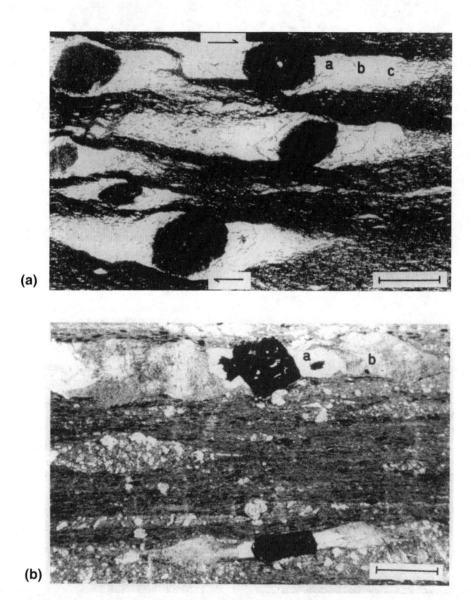

Fig. 10.20   X–Z sections of three natural examples of pressure shadows. In all cases the scale is 1 mm. (a), (c) XPL; (b) PPL. (a) Fine-grained cordierite schist/phyllite with complex deformable pressure shadows of calcite (a), quartz (b) and Qtz + Ms (c) adjacent to the cordierite porphyroblasts. The cordierite appears very dark because it contains extremely abundant and fine-grained inclusions of quartz and opaques. (b) Pyritic tuff with pressure shadows of quartz (a) and chlorite (b) adjacent to euhedral pyrite crystals.

Depending on the way in which they grow, fibres of 'pyrite-type' shadows can be subdivided into two categories. The first of these, 'displacement-controlled fibres', show consistent geometry of progressive growth of fibres along the displacement path, between pressure shadow wall and towards the resistant object (Fig. 10.21). The second type, 'face-controlled fibres', exhibit fibre growth normal to the face(s) of the rigid object, irrespective of the

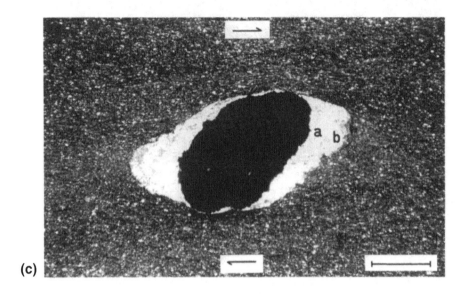

(c) A framboidal pyrite crystal in slate, with deformable pressure shadows of fibrous quartz (a) and calcite (b).

FIG. 10.21 Displacement-controlled fibres around pyrite (after Etchecopar & Malavieille, 1987). Scale = 1 mm (XPL). (Reproduced with permission of Elsevier Science).

Shear-sense indicators

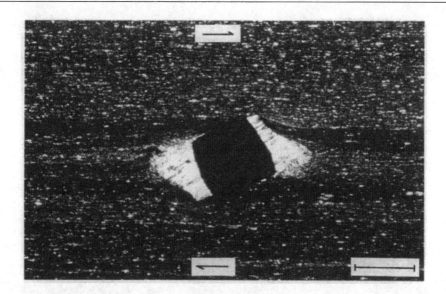

FIG. 10.22   Face-controlled quartz fibres developed in a pressure shadow adjacent to pyrite. Slate, south Cornwall, England. Scale = 1 mm (XPL).

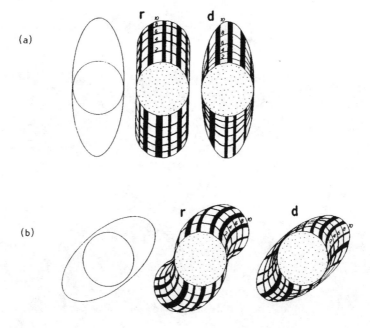

FIG. 10.23   Geometric differences between rigid fibre (r) and deformable fibre (d) pressure shadows in situations of coaxial (a) and non-coaxial (b) deformation (after Ramsay & Huber, 1983, Figs 14.15–14.16; courtesy of Academic Press).

displacement directions (Fig. 10.22). Depending on the *P–T* conditions, the fibres/pressure shadows may be deformable or rigid. At high temperatures and in more ductile conditions, pressure shadows are generally deformable with strong recrystallisation while under lower-temperature conditions they are usually rigid and fibrous. The geometric contrasts between 'rigid fibre' (r) and 'deformable fibre' (d) pressure shadows in the situation of (a) coaxial and (b) non-coaxial deformation are visually summarised in Fig. 10.23.

Studies by Etchecopar & Malavieille (1987) compare various computer-generated 'pyrite-type' pressure shadows with examples taken from natural rocks. From this they are able to estimate the bulk strain and shear sense in the deformed rocks. They consider pressure shadow asymmetry in *X–Z* sections to be one of the best criteria for determining sense of shearing in rocks. Such analysis clearly has wide application in the study of metamorphic terrains. Most obviously, pressure shadows around pyrite porphyroblasts from low-grade 'slate belts' can be studied, but pressure shadows around porphyroblasts, such as garnet in higher-grade schists, can also be assessed in order to determine the sense and amount of shear.

The types of pressure shadow fibres predicted for successive amounts of shear (γ = 0–4; all other variables constant), for a model based on 'deformable and face-controlled' fibres, are shown in Fig. 10.24. In contrast, the nature of pressure shadow fibres predicted for 'rigid and displacement controlled' fibres over the same range of incremental shear strain is illustrated in Fig. 10.25. In Fig. 10.26, the predicted developments in situations of pure

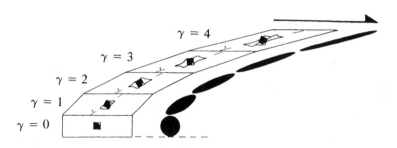

FIG. 10.24 A computer simulation of deformable, face controlled pressure shadow fibres developed after various increments of shear strain (i.e. γ = 0–4) (modified after Fig. 5 of Etchecopar & Malaveille, 1987).

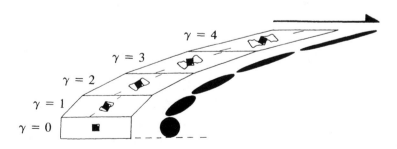

FIG. 10.25 A computer simulation of rigid, displacement-controlled pressure shadow fibres after various increments of shear strain (i.e. γ = 0–4) (modified after Fig. 6 of Etchecopar & Malavieille, 1987).

Shear-sense indicators

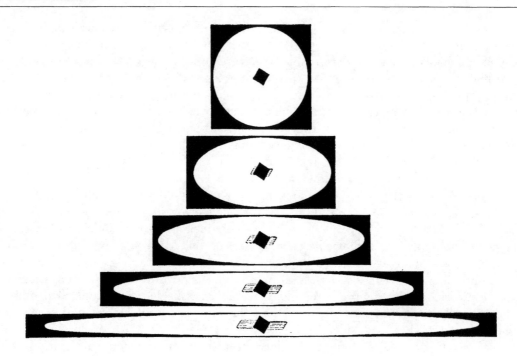

FIG. 10.26 A computer simulation of pressure shadow fibres expected in cases of pure shear flattening, assuming the fibres to be rigid and displacement-controlled (after Fig. 7 of Etchecopar & Malavieille, 1987). (Reproduced with permission of Elsevier Science).

shear flattening are illustrated. By comparison with Fig. 10.24, it is apparent that the natural example of pressure shadow fibres illustrated in Fig. 10.22 is face-controlled, and indicative of bulk shear of the order of $\gamma = 1$. X–Z sections for some more natural examples of pressure shadows are shown in Fig. 10.20. By comparison with Figs 10.24–10.26, a rough visual estimate of the dominant fibre type and amount and sense of shear can be made for each (see the figure caption for details).

10.8 Grain-shape fabrics and crystallographic preferred orientations

It is common, especially in quartzofeldspathic mylonites and highly sheared carbonates, to observe distinct grain-shape fabrics (Fig. 10.27) oblique to the shear-zone margins and C surfaces (e.g. Simpson & Schmid, 1983; Lister & Snoke, 1984; Schmid et al., 1987; De Bresser, 1989; Shelley, 1995). Such fabrics are especially common in monomineralic greenschist facies mylonites, and usually show an obliquity of 25–40° to the shear-zone margin. Oblique shape fabrics are defined by alignment of the long axes of dynamically recrystallised grains, especially of quartz or calcite, depending on the lithology. Since the rock, and grains, may recrystallise more than once during the deformation and metamorphic history, the particular shape fabric will probably relate to the latest stages of deformation only. Nevertheless, this may be a useful kinematic indicator, since the flattening plane of the new grains will lie approximately perpendicular to the direction of maximum compression during the final increment of deformation (Fig. 10.27). Lister & Snoke (1984) illustrate how in mylonites, the atti-

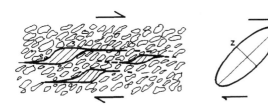

FIG. 10.27 A schematic illustration of oblique grain shape fabrics in relation to mica-fish and shear sense.

tude of quartz grain fabrics agrees with the sense of shear indicated by mica-fish. More recently, Schmid et al. (1987) give an account of results from simple shear experiments on calcite rocks. In particular, their study describes the nature of oblique calcite grain-shape fabrics produced by dextral shearing of Carrara marble and Solenhofen limestone at various temperatures.

As well as grain shape fabrics, many minerals (e.g. quartz, calcite and olivine) are known to have a crystallographic preferred orientation (CPO) in deformed rocks. Ultramylonites developed in quartz-rich rocks can superficially resemble fine-grained quartzites (Fig. 8.13(c)). However, the insertion of the 'sensitive tint plate' and rotation of the microscope stage will reveal how well aligned (crystallographically) the quartz grains are. Highly deformed mylonitic rocks have a strong CPO such that in one position the bulk of grains in the field of view should appear blue, whereas when rotated through 90° they will be largely yellow–orange (Plates 7(a) & (b)).

Following early work by Nicholas et al. (1971) on peridotites, many researchers have analysed CPOs when studying the kinematics of ductile deformation. The asymmetric nature of quartz c-axes with respect to the principal foliation(s) of shear zones has been extensively researched and discussed. It is now standard practice to study such fabrics when interpreting the shear sense in mylonitic rocks (e.g. Lister & Williams, 1979; Simpson & Schmid, 1983; Lister & Snoke, 1984). Similarly, the obliquity of lattice preferred orientation in olivine with respect to a given foliation is widely accepted as an expression of rotational deformation and may be used to determine sense of shear (Etchecopar & Vasseur, 1987). Wenk et al. (1987) give a detailed account of how calcite CPOs in marble can be used in distinguishing pure shear from simple shear, assessing the sense of shear and estimating finite strain.

Despite the undoubted value of CPOs as shear-sense indicators, a degree of caution should always be exercised because, in some cases, opposing senses of shear may be recorded within different samples from the same thrust zone (e.g. Bell & Johnson, 1992). There are a number of reasons why this may occur. One reason is that the presence of other phases in the matrix aggregate will affect the nature of the CPO recorded and, second, recrystallisation may cause weakening of the CPO, such that the pattern is less distinct and shear-sense evaluation becomes less clear cut.

References

Arboleya, M.L. & Engelder, T. (1995) Concentrated slip zones with subsidiary shears: their development on three scales in the Cerro Brass fault zone, Appalachian valley and ridge. *Journal of Structural Geology*, 17, 519–532.

Barker, A.J. (1994) Interpretation of porphyroblast inclusion trails: limitations imposed by growth kinetics and strain rates. *Journal of Metamorphic Geology*, 12, 681–694.

Bell, T.H. (1985) Deformation partitioning and porphyroblast rotation in metamorphic rocks: a radical reinterpretation. *Journal of Metamorphic Geology*, 3, 109–118.

Bell, T.H. & Johnson, S.E. (1992) Shear sense: a new approach that resolves conflicts between criteria in metamorphic rocks. *Journal of Metamorphic Geology*, 10, 99–124.

Berthé, D. Choukroune, P. & Jegouzo, P. (1979) Orthogneiss, mylonite and non-coaxial deformation of granites: the example of the south Armorican shear-zone. *Journal of Structural Geology*, 1, 31–42.

Blenkinsop, T.G. & Treloar, P.J. (1995) Geometry, classification and kinematics of S–C fabrics. *Journal of Structural Geology*, **17**, 397–408.

Chester, F.M., Friedman, M. & Logan, J.M. (1985) Foliated cataclasites. *Tectonophysics*, **111**, 139–146.

De Bresser, J.H.P. (1989) Calcite *c*-axis textures along the Gavarnie thrust zone, central Pyrenees. *Geologie en Mijnbouw*, **68**, 367–376.

Dennis, A.J. & Secor, D.T. (1987) A model for the development of crenulations in shear zones with applications from the Southern Appalachians Piemont. *Journal of Structural Geology*, **9**, 809–817.

Etchecopar, A. & Malavieille, J. (1987) Computer models of pressure shadows: a method for strain measurement and shear-sense determination. *Journal of Structural Geology*, **9**, 667–677.

Etchecopar, A. & Vasseur, G. (1987) A 3-D kinematic model of fabric development in polycrystalline aggregates: comparisons with experimental and natural examples. *Journal of Structural Geology*, **9**, 705–717.

Johnson, S.E. (1993) Testing models for the development of spiral-shaped inclusion trails in garnet porphyroblasts: to rotate or not to rotate, that is the question. *Journal of Metamorphic Geology*, **11**, 635–659.

Lister, G.S. & Snoke, A.W. (1984) S–C mylonites. *Journal of Structural Geology*, **6**, 617–638.

Lister, G.S. & Williams, P.F. (1979) Fabric development in shear zones: theoretical controls and observed phenomena. *Journal of Structural Geology*, **1**, 283–297.

Logan, J.M., Friedman, M., Higgs, N., Dengo, C. & Shimanto, T. (1979) Experimental studies of simulated gouge and their application to studies of natural fault zones, in *Analysis of actual fault zones in bedrock*. U.S. Geological Survey, Open-file Report, 79- **1239**, 305–343.

Mawer, C.K. (1987) Shear criteria in the Grenville Province, Ontario, Canada. *Journal of Structural Geology*, **9**, 531–539.

Nicholas, A., Bouchez, J.L., Boudier, F. & Mercier, J.C. (1971) Textures, structures and fabrics due to solid state flow in some European lherzolites. *Tectonophysics*, **12**, 55–85.

Passchier, C.W. (1991) Geometric constraints on the development of shear bands in rocks. *Geologie en Mijnbouw*, **70**, 203–211.

Passchier, C.W. (1994) Mixing in flow perturbations: a model for development of mantled porphyroclasts in mylonites. *Journal of Structural Geology*, **16**, 733–736.

Passchier, C.W. & Simpson, C. (1986) Porphyroclast systems as kinematic indicators. *Journal of Structural Geology*, **8**, 831–844.

Passchier, C.W., ten Brink, C.E., Bons, P.D. & Sokoutis, D. (1993) Delta-objects as a gauge for stress sensitivity of strain rate in mylonites. *Earth and Planetary Science Letters*, **120**, 239–245.

Platt, J.P. (1984) Secondary cleavages in ductile shear zones. *Journal of Structural Geology*, **6**, 439–442.

Platt, J.P. & Vissers, R.L.M. (1980) Extensional structures in anisotropic rocks. *Journal of Structural Geology*, **2**, 387–410.

Ponce de Leon, M.I. & Choukroune, P. (1980) Shear zones in the Iberian Arc. *Journal of Structural Geology*, **2**, 63–68.

Ramsay, J.G. & Huber, M.I. (1983) *The techniques of modern structural geology; Volume 1: Strain analysis*. Academic Press, London.

Schmid, S.M., Panozzo, R. & Bauer, S. (1987) Simple shear experiments on calcite rocks: rheology and microfabric. *Journal of Structural Geology*, **9**, 747–778.

Shelley, D. (1995) Asymmetric shape preferred orientations as shear-sense indicators. *Journal of Structural Geology*, **17**, 509–517.

Sibson, R.H. (1990) Faulting and fluid flow, in *Fluids in tectonically active regimes* (ed. B.E. Nesbitt). Mineralogical Association of Canada, Short Course No. 18, 93–132.

Simpson, C. & Schmid, S.M. (1983) An evaluation of criteria to deduce the sense of movement in sheared rocks. *Bulletin of the Geological Society of America*, **94**, 1281–1288.

van den Driessche, J. & Brun, J.P. (1987) Rolling structures at large shear strain. *Journal of Structural Geology*, **9**, 691–704.

Wenk, H.-R., Takeshita, T., Bechler, E., Erskine, B.G. & Matthies, S. (1987) Pure shear and simple shear calcite textures. Comparison of experimental, theoretical and natural data. *Journal of Structural Geology*, **9**, 731–745.

Chapter eleven

Veins and fluid inclusions

11.1 Controls on fluid migration and veining

Most prograde metamorphic reactions involve devolatilisation. The main volatile components produced are H_2O and CO_2, although the relative proportions of these varies considerably from one rock to the next. Once produced, volatiles tend to migrate upwards through the crust, the mechanism and rate of flow being largely a function of temperature, fluid pressure (P_f) gradients, and bulk rock permeability (e.g. Fyfe *et al.*, 1978; Ferry, 1980; Etheridge *et al.*, 1983; Wood & Walther, 1986; Yardley, 1986; Thompson, 1987).

Fluids become focused into zones of high permeability and lower P_f. Penetratively foliated schists, fault zones and thrust zones are particularly favoured, whereas massive quartzites and marble (unless highly fractured) have low permeabilities and experience much lower fluid flux. For a given rock, the degree of permeability relates to the effectiveness of fluid migration via grain boundaries, and discrete fractures or cleavage. On the local scale, segregations of new mineral growth are commonly observed in strain shadows of porphyroblasts (e.g. Figs 10.20–10.22) and in boudin necks on both the meso- and the microscale (Fig. 11.1).

While veining is extensive in many metamorphic rocks and terrains, in other cases it is extremely scarce. This can be interpreted either in terms of differences in bulk fluid flow (the most veined rocks representing highest fluid flux), or in terms of permeability differences. In cases with little veining the permeability of the rocks may have been sufficient to permit fluid flow through the rock in a pervasive fashion, maintaining P_f below the value for fracturing. However, in intensely veined examples it is likely that the permeability of the unfractured rock was insufficient to allow fluid escape at the rate at which it was being produced by devolatilisation reactions. This would give rise to an increase in P_f until ultimately the rock strength would be exceeded and a series of discrete fractures would develop to increase the rock's bulk permeability. Healing of fractures by precipitation of minerals out of solution competes with hydraulic fracturing, but progressive upward migration of fluids maintains high P_f and favours further microcrack propagation. Because SiO_2 solubility in aqueous solutions diminishes with decreasing P and T, SiO_2 saturated fluids will precipitate quartz as they ascend through the crust.

11.2 Initial description and interpretation of veins

When considering the characteristics of veins and the nature of their formation, field studies provide key information concerning the mineral assemblage, orientation and timing of each vein-set, both in relation to other veins and to specific deformation and metamorphic

Veins and fluid inclusions

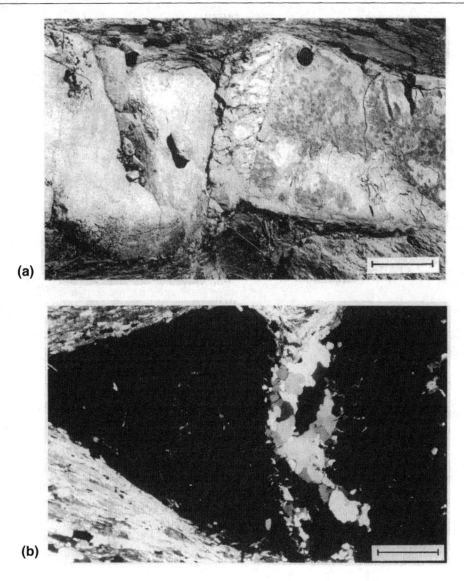

FIG. 11.1 (a) Mesoscale vein quartz segregation in a boudin neck: a boudinaged psammite layer in an interbedded psammite–pelite sequence. Snake Creek, Queensland, Australia. The lens cap is 50 mm in diameter. (b) A microscale quartz segregation in a neck of fractured hornblende porphyroblast: Hbl–mica schist ('*Garbenschiefer*') from Troms, Norway. Scale bar = 1 mm (XPL).

events. The use of veins for shear-sense determinations has already been covered in Section 10.2.

Since the majority of veins and segregations have sharply defined margins, and are of different colour to the host rock, they are easily identified on all scales. At moderate and high-temperature metamorphism, most veins are strongly recrystallised polygonal aggregates, but at sub-greenschist facies, fibrous veins of quartz and calcite are common. In relation to certain high-temperature metamorphic

Initial description and interpretation of veins

processes such as charnockitic alteration of granulites (e.g. Newton, 1992) and development of 'patch' and 'diktyonitic' migmatites (e.g. McLellan, 1988), diffuse veins and segregations may be encountered, with less well defined boundaries.

Monomineralic veins such as quartz, calcite and epidote are especially common, but veins that are bi-mineralic (e.g. qtz–cal and qtz–chl veins) or polymineralic (e.g. mineralised veins) are also seen in metamorphic rocks. Some basic observations about the degree of interaction with the wall-rock are important. For example, many quartz veins draw silica from the adjacent wall-rock to leave darkened, quartz-depleted margins. In slates and schists this is often quite obvious in the field (Fig. 11.2). In other cases fluids infiltrating the wall-rock adjacent to veins may cause metasomatic alteration and new mineral growth in the wall-rock. The interaction of wall-rock and fluid passing through a fracture commonly leads to nucleation of one or more mineral phases at the vein margin, whereas the central portion of the vein often consists of one phase only (e.g. quartz). In Fig. 11.3 actinolite is developed at the margin of a quartz vein passing through a metabasite. Although best observed in thin section, such features may be observable in the field in coarse-grained veins.

Although many features can be identified in the field, for a more complete understanding of veins and mineral segregations, thin-section studies are required. Depending on the mineral assemblage, it may be possible to gain a general understanding of the P–T conditions at the time of vein formation. Clearly, quartz and calcite are of little use here, since they are stable over almost the full range of metamorphic conditions. However, veins containing minerals with more restricted stability fields (e.g. jadeite, andalusite and prehnite) can provide important information. In recent decades, the use of cathodoluminescence

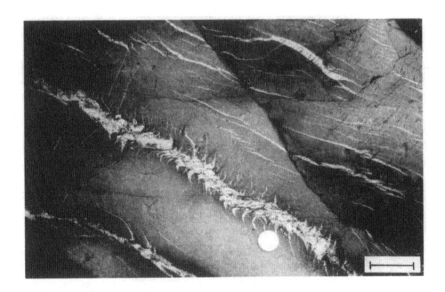

FIG. 11.2 Multiple quartz veins intruding slate. Bude, England. Scale = 5 cm. Note the darkened area of silica depletion immediately adjacent to the main vein. Also note the fact that this 3 cm wide vein initiated as a zone of weakness defined by an *en echelon* array of thin tension gashes. These have subsequently experienced a component of right-lateral (dextral) shear, to give the sigmoidal traces now observed.

Veins and fluid inclusions

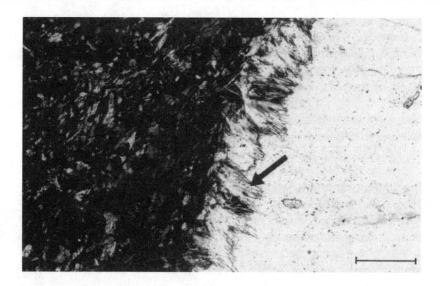

FIG. 11.3 Actinolite (arrowed) developed at the margin of a quartz vein (right) passing through a greenschist facies metabasite (left). Pyrite Belt, Spain. Scale = 0.1 mm (XPL).

(Marshall, 1988) has proved highly revealing in the study of vein mineralogy and genesis (Plates 7(c) & (d)). In addition, it has been usefully employed for the recognition of sealed microfractures through crystals, which are either unrecognisable or else poorly defined by standard microscopy. Since the common carbonate minerals luminesce very differently, it has proved particularly useful for their study (Plates 7(c) & (d)), as well as for the study of overgrowth textures on quartz.

With regard to the mechanism of vein development, it is necessary to consider (a) whether a fracture already existed and fluid passed through and precipitated out of solution, or (b) whether increased fluid pressure induced mechanical failure of the rock and promoted fracture propagation, into which vein material precipitated. Open fissures can be maintained in the upper 5–10 km of crust, since fluids passing through these fractures are generally at pressures close to hydrostatic pressure and considerably less than lithostatic pressure. Such fractures are not always completely sealed by the precipitating phase, and minerals growing from the walls commonly develop large crystals with perfect form. In situations in which increased fluid pressure (often due to local reaction) exceeds rock strength, hydraulic fracturing will occur. Once such fracturing has initiated, fractures will propagate until such a time as fluid pressure falls below rock strength. The propagating tip is characteristically tapered or splayed. At deeper levels in the crust it is a widely held view that P_f approximates to P_l and open fractures only develop transiently, and do not usually exist for any significant length of time. However, Ague (1995) presents convincing evidence for Ky + Grt + Qtz veins, where the well-formed nature of the minerals suggests growth into large aperture fractures at P–T conditions corresponding to c. 30 km depth within continental crust. Analogue modelling of single-stage fracture opening events and subsequent crystallisation and vein infill have been undertaken by Wilson (1994) using ice and an aqueous solution. The microstructures recorded have been used to evaluate the likely processes operating during single-stage fracture-opening and vein development in metamorphic rocks. Wilson (1994) concludes that the presence of a deviatoric stress plays a crucial role, influencing both crystallisation and the resultant microstructure of veins. Static growth in a fluid-filled cavity in equilibrium with the local stress produces euhedral prismatic crystals (e.g. 'drusy' or 'comb' quartz cavity infill) by a mechanism of free-face growth. In contrast, syntectonic growth from a fluid-filled cavity in the presence of a deviatoric stress produces veins comprising anhedral polygonised aggregates, as a result of contact growth and recrystallisation. Where vein formation occurs in a solid grain aggregate in the presence of a deviatoric stress, Wilson (1994) shows that fibrous veins will develop by incremental microcracking and sealing (the 'crack-seal mechanism').

11.3 The 'crack–seal' mechanism of vein formation

The 'crack–seal' mechanism (Ramsay, 1980; and see Fig. 11.4) is an important process responsible for the formation of many veins under low-grade conditions (e.g. greenschist and sub-greenschist facies). It is often difficult to assess the mechanism responsible for vein formation under medium- and high-grade conditions, since the veins are usually strongly recrystallised (Fig. 11.5). The 'crack–seal' mechanism involves multistage crack opening and vein infill in response to fluctuations in P_f. By successive increments it can lead to veins of considerable width (typically mm–cm-scale; Plates 8(a)–(c)). Due to build-up of elastic strains in the rock in response to rising P_f, often induced by poor or limited rock permeability (see above), a fracture is formed. This will propagate until such a time as P_f diminishes to a value less than rock strength. Judging by incremental growth stages observed in veins (Plates 8(a)–(c)) it appears that such fractures typically

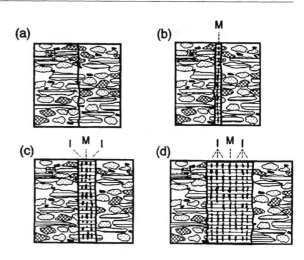

FIG. 11.4 Successive stages in the development of a vein by the crack–seal process (see text for details). M, median line; I, inclusion bands.

open by 5–100 μm at a time. Precipitation of solute species from the fluid leads to fracture infill, until eventually the fracture is more or less completely sealed. Once sealed, the permeability of the rock becomes reduced once again, and assuming continual replenishment of fluid, P_f will rise once more until the critical value at which the rock fails again and further cracking is induced. The initial vein formation strongly influences the location of subsequent veins due to the planar anisotropy set-up. Subsequent fractures will commonly split earlier veins in half or form along one of the pre-existing vein–wall contacts. The whole process of 'crack-seal' and vein-widening continues by a number of cycles such as described above. It is the regular growth increments shown by 'crack-seal' veins that gives strong evidence for hydraulic fracturing related to changing P_f as the vein-forming process. Characteristically, such veins record successive increments of growth by thin 'inclusion bands' parallel to the vein walls (Plate 8(c)) and which generally represent detached screens of wall rock now incorporated in the vein. Even within veins that

Veins and fluid inclusions

FIG. 11.5 A recrystallised quartz veinlet cutting amphibolite. Troms, Norway. Scale = 0.5 mm (XPL).

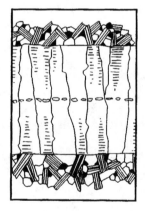

FIG. 11.6 A schematic illustration showing inclusion trails developed perpendicular to the walls of a fibrous vein.

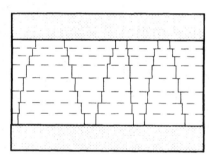

FIG. 11.7 A schematic illustration of the sawtooth form of the contacts between individual crystals in a fibrous vein.

have recrystallised at higher grades of metamorphism, it may be possible to recognise these bands (Plate 8(d)) and thus understand how the vein formed. In addition to 'inclusion bands' approximately parallel to vein walls, fibrous veins may also show inclusions arranged in trails sub-perpendicular to the wall (Fig. 11.6). These inclusion trails normally comprise separate optically parallel small inclusions of the same mineral, and are usually entirely enclosed within coarse fibrous crystals of the main vein phase or phases (typically quartz or calcite). The contact between individual fibres is sub-perpendicular to the vein-wall, and commonly has a stepped saw-toothed form (Fig. 11.7). Planes of trapped fluid inclusions sub-parallel to the vein wall, and representing sealed surfaces, are an additional feature of such veins.

11.4 Interpretation of fibrous veins

Four main types of fibrous vein system can be distinguished (Fig. 11.8), and are common in rocks undergoing deformation at sub-greenschist to mid-greenschist facies conditions. At higher temperatures the fibrous form of the crystal is thermodynamically unstable, and is superseded by equidimensional polygonal crystals; compare fibrous quartz pressure shadows of anchimetamorphic slate (Fig. 10.22) with polygonal quartz pressure shadows in garnet–mica schist (Plate 5(c)). The various types of fibrous veins are now described and discussed.

11.4.1 Syntaxial fibre veins

These veins involve progressive fibre growth from each wall of the fracture towards the centre. The fibres thus fall into two halves either side of a centrally located suture (Fig. 11.8(a)). The crystals forming the fibres show a close compositional link with the rock through which the vein cuts (e.g. calcite veins in marble or quartz veins in sandstone). In the case of curved syntaxial fibres (Fig. 11.8(a)), the fibres at the vein–wall contact are always perpendicular to the wall, whereas in the next type of veins to be considered (antitaxial fibre veins) the contact is oblique. Curved fibres reflect changes in orientation of the line of maximum incremental longitudinal strain during the deformation history.

11.4.2 Antitaxial fibre veins

Antitaxial fibre veins involve growth towards the wall-rock rather than from it. The fibres show crystallographic continuity from one wall to the other and the mineral phase(s) involved are often unrelated to the wall-rock mineralogy, but are derived from some local source. The centre of the vein is usually marked by a screen of small wall-rock fragments to define what is termed the 'median line'. Although changes in orientation of the line of maximum incremental longitudinal strain may give rise to curved antitaxial fibres, at the 'median line' such fibres are always perpendicular to the rock walls (Fig. 11.8(b)), since this coincides with initial fracture opening. Some antitaxial veins show several or many 'inclusion bands' parallel to the median line. Such trails indicate successive vein openings by the 'crack–seal' mechanism (see above).

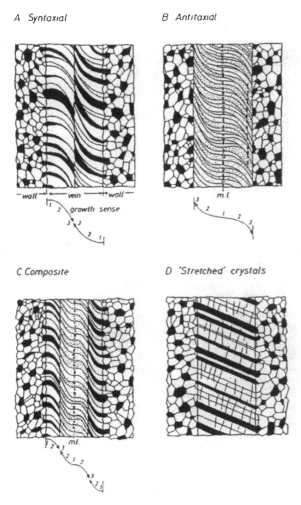

FIG. 11.8 (a) A syntaxial fibre vein. (b) An antitaxial fibre vein. (c) A composite fibre vein. (d) A stretched crystal fibre vein. (After Fig. 13.9 of Ramsay & Huber, 1983; courtesy of Academic Press.) m.l. is median line.

11.4.3 Composite fibre veins

Such veins comprise a central zone (with median line) of one mineral phase, bounded on each side by marginal zones consisting of another crystal species. The crystal species of the marginal zones have compositional similarity to crystals of the wall-rock (on to which they root). In examples from the Helvetic Alps, Ramsay & Huber (1983) observe that the central portion of composite veins has geometric characteristics identical to that of an antitaxial vein, while the marginal zones have characteristics of syntaxial veins (Fig. 11.8(c)).

FIG. 11.9 A stretched quartz fibre vein. Croyde Bay, Devon, England. Scale = 2 cm.

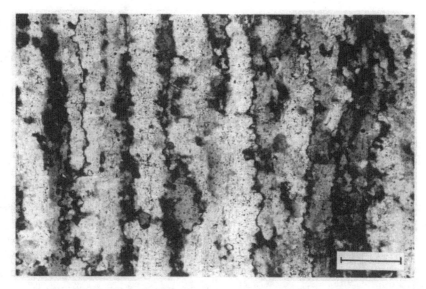

FIG. 11.10 Detail of the serrated contacts between individual fibres of a stretched crystal fibre vein. Vein quartz, Croyde Bay, Devon, England. Scale = 0.1 mm (XPL).

11.4.4 'Stretched' (or 'ataxial') crystal fibre veins

This fourth category of fibrous veins involves crystal fibres of the same or similar composition to those minerals found in the wall-rock. The fibres are typically perpendicular or at a high angle to the vein margins (Fig. 11.9), and always show crystallographic continuity from one wall to the other (Fig. 11.8(d)). In the case of quartz, thin-section studies usually reveal that contacts between adjacent fibres are serrated (Fig. 11.10). In detail, the serrations show some regularity suggestive of a crack–seal mechanism for vein formation.

The orientation of fibres in 'stretched' crystal fibre veins is commonly used as an indication of the vein opening direction, but this is not strictly true, because fibrous crystals have a tendency to grow normal to the fracture surface (i.e. face-controlled growth rather than displacement-controlled growth). Urai *et al.* (1991) suggest that displacement-controlled fibres only develop under conditions with irregular growth surfaces, small growth increments and isotropic crystal growth.

11.5 Veins and melt segregations at high metamorphic grades

In migmatites, and upper amphibolite facies schists and gneisses, quartzofeldspathic and granitoid veins are common. Some represent melt segregations, whereas others are veins and segregations that have precipitated from an aqueous fluid. In many cases, the veins and segregations are demonstrably of local origin (e.g. Vidale, 1974), and have developed due to metamorphic differentiation, involving diffusion of elements in different directions along concentration gradients. Commonly, it is found that veins are richer in K-feldspar and poorer in biotite relative to surrounding gneisses, suggesting an exchange process involving transfer of K into the vein area, and removal of Mg and Fe from the vein area (e.g. Kretz, 1994). If the vein is forming by local differentiation, this process will lead to depletion of biotite in the vein and a build-up of biotite in a selvage around the vein or segregation. Similar diffusive processes in amphibolite may produce irregular veins or segregations of Pl + Qtz and an area of surrounding host rock enriched in hornblende due to this localised subsolidus metamorphic differentiation, perhaps induced by a localised pressure gradient (Kretz, 1994).

During the development of stromatic migmatites by partial melting of specific layers in a metasedimentary sequence, the biotite of the protolith may do one of several things. It may break down to form minerals such as K-feldspar, it may react out with plagioclase to produce hornblende, or it may crystallise from the partial melt into coarse biotite segregations within the leucosome, but more commonly as concentrated melanosome at the margins of the leucosome (Johannes, 1983). Such biotite selvages a few millimetres thick, separating leucosome from mesosome (Fig. 4.12), are a characteristic feature of stromatic migmatites (e.g. Mehnert, 1968; Johannes, 1983, 1988). Unlike the subsolidus metamorphic differentiation discussed above in relation to certain veins and segregations of gneisses, the biotite selvages at the margins of leucosome are commonly interpreted to form by anatectic differentiation (segregation from a melt) in the presence of H_2O (e.g. Johannes, 1988). However, an alternative explanation presented by Maaløe (1992) is that contraction of melting mesosome causes concentration of refractory minerals such as biotite and hornblende in a selvage at the margins of the leucosome.

11.6 Fluid inclusions

Minute fluid inclusions are trapped in various minerals from a range of metamorphic environments. The inclusions represent small quantities of liquid or vapour trapped inside individual crystals or healed microcracks at some time during the rock's evolution. Although the importance and usefulness of fluid inclusions was realised by Sorby (1858), the general appreciation of such studies was slow to develop. However, in recent decades fluid inclusion studies have received much greater attention, and have been realised as a valuable tool for gaining insight into *P–T* conditions and changing fluid chemistry during a rock's metamorphic history. For details of methods and applications relating to fluid inclusion studies the publications of Roedder (1984), Shepherd *et al.* (1985) and De Vivo & Frezzotti (1994) provide a comprehensive insight.

It is rare to observe fluid inclusions greater than about 150 μm (0.15 mm) in diameter, and more commonly they are less than 30 μm (0.03 mm) in size, with the majority being less than 15 μm. As a general approximation, the larger the host crystal is, the larger is the inclusion trapped. For this reason, vein minerals have received the greatest attention, and have proved most fruitful for fluid inclusion studies. This is also due to the fact that vein minerals have formed in a very fluid-rich environment, and are thus more likely to trap more fluid.

Useful studies can be made by standard transmitted light microscopy, using the high-power objective of the microscope and by adjusting the positions of the substage condenser and aperture diaphragm to obtain optimum illumination. It is advisable to use lower magnifications initially in order to locate inclusions, and to examine their distribution and morphological characteristics (Figs 11.11(a) & (b)). Fluid inclusions are reported in various minerals, but those in quartz, fluorite and calcite have been studied most widely. Of these, it is inclusions in quartz which are of most general interest and use to the metamorphic petrologist. There are a number of reasons why this should be the case. First, quartz is a common mineral, which often forms veins, and is stable over the full range of metamorphic conditions. Second – and very importantly – it is generally considered that quartz shows no significant reaction (exchange of ions) with the trapped fluid, whereas other minerals such as calcite may well exchange cations with solute species.

In terms of their genesis there are two fundamental types of fluid inclusions, namely, PRIMARY and SECONDARY. Primary (P) inclusions become trapped within minerals as they crystallise within a fluid-rich environment. Secondary (S) inclusions represent small pockets of fluids that have become trapped along post-crystallisation microcracks as they have sealed. Planar surfaces of secondary inclusions are characteristic of healed cracks and are seen as linear inclusion trails in thin section (Figs 11.11(a) & (b)). A third genetic class of fluid inclusions, termed pseudo-secondary (PS) inclusions, develop in a similar way to S-inclusions, but differ by the fact that the fracture heals before crystal growth is complete.

During prograde metamorphism, general recrystallisation and grain coarsening leads to crystal lattice modifications, and the destruction of fluid inclusions trapped at lower temperatures. Overpressuring of early inclusions due to the thermal expansion of the fluid being greater than the surrounding solid commonly leads to 'decrepitation' (sudden and abrupt leakage). In natural examples of decrepitated inclusions, the inclusions often appear dark, may have several sharply pointed corners, or exhibit a halo of daughter inclusions. Experimental work with synthetic fluid inclusions (e.g. Sterner & Bodnar, 1989) has shown that identical features form by implosion, when a fluid inclusion is substantially underpressured – that is, when the confining pressure is substantially increased above internal fluid pressure. In natural samples of granulites, such decrepitated inclusions with a halo of microscopic satellite inclusions are commonly encountered, and are interpreted in terms of the rock experi-

Fluid inclusions

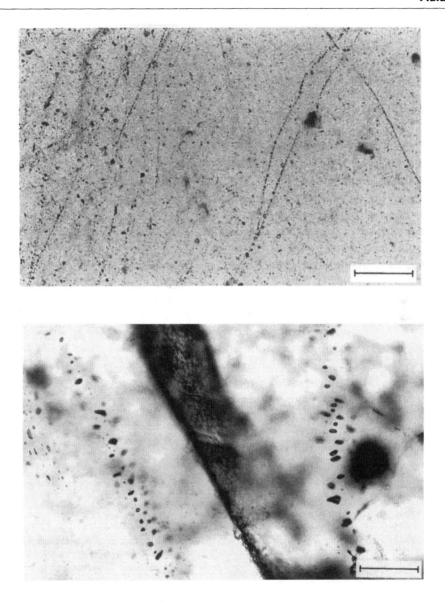

FIG. 11.11 (a) A typical example of a linear array of secondary fluid inclusions along healed microcracks in vein quartz from Troms, Norway. Scale = 0.5 mm (PPL). (b) Detail of the inclusions shown in (a). Scale = 100 μm (PPL).

encing increasing pressure accompanying cooling.

Because of the problems of recrystallisation and decrepitation associated with prograde metamorphism, genuine primary inclusions related to the early history of the rock are infrequently encountered. It is fluid inclusions formed synchronous with, or after, peak metamorphic temperatures that are most frequently preserved in veins and sealed microcracks. In consequence, fluid inclusion studies are most useful for assessing conditions during the uplift trajectory of a

Veins and fluid inclusions

given rock. Careful petrographic study may allow various generations of fluid inclusions to be recognised. Non-planar clusters are most likely to be the earliest inclusions, while linear trails representing distinct inclusion planes relate to later healed microcracks. In practice, it is never easy to make definite identification of primary inclusions; nor is it always easy to determine the relative ages of secondary inclusion trails by petrographic observation alone.

Fluid inclusions show a variety of shapes, which in part often relate to the structure of the host mineral. Inclusions mirroring the symmetry of the host and forming perfect 'negative crystals' are common in some minerals. Cubic inclusions in halite (Fig. 11.12(a)) are a good example of this. Inclusions in quartz may also have a form relating to the symmetry of the host (Fig. 11.12(b)), but invariably are not perfect, and have an irregular, or rounded to oval shape (Fig. 11.12(c)). It is possible to construct various elaborate schemes to classify fluid inclusions in terms of the differing proportions and types of solids, liquids and vapour observed in inclusions at room temperature. However, the vast majority of fluid inclusions observed in metamorphic rocks fall into one of five categories (Fig. 11.13):

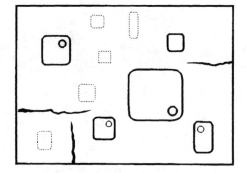

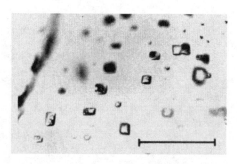

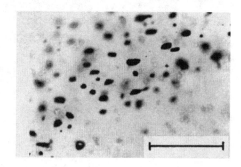

(a) Monophase, liquid (usually aqueous) (Figs 11.13(a) & (f)).
(b) Two-phase, liquid-rich (i.e. liquid + vapour, where L > V) (Figs 11.13(b) & (g)).
(c) Vapour-rich (vapour dominates, but in many cases a thin rim of liquid may still be present) (Figs 11.13(c) & (h)).
(d) CO_2-bearing (Figs 11.13(d) & (i)). Such inclusions usually consist of CO_2 + H_2O as immiscible liquids. If required, they can be subdivided according to which liquid dominates. At room temperature (e.g. Fig. 11.13(i)), the inclusion often appears to be three-phase ($CO_{2(V)}$ + $CO_{2(L)}$ + H_2O). The CO_2-rich phase may contain significant proportions of CH_4 and N_2 (Shepherd et al., 1985).

FIG. 11.12 (a) A schematic example of fluid inclusions in halite, forming perfect negative crystals. (b) Well-formed fluid inclusions in quartz, their shape being controlled by the crystal structure of the host. Scale = 100 μm (PPL). (c) An array of rounded to oval fluid inclusions in quartz. Scale = 50 μm (XPL).

(e) Solid-phase bearing – containing one or more solid phases; cubes of NaCl or KCl being most common and having precipitated from the trapped fluid ('daughter minerals') (Figs 11.13(e) & (j); Plate 8(f)). It is also possible that solid phases suspended in the fluid (e.g. mica) may become trapped in the inclusion ('captive minerals').

Fluid inclusions

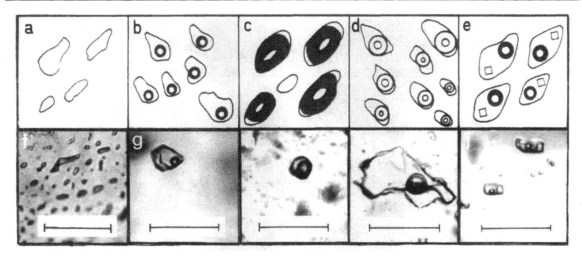

FIG. 11.13 Schematic and natural examples of the five main types of fluid inclusions: (a, f) monophase; (b, g) two-phase, liquid-rich; (c, h) two-phase, vapour-rich; (d, i) CO_2-bearing; (e, j) solid phase bearing. The natural example shown in (j) is an aqueous fluid with 'daughter minerals' of halite (cubes). Scale bars are 20 μm.

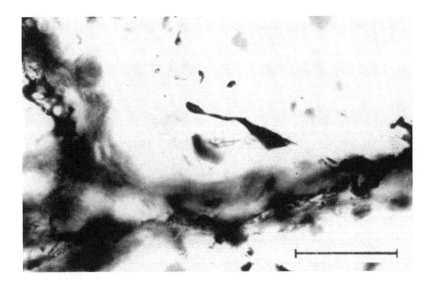

FIG. 11.14 The necking-down of a fluid inclusion. Scale = 50 μm (PPL).

Variable phase ratios in fluid inclusions can be used as evidence for immiscible (phase-separated) fluids at the time of trapping. Immiscible CO_2–H_2O and CH_4–H_2O fluids have been reported from many metamorphic terrains (e.g. Yardley & Bottrell, 1988), and are significant in that the physical flow characteristics of two-phase fluids are quite different from those of single fluids. When one fluid is significantly in the minority, it will either occur as isolated globules ('non-wetting phase') suspended in the majority phase or, alternatively, will adhere to grain surfaces

Veins and fluid inclusions

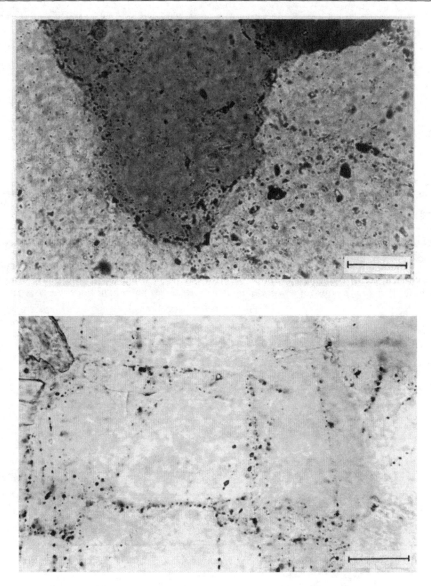

FIG. 11.15 Micro-inclusions ('bubbles') along grain boundaries (vein quartz, Troms, Norway): (a) scale bar = 50 μm, partial XPL; (b) scale bar = 125 μm, partial XPL.

('wetting phase'). The separation into two distinct phases generally leads to one phase flowing more rapidly than the other and may also lead to a reduction in effective permeability. These effects can have a profound influence on reaction pathways and the progress of metamorphism, especially in the case of carbonate rocks.

Having decided whether an inclusion or set of inclusions is primary or secondary, and having classified in terms of the relative proportions of solid, liquid and vapour, it is also necessary to determine whether or not the inclusion has leaked. The tell-tale signs of necking-down and leakage are variable phase ratios in a group of inclusions, and inclusions connected by a thin tube or with protruding tails (Fig. 11.14). However, recent experimental and TEM studies (e.g. Bakker & Jansen, 1994) have established that even inclusions that show none of the tell-tale signs of leakage when examined during transmitted light microscopy may in fact have experienced dislocation-controlled diffusional leakage. This can have important consequences when interpreting fluid chemistry and trapping conditions, because phase ratios can change during leakage. For example, Crawford & Hollister (1986), and Bakker & Jansen (1991, 1994) have shown that due to differences in wetting characteristics, H_2O preferentially leaks out of H_2O-CO_2 inclusions. Fluid inclusions are particularly prone to leakage or else decrepitation if they experience an underpressure or overpressure in excess of 1 kbar relative to the conditions under which they were trapped (Bodnar et al., 1989; Sterner & Bodnar, 1989), or if there is extensive recrystallisation of the mineral aggregate in which the fluid inclusions are contained. The effects of recovery and recrystallisation on fluid inclusions were considered in Section 8.2, but one of the most distinctive features is to see residual very fine (< 10 μm) fluid inclusions decorating grain boundaries (Figs 11.15(a) & (b)), the fluid having been swept to these areas facilitated by dislocation climb and other intracrystalline processes (Kerrich, 1976; Wilkins & Barkas, 1978; O'Hara & Haak, 1992). From TEM studies it has been confirmed that dislocations are often decorated with minute nanometre-scale fluid inclusions, termed 'bubbles', and that such dislocations lined with an array of fluid bubbles commonly emanate from larger fluid inclusions (Fig. 11.16), thus suggesting leakage (e.g. Bakker & Jansen, 1991, 1994; Reeder, 1992; Wang et al., 1993).

Having identified the different sets of inclusions present in a rock by optical studies, it is possible by using a heating-freezing stage to go

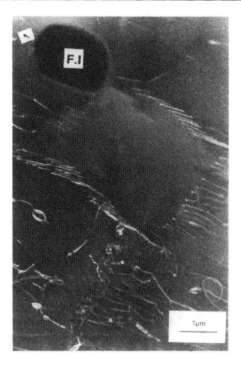

FIG. 11.16 A TEM (dark field) image showing a sub-grain (in quartz) bordered by low-angle tilt boundaries. Adjacent sub-grains consist of two or more sets of dislocations. Note the extrinsic dislocation loop emanating from the fluid inclusion (FI) bubble projecting into a sub-grain interior (top left, arrowed) (unpublished photograph courtesy of L. Hopkinson).

further, and obtain quantitative information about aspects such as the salinity of aqueous fluids and the homogenisation temperatures of two-phase inclusions. In metamorphic rocks salinities commonly range from 0 wt% NaCl equivalent up to 30–40 wt% NaCl equivalent. Inclusions with >26 wt% NaCl characteristically show a daughter mineral (halite cube) at room temperature (e.g. Fig. 11.13(j)). It is not really possible to say too much about the environment or conditions of metamorphism simply on the basis of fluid chemistry. However, it is a common feature to find that peak metamorphic amphibolite facies fluid inclusions have a low-salinity H_2O-CO_2 composition, while many greenschist facies fluids associated with retrogression of metapelites and metavolcanics have a highly saline aqueous character, with one or more daughter minerals (e.g. Bennett & Barker, 1992). Under granulite facies conditions, one of the most characteristic features is to find CO_2-rich fluid inclusions (e.g. Touret, 1971, 1977). This agrees well with the theoretical considerations that suggests granulite facies assemblages require low aH_2O to exist.

More recently, the use of laser Raman spectroscopy (Burke, 1994) has enabled microanalysis of daughter minerals and certain species within individual fluid inclusions. This technique involves laser beam excitation of the various molecules and induces stretching and vibration of bonds between atoms, to produce Raman radiation. The Raman spectra detected have enabled identification and quantification of various polyatomic species (e.g. SO_4, CH_4, CO_2, N_2, NaCl, H_2O and so on) present in the inclusions (either as solid, vapour or fluid).

References

Ague, J.J. (1995) Deep crustal growth of quartz, kyanite and garnet into large-aperture, fluid-filled fractures, north-eastern Connecticut, USA. *Journal of Metamorphic Geology*, 13, 299–314.

Bakker, R.J. & Jansen, J.B.H. (1991) Experimental post-entrapment water loss from synthetic CO_2–H_2O inclusions in natural quartz. *Geochimica et Cosmochimica Acta*, 55, 2215–2230.

Bakker, R.J. & Jansen, J.B. (1994) A mechanism for preferential H_2O leakage from fluid inclusions in quartz, based on TEM observations. *Contributions to Mineralogy and Petrology*, 116, 7–20.

Bennett, D.G. & Barker, A.J. (1992) High salinity fluids: the result of retrograde metamorphism in thrust zones. *Geochimica et Cosmochimica Acta*, 56, 81–95.

Bodnar, R.J., Binns, P.R. & Hall, D.L. (1989) Synthetic fluid inclusions – VI. Quantitative evaluation of the decrepitation behaviour of fluid inclusions in quartz at one atmosphere confining pressure. *Journal of Metamorphic Geology*, 7, 229–242.

Burke, E.A.J. (1994) Raman microspectrometry of fluid inclusions: the daily practice, in *Fluid inclusions in minerals: methods and applications. Short Course of the Working Group (IMA) 'Inclusions in Minerals' (Pontignano – Siena, 1–4 September 1994)* (eds B. De Vivo & M.L. Frezzotti). Virginia Tech, USA, 25–44.

Crawford, M.L. & Hollister, L.S. (1986) Metamorphic fluids: the evidence from fluid inclusions, in *Fluid–rock interactions during metamorphism* (eds J.V. Walther & B.J. Wood). *Advances in Physical Geochemistry*, 5, 1–35. Springer-Verlag, New York.

De Vivo, B. & Frezzotti, M.L. (eds) (1994) Fluid inclusions in minerals: methods and applications. *Short Course of the working group (IMA) 'Inclusions in Minerals' (Pontignano – Siena, 1–4 September 1994)*. Virginia Tech, USA, 376 pp.

Etheridge, M.A., Wall, V.J. & Vernon, R.H. (1983) The role of the fluid phase during regional metamorphism and deformation. *Journal of Metamorphic Geology*, 1, 205–226.

Ferry, J.M. (1980) A case study of the amount and distribution of heat and fluid during metamorphism. *Contributions to Mineralogy and Petrology*, 71, 373–385.

Fyfe, W.S., Price, N.J. & Thompson, A.B. (1978) *Fluids in the Earth's crust*. Elsevier, Amsterdam, 383 pp.

Johannes, W. (1983) On the origin of layered migmatites, in *Migmatites, melting and metamorphism* (eds M.P. Atherton & C.D. Gribble). Shiva, Nantwich, 234–248.

Johannes, W. (1988) What controls partial melting in migmatites? *Journal of Metamorphic Geology*, 6, 451–465.

Kerrich, R. (1976) Some effects of tectonic crystallisation on fluid inclusions in vein quartz. *Contributions to Mineralogy and Petrology*, 59, 195–202.

Kretz, R. (1994) *Metamorphic crystallization*. John Wiley, Chichester, 507 pp.

McLellan, E.L. (1988) Migmatite structures in the Central Gneiss Complex, Boca de Quadra, Alaska. *Journal of Metamorphic Geology*, 6, 517–542.

Maaløe, S. (1992) Melting and diffusion processes in closed-system migmatization. *Journal of Metamorphic Geology*, 10, 503–516.

Marshall, D.J. (1988) *Cathodoluminescence of geological materials*. Unwin Hyman, London, 172 pp.

Mehnert, K.R. (1968) *Migmatites and the origin of granitic rocks*. Elsevier, Amsterdam.

Newton, R.C. (1992) Charnockitic alteration: evidence for CO_2 infiltration in granulite facies metamorphism. *Journal of Metamorphic Geology*, 10, 383–400.

O'Hara, K. & Haak, A. (1992) A fluid inclusion study of fluid pressure and salinity variations in the footwall of the Rector Branch thrust, North

References

Carolina, USA. *Journal of Structural Geology*, **14**, 579–589.

Ramsay, J.G. (1980) The crack seal mechanism of rock deformation. *Nature*, **284**, 135–139.

Ramsay, J.G. & Huber, M.I. (1983) *The techniques of modern structural geology; Volume 1: Strain analysis*. Academic Press, London, 307 pp.

Reeder, R.J. (1992) Carbonates: growth and alteration microstructures, in *Minerals and reactions at the atomic scale: Transmission electron microscopy* (ed. P.R. Buseck). Mineralogical Society of America, Reviews in Mineralogy, No. 27, Ch. 10, 381–424.

Roedder, E. (1984) *Fluid inclusions*. Reviews in Mineralogy 12, Mineralogical Society of America.

Shepherd, T.J., Rankin, A.H. & Alderton, D.H.M. (1985) *A practical guide to fluid inclusion studies*. Blackie, Glasgow, 239 pp.

Sorby, H.C. (1858) On the microscopical structure of crystals indicating the origin of rocks and minerals. *Quarterly Journal of the Geological Society*, **14**, 453–500.

Sterner, S.M. & Bodnar, R.J. (1989) Synthetic fluid inclusions – VII. Re-equilibration of fluid inclusions in quartz during laboratory simulated metamorphic burial and uplift. *Journal of Metamorphic Geology*, 7, 243–260.

Thompson, A.B. (1987) Some aspects of fluid motion during metamorphism. *Journal of the Geological Society*, **144**, 309–312.

Touret, J. (1971) Les facies granulite en Norvège méridionale II: les inclusions fluides. *Lithos*, **4**, 423–436.

Touret, J. (1977) The significance of fluid inclusions in metamorphic rocks, in *Thermodynamics in Geology* (ed. D.G. Fraser). D. Reidel, Dordrecht, 203–227.

Urai, J.L., Williams, P.F. & van Roermund, H.L.M. (1991) Kinematics of crystal growth in syntectonic fibrous veins. *Journal of Structural Geology*, **13**, 823–836.

Vidale, R.J. (1974) Vein assemblages and metamorphism in Dutchess County, New York. *Bulletin of the Geological Society of America*, **85**, 303–306.

Wang, J.N., Boland, J.N., Ord, A. & Hobbs, B.E. (1993) Microstructural and defect development in heat treated Heavitree Quartzite, in *Defects and processes in the solid state: geoscience applications* (eds J.N. Boland & J.D. FitzGerald). Developments in Petrology No. 14 ('The McLaren Volume'). Elsevier, Amsterdam, 359–381.

Wilkins, R.W.T. & Barkas, J.P. (1978) Fluid inclusions, deformation and recrystallization in granite tectonites. *Contributions to Mineralogy and Petrology*, **65**, 293–299.

Wilson, C.J.L. (1994) Crystal growth during a single-stage opening event and its implications for syntectonic veins. *Journal of Structural Geology*, **16**, 1283–1296.

Wood, B.J. & Walther, J.V. (1986) Fluid flow during metamorphism and its implications for fluid–rock ratios, in *Fluid–rock interactions during metamorphism* (eds J.V. Walther & B.J. Wood). Advances in Physical Geochemistry, **5**, 89–108. Springer-Verlag, New York.

Yardley, B.W.D. (1986) Fluid migration and veining in the Connemara Schists, Ireland, in *Fluid–rock interactions during metamorphism* (eds J.V. Walther & B.J. Wood). Advances in Physical Geochemistry, **5**, 109–131. Springer-Verlag, New York.

Yardley, B.W.D. & Bottrell, S.H. (1988) Immiscible fluids in metamorphism: implications of two-phase flow for reaction history. *Geology*, **16**, 199–202.

Chapter twelve

Deciphering polydeformed and polymetamorphosed rocks

12.1 Polymetamorphism

The majority of metamorphic rocks show evidence of more than one phase of deformation, and often multiple stages of metamorphism. By careful study of the mineralogical and textural features of a rock, it should be possible to build up a detailed understanding of the interrelationships between the various mineral phases, and how they relate to specific metamorphic and deformation events.

Some porphyroblasts (especially garnets) may show chemical or else textural zonation resulting from more than one phase of growth (Plates 6(d) & (e)). The included minerals and various growth stages can give an important insight into precursor assemblages and the metamorphic conditions experienced by the rock. In many cases, they may provide a key to some of the reactions that have occurred during the early evolution of the rock. Unless there is extensive volume diffusion, included phases in porphyroblasts become shielded from changing conditions in the matrix of the rock as metamorphism proceeds. In many cases, the inclusions are of minerals such as quartz which, being stable over such a wide range of P–T conditions, are of little use in establishing much about the rock's earlier history. However, in other cases, minerals such as chloritoid, staurolite and glaucophane may be included. Because of their more restricted, and characteristic, stability fields, they provide a useful indication of previous P–T conditions experienced by the rock. In certain cases it may be possible to establish with some confidence the nature of some of the prograde reactions. In the case of garnet–staurolite schists for example, it is common to observe chloritoid inclusions in the garnet but no chloritoid in the matrix of the rock (Plate 5(a)). This provides evidence for chloritoid being part of the earlier assemblage of the rock, but having reacted-out to form staurolite during prograde metamorphism by a reaction in the KFMASH system, such as

$$\text{Ctd} \rightleftharpoons \text{Grt} + \text{Chl} + \text{Stt} + H_2O. \quad (12.1)$$

Without the inclusion evidence, the nature of the staurolite-forming reaction would be much more speculative.

The inclusion trails common to many porphyroblasts can also provide valuable information relating to the timing and style of previous deformation events. The detailed

Deciphering polydeformed and polymetamorphosed rocks

interpretations of porphyroblast–foliation relationships were described in Chapter 9, and consequently will not be repeated here. The conclusions regarding the timing of growth of various minerals with respect to specific deformation and metamorphic events can be usefully summarised using 'mineral growth-deformation sequence' diagrams. The example shown in Fig. 12.1 uses a bar of variable length to represent the times at which various minerals of the assemblage grew in relation to different tectonic fabrics (dashed lines represent less significant mineral growth or uncertainty). Figure 12.2 is a slightly more refined version of Fig. 12.1, and illustrates the time(s) of maximum nucleation and growth of particular phases by bulges in the horizontal bar. In each of these diagrams, the standard notation of S_1, S_2 and S_3 is used to denote planar fabrics associated with successive deformation events D_1, D_2 and D_3. M_1 and M_2 denote specific metamorphic events.

12.2 Local and regional complications

Across a given tectonometamorphic terrain (especially one of orogenic metamorphism), diachroneity of metamorphism in relation to deformation may be observed. In orogenic belts not only will the timing of specific deformation phases be diachronous across the region (e.g. as a given deformation event propagates from hinterland to foreland), but the timing of these events with respect to metamorphism may also show significant variation. If this is the case, a systematic variation in porphyroblast–foliation relationships may be observed on a regional scale. This grossly oversimplified scenario is shown schematically in Fig. 12.3, to emphasise that deformation and metamorphic events are not intrinsically linked, and are likely to show both spatial and temporal variation across a given terrain. Notice in Fig. 12.3 that at time t_1 locality A is experiencing D_1 deformation, whereas locality C does not experience this until t_2. Also notice that porphyroblast growth associated with M_1 metamorphism is inter-D_1–D_2 (S_1–S_2) at A, and shows syn-S_2 relationships at B, but post-S_2 relationships at C.

When drawing conclusions from porphyroblast–foliation relationships, it is important to realise that local variations in bulk rock chemistry can have a profound effect on the sequence of metamorphic reactions and precise timing of porphyroblastesis. This means that the observed relationships and textural features may show significant variation between different lithologies. For example, garnets will probably nucleate more readily in pelites than in semi-pelitic or psammitic rocks. Thus, even at the same exposure, garnets from different

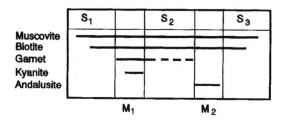

FIG. 12.1 An example of a mineral growth–deformation sequence diagram. The solid lines represent the timing of the mineral growth with respect to the development of different fabrics in the rock. The dashed lines represent less significant mineral growth or uncertainty.

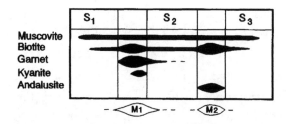

FIG. 12.2 An example of a mineral growth–deformation sequence diagram, where the bars represent the timing of the mineral growth. The acme of the growth is shown by bulges in the bars.

P–T–t paths

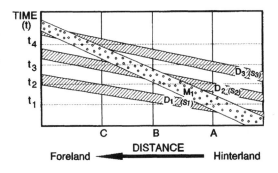

FIG. 12.3 A schematic diagram illustrating diachroneity of deformation and metamorphism from the hinterland to the foreland of an orogenic belt.

lithologies can show different growth relationships and zonation with respect to a given tectonic fabric. Because of this, it is important to compare samples from comparable lithologies.

A further complication in polydeformed areas is that near to major fault or thrust zones a more complex structural and metamorphic history will be experienced. Such differences may be especially apparent in rocks in the footwall and hangingwall to major thrusts during orogenic metamorphism. In such cases, the emplacement of a 'hot' slab on to a 'cold' footwall gives rise to simultaneous heating and prograde reactions in the footwall, and cooling with accompanying retrogression in the hangingwall, promoted by infiltration of fluids liberated during prograde reactions in the footwall. This means that closely associated rocks each side of the thrust may show quite different P–T–t paths until the point at which movement on the thrust terminates, and they then have a common, linked evolution. Documented examples of such histories include the work of Pecher (1989) in the Himalayas, and Anderson et al. (1992) in the north Scandinavian Caledonides. Such differences between hangingwall and footwall evolution next to major thrusts mean that in porphyroblasts from these areas inclusion fabrics and growth zones may in part reflect local features. These cautionary points should always be considered when attempting to interpret regional patterns of metamorphism. Additionally, if more than one tectonic unit is involved (e.g. several nappes) it is essential to treat each unit separately, and determine the tectonometamorphic history of each. Such an approach may reveal differences that give grounds for defining separate terranes.

12.3 P–T–t paths

12.3.1 Introduction

The evaluation of 'peak' metamorphic conditions has traditionally been one of the main objectives of the petrologist studying metamorphic rocks. On the basis of empirical observation, theoretical considerations and, increasingly, by detailed experimental work, it is well established that certain minerals and mineral assemblages characterise a given range of P–T conditions.

The metamorphic facies concept (Chapter 2) is based on equilibrium mineral assemblages that characterise specific P–T conditions. Assigning the equilibrium assemblage of a given rock to a particular metamorphic facies is the first approach to quantifying P–T, and should give a reasonable estimate for the magnitude of P–T at peak metamorphism (see Fig. 2.1 for relative positions of different facies in P–T space, and Appendix III for typical assemblages of each metamorphic facies). A 'metamorphic facies series' denotes the systematic change in peak metamorphic conditions attained by a sequence of rocks across a given metamorphic terrane. For example, the classic Barrovian sequence of Scotland (Barrow, 1893) shows a typical medium-pressure facies series from greenschist facies rising to upper amphibolite facies. However, this should not be taken to represent a palaeogeothermal gradient; nor should it be considered as the prograde P–T path taken by the highest-grade rocks. Spear et

199

Deciphering polydeformed and polymetamorphosed rocks

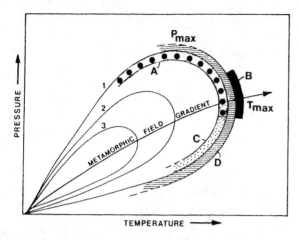

FIG. 12.4 A schematic P–T diagram (modified from Spear et al., 1984; courtesy of the Geological Society of America) showing the different paths taken by three rocks (1, 2 and 3) in a metamorphic sequence (in terms of metamorphic grade, 1 > 2 > 3). Note (1) that the low-grade rock does not represent the earlier part of the high-grade rocks' metamorphic history, and (2) that the change in metamorphic grade seen in the field (the metamorphic field gradient) represents the locus of the peak P–T conditions attained by each rock, and not the palaeogeothermal gradient. The portions of the P–T–t loop that can be constructed from different lines of evidence are also indicated. A, Realm of entrapment of mineral inclusions in porphyroblasts; B, 'peak' P–T conditions from geothermometry/barometry; C, retrograde path from reset mineral equilibria; D, realm of entrapment of fluid inclusions.

al. (1984) emphasise that the locus of 'peak' P–T conditions in samples across a given metamorphic terrain, termed the 'metamorphic field gradient', is rarely coincident with the rocks' P–T path. It is more likely to be representative of the maximum temperature or maximum entropy state achieved by rocks along their individual paths (Fig. 12.4).

It is highly erroneous to build up a P–T loop for an area based on assemblages in several rocks from various grades across the particular metamorphic belt. Instead, it is necessary to build up P–T trajectories for individual samples. Some samples lend themselves to this more readily than others. Ideally suited are porphyroblastic samples with a range of mineral inclusions in the porphyroblasts. With such a rock it is necessary to establish which phases formed the peak metamorphic assemblage, which minerals were present before peak conditions (as determined by inclusions in porphyroblasts) and which phases developed late (often as retrogressive phases). By defining P–T at particular stages in the metamorphic evolution of a given rock, it should be possible to mark on a petrogenetic grid the approximate path that a particular rock has taken through P–T space. On the basis of current knowledge, an inherent assumption of most petrogenetic grids is that $P_{total} = P_{H_2O}$. Taking other fluids such as CO_2 into consideration, or simply lowering aH_2O, can cause significant shifts in the position of univariant reactions and stability fields for specific assemblages. This means that the construction of a P–T path based solely on petrogenetic grids on which $P_{total} = P_{H_2O}$ may have serious limitations.

Geothermometry and geobarometry

In addition to the estimation of metamorphic conditions based on key mineral assemblages and petrogenetic grids, quantitative estimates of P–T conditions may also be obtained using various experimentally, empirically or theoretically calibrated geothermometers and geobarometers. Most of these rely on the accurate measurement of the chemical compositions of mineral phases in the equilibrium assemblage. For this reason, it is only since the electron microprobe has become a standard tool of the metamorphic petrologist that this approach to quantifying P–T conditions has been possible.

Before attempting any geothermometry or geobarometry, it is of fundamental importance to understand fully the mineralogical and textural relationships preserved within the rock. Of primary importance is to establish that the thermobarometry is based on an equilibrium assemblage. Thorough reviews by

Essene (1982, 1989) and Spear (1993, Chapter 15) examine the various geothermometers and geobarometers used by metamorphic petrologists. In these reviews, the positive points of different approaches are identified, but the assumptions and limitations of the different systems and techniques are also critically evaluated. In addition to the use of petrogenetic grids, the main approaches to geothermometry and geobarometry that have developed over the past two decades fall under one of five main headings, namely:

(a) **exchange thermometry** (e.g. Grt–Bt pairs or Grt–Hbl pairs);
(b) **solvus thermometry** (e.g. Cal–Dol pairs and two-feldspar thermometry);
(c) **solution models for multiple systems**, or 'net-transfer equilibria' (e.g. Grt–Rt–Al$_2$SiO$_5$–Ilm–Qtz);
(d) **internally consistent thermodynamic data set** approach; and
(e) **fluid inclusions**.

In addition to these, illite crystallinity, vitrinite reflectance and conodont colour index have all been used in low-grade rocks. It is outside the scope of this book to give an in-depth analysis of the pros and cons of the various approaches to thermobarometry in metamorphic rocks, but instead the reader is referred to the excellent reviews cited above. Essene (1989) gives a particularly useful evaluation of optimal thermobarometers for each metamorphic facies, and gives an extensive list of references. Where possible, it is always advisable to base any evaluation of P–T conditions on more than one approach. If this is done with care, thermobarometry in many metamorphic terrains may be accurate to ± 50°C and ± 1 kbar (Essene, 1989). Of vital importance is the careful characterisation of equilibrium mineral assemblages (including accessory phases), and a detailed evaluation of the microstructural interrelationships of all phases in the rock. When undertaking chemical micro-analysis of individual phases of the system, it is important to establish the nature of any chemical variations (e.g. zonation) within minerals, and to assess whether this reflects prograde or retrograde reactions. Chemical (X-ray) mapping and back-scattered electron imaging (Fig. 5.22) are often employed to evaluate detailed mineralogical relationships prior to microprobe analysis of specific points for thermobarometry.

Having evaluated peak metamorphic conditions based on observed mineral assemblages and, where possible, utilising one or more of the available geothermometers and geobarometers, the position of this event on the P–T trajectory can be accurately defined. Spear *et al.* (1984) give a schematic summary diagram, reproduced here as Fig. 12.4, showing the types of information that can be used to construct different parts of the P–T loop for individual samples. Although temperatures can often be evaluated quite well, accurate pressure estimates are usually more difficult to obtain. A study of porphyroblast–foliation relationships allows assessment of the timing of metamorphic events with respect to given deformation events. Combining all of the above information will give a P–T–t path, but without the absolute quantification of time.

Radiometric dating

Dating of specific metamorphic phases associated with a given metamorphic event will provide a date for a specific part of the P–T–t loop. Similarly, radiometric dating of an intrusion responsible for a certain metamorphic event, or constrained in the tectonometamorphic evolution by cross-cutting relationships will place a specific time on part of the trajectory. Depending on the amount of information accrued, it may in some instances be possible to obtain an indication of uplift rates.

There are various techniques that can be employed to obtain a whole-rock age, or a metamorphic age based on specific mineral separates. The various techniques, and the theory behind them, are dealt with in considerable detail by Faure (1986), and will not be dealt with here. The age equation,

$$\frac{1}{\lambda}\ln\left(\frac{D^*}{N}+1\right)=t \qquad (12.2)$$

gives the basic relationship from which the age of a sample can be determined. In this equation, D^* is the number of radiogenic daughter atoms produced from the parent, N is the number of remaining parent atoms, and λ is the decay constant. Knowing λ, and by measuring D^* and N, the age of the sample can be determined.

At high temperatures, a mineral will lose radiogenic daughter product, at medium temperatures daughter product will start to accumulate, and then at low temperatures daughter product accumulation will take place without loss. Eventually, the rate of increase of daughter product with time becomes more or less constant. The temperature at which this rate of increase of daughter to parent becomes constant is the *closure temperature* (T_c). Closure temperatures of individual minerals provide one of the most useful approaches to understanding the T–t evolution of a given rock, especially those that have cooled slowly. By mathematical analysis, Dodson (1973) devised an expression for T_c incorporating the various factors that have a direct influence; namely, the cooling rate (dT/dt), the chemical diffusivity (a function of activation energy for chemical diffusion and temperature) and the size of the diffusion domain of the crystal (which influences the length scale for diffusion).

Although T_c is mostly dependent on the activation energy of diffusion for specific minerals, it is also a function of cooling rate, and thus it is not possible to define a unique closure temperature for individual minerals. Nevertheless, closure temperature estimates have now been established for various metamorphic minerals (Table 12.1), and by using

TABLE 12.1 Estimated closure temperatures for various geochronological systems that are useful for constraining metamorphic cooling histories (based on Spear, 1993; Zeitler, 1989).

Mineral	Isotope system	T_c(°C)
Garnet	U–Pb	> 800
Zircon	U–Pb	> 700
Allanite	U–Pb	600–750
Monazite	U–Pb	> 700
Sphene (titanite)	U–Pb	500–600
Garnet	Sm–Nd	≈ 600
Hornblende	^{40}Ar/^{39}Ar	470–550
Muscovite	^{40}Ar/^{39}Ar	350–425
Biotite	^{40}Ar/^{39}Ar	260–350
K-feldspar	^{40}Ar/^{39}Ar	125–350
Zircon	Fission tracks	≈ 175–260
Sphene (titanite)	Fission tracks	≈ 250
Apatite	Fission tracks	≈ 100–150

minerals with different values of T_c, it is possible to obtain a series of dates corresponding to specific temperatures during the cooling history of the rock. This enables the cooling histories of individual rocks to be defined on an absolute rather than just a relative timescale. In Table 12.1 it is shown that U–Pb dating is most useful for rocks that have attained high-temperature granulite facies conditions, and ^{40}Ar/^{39}Ar step-heating dates from micas and hornblende are most useful for rocks that have experienced moderate- and high-temperature Barrovian-style orogenic metamorphism, while fission track dates are of greatest use at the lowest temperatures in the final stages of metamorphic cooling, and in studies of basin evolution. Having established the detailed P–T–t evolution of a given rock, this can in turn be related to crustal-scale processes such as uplift. This has proved increasingly useful towards our understanding of crustal processes and the tectonometamorphic evolution of individual rocks (e.g. Parrish *et al.*, 1988; Mezger *et al.*, 1991; Anderson *et al.*, 1992). As with geothermobarometry, it is crucial when undertaking thermochronology to be able to integrate it with petrological observations, and to be sure of the precise mineralogical and microstructural features of the rock being dated, and the particular metamorphic event to which the dated material relates.

In the following sections, examples are discussed that represent the variety and complexity of P–T trajectories from rocks of different metamorphic environments. In particular, those P–T–t paths that are most diagnostic of a particular style of metamorphism are highlighted. In all cases, the emphasis is on the mineralogical, microstructural and thermochronological evidence used to establish a particular P–T–t path.

12.3.2 Orogenic metamorphism

Orogenesis related to continental collision is associated with extensive crustal thickening. Large areas of crust become deeply buried, and the rocks involved experience an accompanying pressure increase (England & Thompson, 1984). This buried crust then experiences a period of heating as the perturbed crustal geotherm evolves towards a new equilibrium geotherm. Thrusting on various scales can give rise to complex 'saw-toothed' geotherms (e.g. Oxburgh & Turcotte, 1974; England & Richardson, 1977), but with time the conduction of heat causes heating of the footwall and cooling of the hangingwall, thus smoothing the geotherm. As time progresses, erosion and/or extensional thinning of the mountain belt causes unroofing (exhumation) of the deeply buried and metamorphosed core of the orogen. During this stage, the rocks in question experience both pressure and temperature decrease, such that the most characteristic feature of rocks displaying Barrovian-style orogenic metamorphism is one of a 'clockwise' P–T–t trajectory (Fig. 12.5). The decompression stage of such P–T–t trajectories is often referred to as 'uplift', but terms such as 'unroofing' or 'exhumation' are a more accurate description, since such decompression may not only be due to rigid-block uplift and erosion, but can also result from extensional deformation along low-angle normal faults, and ductile thinning of the crust (Hames et al., 1989; Haugerud & Zen, 1989). The high-grade rocks of the Tauern Window, Eastern Alps provide a good example of a situation in which the two differing interpretations of the decompression path have been presented. Until the work of Selverstone and colleagues (e.g. Selverstone et al., 1984; Selverstone & Spear, 1985; Selverstone, 1988), the Tauern Window rocks had largely been considered as a classic example of high-grade metamorphism as a result of overthrusting and subsequent erosion-controlled exhumation. However, Selverstone (1988) presented evidence to suggest that the western margin of the Tauern Window is a low-angle normal fault, and that a prolonged period of Oligocene–Miocene extension (after the main phase of nappe emplacement) involved ductile stretching and low-angle normal faulting to exhume the high-grade Tauern Window rocks.

Although the peak metamorphic and retrograde (exhumation) stage of a typical 'clockwise' P–T trajectory is usually well defined, the

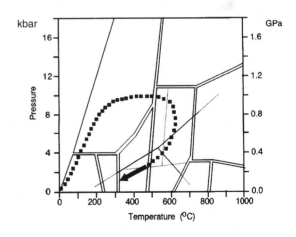

FIG. 12.5 A clockwise P–T trajectory of orogenic metamorphism. Boxed areas shown on the diagram are the metamorphic facies designated in Fig. 2.1. Al$_2$SiO$_5$ triple-point and univariant curves are also shown.

evidence for the early *P–T* evolution is usually fairly limited. Inclusions of mineral phases in porphyroblasts provide some of the most useful evidence for constraining early stages of the evolution (Plate 5(a)), but in most cases there are few diagnostic minerals. Alternatively (e.g. Spear, 1986; Burton *et al.*, 1989), sections of the *P–T* trajectory can be modelled by use of chemical variations in zoned garnet porphyroblasts and inclusions of biotite and plagioclase contained within the individual porphyroblasts (Fig. 12.6). Numerical modelling by England & Thompson (1984), Thompson & England (1984) and Ridley (1989) has also shown that crustal thickening during orogenesis produces Ky (± Sil) grade orogenic metamorphism and characteristic 'clockwise' *P–T–t* trajectories.

On the basis of various lines of evidence, 'clockwise' *P–T–t* paths associated with collisional orogenesis and Barrovian-style metamorphism have been described for many orogenic belts by numerous workers. For an area in the north Scandinavian Caledonides, Barker (1989), Barker & Anderson (1989) and Anderson *et al.* (1992) used a combination of $^{40}Ar/^{39}Ar$ closure ages for hornblende and muscovite, coupled with estimates of peak metamorphism based on geothermobarometry, to constrain the *P–T–t* evolution of individual nappes. Included mineral phases were used as evidence for some of the prograde reactions, and closure temperatures for muscovite, coupled with fluid inclusion studies of retrogression-related veins (Bennett & Barker, 1992), allowed the retrograde stage of the *P–T–t* evolution to be constrained to some degree. Two photographs of a garnet–mica schist from the Troms region of north Norway are shown in Fig. 12.7. The sample lies within one of the higher nappes of the Scandinavian Caledonides, and has been metamorphosed

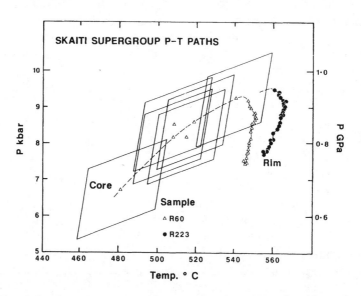

FIG. 12.6 A clockwise *P–T* trajectory based on porphyroblast chemical zonation (after Burton *et al.*, 1989; courtesy of the Geological Society). The *P–T* paths were determined from modelling of the inclusion-free rims of garnets in samples R60 (triangles) and R223 (circles) from the Skaiti Supergroup, Sulitjelma, Norway. The dashed portion of the R60 curve was drawn using six *P–T* points (shown with associated error boxes) calculated from biotite and plagioclase inclusions in the inclusion-rich core at varying distances from the centre. The modelling shows two distinct periods of growth: cores growing during increasing *P–T* and the rims growing during decompression.

P—T—t paths

during the Scandian phase of the Caledonian Orogeny. The garnets have two distinct growth stages; an early rather rounded core region, followed by a later rim growth producing subhedral to euhedral form. The early stage of growth is relatively inclusion-free, although in some porphyroblasts (Fig. 12.7(b)) the core exhibits a 'straight' inclusion fabric defined by small crystals of quartz. This is interpreted as overgrowth of an early (S_1) schistosity, the main fabric in the rock being the regional (S_2) schistosity, defined by aligned

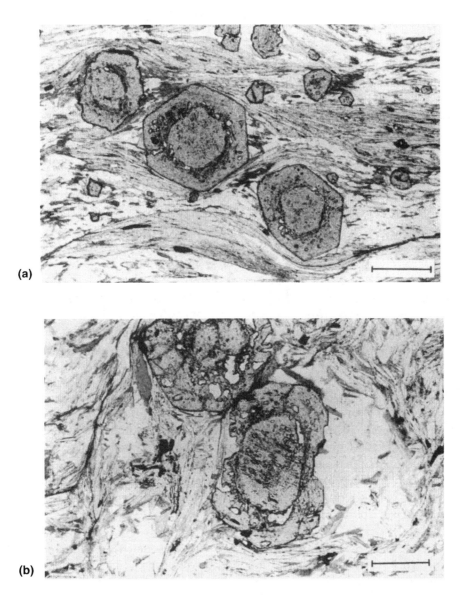

FIG. 12.7 A garnet–mica schist from a Caledonian nappe of the Troms region, Norway (see the text for details of the polymetamorphic history). Scale = 0.5 mm (PPL).

muscovite and biotite (Fig. 12.7(a)). The boundary between the first and second growth stages is marked by a distinct line, followed by abundant quartz inclusions in the innermost part of the second growth stage. This second growth zone is only observed in a few other samples from the nappe, and there is no correlation with proximity to the basal or roof thrusts. Because of this, possible frictional or down-heating effects are ruled out as the cause of the second growth stage. It is more likely that bulk rock chemistry is the main control, and that the outer growth zone developed in response to a specific discontinuous reaction within the rock, such as chloritoid breakdown. The dense zone of quartz inclusions probably reflects rapid initial growth of the second phase of garnet. This phase of growth occurred synchronous with S_2, but terminated before the S_2 schistosity was completely developed. In consequence, the porphyroblasts are wrapped by the schistosity (Fig. 12.7(a)). Plate 5(a) is from the same nappe, and shows inclusions of chloritoid in a garnet porphyroblast, whereas elsewhere in the matrix staurolite occurs and chloritoid is absent. This provides evidence that staurolite formed from a chloritoid breakdown reaction and, coupled with garnet 'core–rim' thermobarometry, helps to constrain part of the P–T trajectory (Fig. 12.8).

In the Central Alps, Italy, Diella et al. (1992) studied the metamorphic evolution of South Alpine metamorphic basement and determined a 'clockwise' P–T trajectory on the basis of detailed microstructural and petrological observations from metapelites. They established the nature of key univariant reaction curves that had been crossed during uplift. For example, in the Val Vedello basement (Fig. 12.9), they showed how, during decompression, late kyanite grew and rutile became destabilised, as indicated by the development of ilmenite rims. This is interpreted in terms of the reaction (Bohlen et al., 1983)

$$Alm + Rt \rightleftharpoons Ilm + Ky + Qtz.$$

For units north of the Main Karakorum Thrust, northern Pakistan (Himalayas), Allen & Chamberlain (1991) were able to infer 'clockwise' P–T trajectories on the basis of reaction histories based on observed microstructural and textural relationships. For example, in samples from the Braldu Nappe, they frequently recorded a corona of Qtz + Bt + Sil after kyanite, and replacing Grt + Ms. Additionally, they noted the common case of kyanite incipiently replaced by sillimanite. Integrating these petrographic observations with the data obtained from geothermobarometry, they deduced the P–T path shown in Fig. 12.10. In a study of metamorphism in the Central Himalaya, Pecher (1989) discussed the P–T–t trajectories in units either side of the Main Central Thrust (MCT). Pecher (1989) concluded that the Midland Formation in the footwall to the MCT experienced prograde metamorphism due to downward transfer of heat during the 'hot' emplacement of the Tibetan Slab (also known as the 'High Himalayan Crystalline Sequence') which forms the hangingwall to the MCT. The Tibetan slab is interpreted to have experienced simultaneous cooling as heat was transferred to the footwall. This example usefully illustrates that while the 'clockwise' P–T trajectories may have similar forms for a range of rocks from different nappes in a given terrain, in detail the trajectories are more complex. When time is taken into consideration there may be considerable diachroneity between seemingly similar P–T trajectories. This point was also emphasised in the work of Anderson et al. (1992) and Barker (1995) in a study of part of the Scandinavian Caledonides.

The Taconian (Middle to Late Ordovician) and Western Acadian (late Silurian to middle Devonian) metamorphism of New England, USA, also produced 'clockwise' P–T paths characteristic of collisional orogenesis involving

P—T—t paths

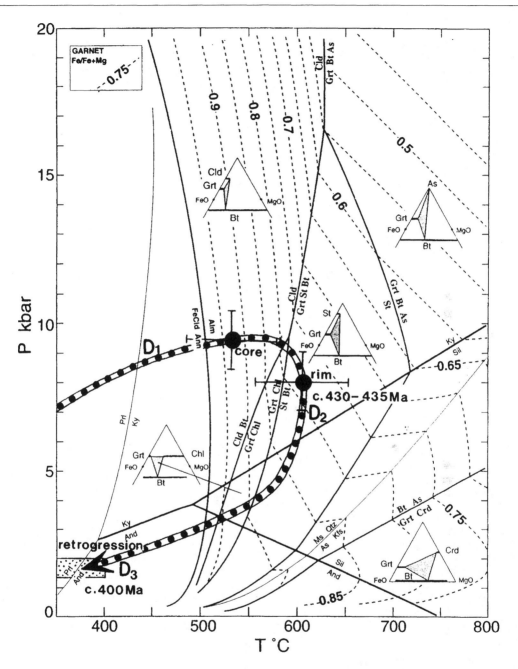

FIG. 12.8 A *P–T* trajectory for the Høgtind Nappe, Troms, Norway (based on Barker, 1989; Barker & Anderson, 1989; Anderson *et al.*, 1992; Bennett & Barker, 1992), superimposed on the calculated *P–T* grid of Spear & Cheney (1989) and Spear (1993), for the KFMASH system. Only those assemblages containing Grt + Bt are shown. The dashed contours show the Fe/(Fe + Mg) in garnet as a function of *P* and *T* in each assemblage. Note that the Fe/(Fe + Mg) in garnet changes along the univariant curves.

Deciphering polydeformed and polymetamorphosed rocks

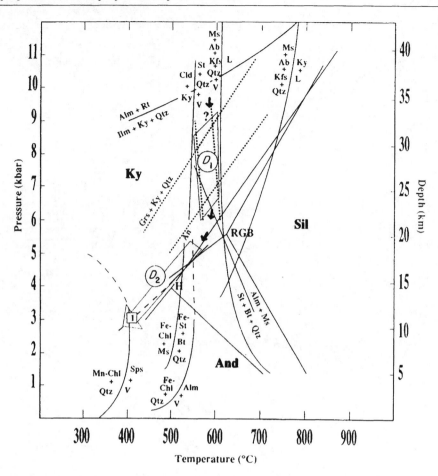

Fig. 12.9 A modified version of Fig. 9 of Diella *et al.* (1992), courtesy of Blackwell Science, showing key univariant reactions crossed during the uplift history of metapelites from the Val Vedello basement, Central Alps, Italy. D_1, D_2 = P–T conditions for assemblages formed during D_1 and D_2 structural events. Al_2SiO_5 triple-points: H = Holdaway (1971); RGB = Richardson *et al.* (1969). Initial melting of metapelite (wet and dry) after Thompson & Tracy (1979). Alm + Rut = Ilm + Ky + Qtz after Bohlen *et al.* (1983). Staurolite equilibria after Hoschek (1969), Chatterjee (1972) and Rao & Johannes (1979). Alm + V = Fe–Chl + Qtz after Hsu (1968) and Naggar & Atherton (1970). Sps + V = Mn–Chl + Qtz after Hsu (1968); 1 = Bt + Grt + Al_2SiO_5 = Chl + Ms + Qtz from Hirschberg & Winkler (1968). The dotted lines represent the minimum and maximum calculated ln K for Grs + Ky + Qtz = An (Ghent, 1976) and the K_d for Grt–Bt equilibrium (Perchuk & Lavrent'eva, 1983).

crustal thickening, followed by a period of heating then rapid decompression during unroofing. This has been described by Hames *et al.* (1989, 1991) and Armstrong *et al.* (1992), by a combination of petrological work, microstructural relationships, geochronology and thermal modelling (Fig. 12.11, path-A). Comparing the theoretical models with the observed P–T trajectory (constrained by mineral equilibrium data and fluid inclusion data), Hames *et al.* (1989) concluded that the rapid decompression following peak Acadian metamorphism is best explained in terms of rapid uplift coupled with erosion and tectonic denudation.

However, not all cases of orogenic (regional)

P—T—t paths

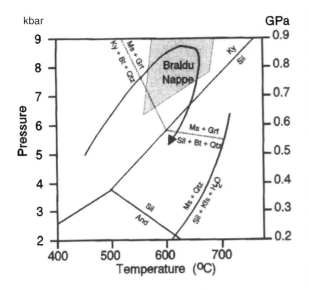

FIG. 12.10 The *P–T* path for the Braldu Nappe, northern Pakistan, Himalayas (modified after Allen & Chamberlain, 1991). The Al_2SiO_5 phase diagram is after Holdaway (1971) and the muscovite breakdown curve is from Chatterjee & Johannes (1974). The dashed line gives slopes calculated from entropy and volume data for the reactions Ms + Grt $\rightleftharpoons$ 2Sil/Ky + Qtz + Bt. The stippled area represents an error box of ± 2 kbar and ± 50°C for the average of the peak *P–T* determinations *for* the Braldu Nappe.

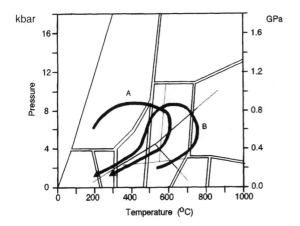

FIG. 12.11 Path A is the Taconian clockwise *P–T–t* trajectory, New England, USA; path B is the Eastern Acadian anticlockwise P–T–t trajectory from an adjacent metamorphic belt (based on Armstrong et al., 1992).

metamorphism record 'clockwise' *P–T–t* trajectories. In particular, some areas that have experienced high-*T*, low-*P* (Buchan-type) metamorphism have been shown to exhibit an 'anticlockwise' *P–T* trajectory. For example, on the basis of petrological, geochronological and fluid inclusion evidence, various authors (e.g. Schumacher *et al.*, 1989; Armstrong *et al.*, 1992; Winslow *et al.*, 1994) have described an 'anticlockwise' path associated with Buchan-type metamorphism in the Eastern Acadian belt, Massachusetts, USA (Fig. 12.11, path-B). The evidence for this path includes sillimanite pseudomorphs after andalusite, and late-stage development of higher-pressure garnet-bearing assemblages in place of low-*P* cordierite assemblages (Winslow *et al.*, 1994). Fluid inclusion studies also support an interpretation in terms of initial low-*P* heating, and then thickening at high temperature followed by isobaric cooling, before final decompressional unroofing. The early evolution is interpreted in terms of widespread melting in the lower crust of a back-arc environment in response to a build-up of heat in basin-fill sediments enriched in radioactive heat-producing elements. Back-arc extension led to widespread emplacement of intermediate and acidic plutons coupled with regional-scale, low-*P*/high-*T*, Buchan-type metamorphism (Armstrong *et al.*, 1992; Winslow *et al.*, 1994). Subsequent east–west shortening during the Acadian orogeny caused crustal thickening. This caused pressure to increase while the rocks were still at high temperature (but starting to cool), and later unroofing completed the 'anticlockwise' *P–T–t* trajectory (Fig. 12.11, path-B).

12.3.3 Orogenic metamorphism with a subsequent thermal overprint

The previous section described the characteristic 'clockwise' *P–T–t* trajectories that typify Barrovian-style orogenic metamorphism, but in cases in which there is more than one metamorphic event, it is quite common to observe an earlier orogenic (regional) metamorphic event overprinted by a late-stage thermal (or 'contact') metamorphic event, when granitoid melt generated in the lower crust ascends and comes to rest at upper crustal levels. Modelling by De Yoreo et al. (1989) examined the role of crustal anatexis and magma migration in relation to the thermal evolution of thickened continental crust. When a single granitoid sheet 2 km thick is emplaced into the upper crust it causes a short-lived and localised perturbation or 'spike' to sillimanite grade conditions, the value for pressure depending on the level at which the intrusion is emplaced. However, with widespread melt generation and melt migration to higher crustal levels, a regionally extensive high-T/low-P metamorphic event may be initiated.

Inger & Harris (1992) described a 17–20 Ma, high-T, metamorphic overprint (M2) in metasediments of the upper part of the High Himalayan Crystalline Sequence in the Langtang Valley, northern Nepal, that had previously experienced an M1 (pre 34 Ma) Barrovian (Ky–St grade) metamorphism. Although this thermal overprint is recognised elsewhere in the same sequence above the MCT, there is still considerable debate concerning its cause.

A more straightforward and common example of a thermal overprint is the case in which metasediments that have experienced sub-greenschist or greenschist facies orogenic (regional) metamorphism in the upper crust experience heating from a late-stage intrusion during their uplift trajectory (Fig. 12.12). The chiastolite slates from the metamorphic aureole of the Skiddaw Granite (Lake District, England) document such a history. Figs 12.13(a) & (b) show andalusite (var. chiastolite) formed at the time of granite intrusion (early Devonian), and subsequently pseudomorphed by an aggregate of sericite rimmed by chlorite (for unaltered chiastolite, see Plate

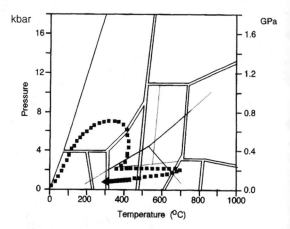

FIG. 12.12 A schematic illustration of a greenschist facies P–T trajectory with a late And/Sil overprint. This is the typical trajectory of low-grade metasediments intruded by late-/post-orogenic granitoids (e.g. Skiddaw Granite aureole, Lake District, England; see Fig. 12.13).

1(c)). The distinctive chiastolite cross is still clearly recognisable (Fig. 12.13(b)), despite the pseudomorphing. The pseudomorphs show clear relationships to indicate that the original porphyroblasts statically overprinted a pre-existing slightly crenulated slaty cleavage: that is, in both cases there is no deflection of the cleavage around the porphyroblasts, and there are no pressure shadows. In Fig. 12.13(a), compositional layering trends from bottom left to top right, and has a continuous (or pervasive) slaty cleavage (S_1) sub-parallel to it. This is overprinted by a close-spaced crenulation cleavage (S_2) trending from bottom right to top left. In Fig. 12.13(b), the S_2 cleavage is not represented, but S_1 (more highly magnified) passes from top to bottom, cross-cutting the prominent compositional layering trending bottom left to top right. The earlier cleavages and greenschist facies fine-grained matrix assemblages formed during low-grade regional metamorphism in the end-Silurian stage of the Caledonian Orogeny. The example in Fig. 12.14(a) is a garnetiferous slate from the Isle of Man, UK. It has a fine-grained matrix

FIG. 12.13 Chiastolite slates from the metamorphic aureole of the Skiddaw Granite, Lake District, England (see the text for details of the polymetamorphic history). Scale = 1 mm in (a), 0.5 mm in (b) (PPL).

comprising Qtz–Ms–Chl–Bt–Ilm, which defines a continuous slaty cleavage. In detail, this is seen to have experienced a later fine-scale crenulation (Fig. 12.14(b)). These Caledonian structures are cross-cut by thin, variably oriented chlorite–muscovite veins, which are in turn overprinted by euhedral spessartine–almandine porphyroblasts (Gillott, 1955). This phase of porphroblastesis occurred during the late-thermal overprint associated with a mid-Devonian granite intrusion.

Deciphering polydeformed and polymetamorphosed rocks

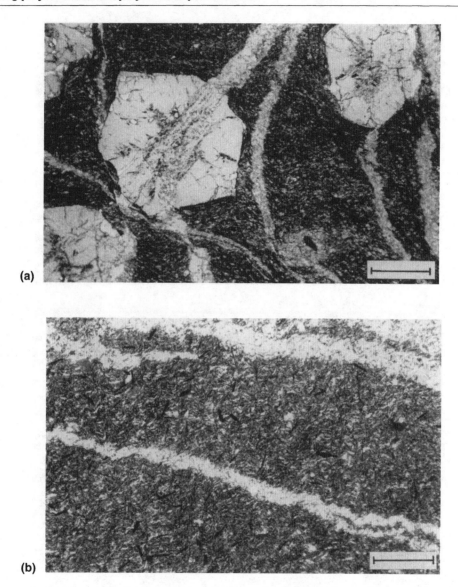

FIG. 12.14 Garnetiferous slate from the Isle of Man, England (see the text for details of the polymetamorphic history). Scale = 1 mm in (a), 0.5 mm in (b) (PPL).

12.3.4 Granulite facies P–T–t paths

Bohlen (1987), Harley (1989) and Spear (1993) have reviewed the characteristic P–T–t paths associated with the evolution of granulite facies rocks. There are two basic types of path to consider, namely, paths showing a stage of near 'isothermal decompression' (ITD), and paths that document a significant period of near 'isobaric cooling' (IBC). The evidence for the two fundamentally different trajectories (Fig. 12.15) comes from the interpretation of reaction features, geothermobarometry and geochronology. The ability to

P–T–t paths

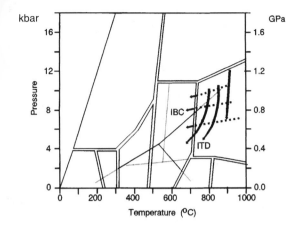

FIG. 12.15 Typical isobaric cooling (IBC) and isothermal decompression (ITD) paths recorded in granulites based on reaction textures and geothermobarometry (simplified from data in Harley, 1989). The boxed areas shown on the diagram are the metamorphic facies designated in Fig. 2.1. Al_2SiO_5 triple-point and univariant curves are also shown.

distinguish between the features indicative of an ITD path as opposed to features of an IBC path is crucial, since the different paths imply fundamentally different crustal processes. In practice, this distinction may not always be straightforward, as Frost & Chacko (1989) and Spear (1993) point out. Initial tectonic thickening of the crust and subsequent rapid exhumation by extensional tectonism, erosional processes or some combination of the two, is the most realistic interpretation of an ITD path.

Bohlen (1987) and Ellis (1987) have described 'anticlockwise' P–T paths with an IBC stage following an earlier high-T, low-P thermal peak induced by magma accretion at the base of existing continental crust. However, other types of IBC path are also possible. One of the problems when interpreting the cause of IBC paths is that while the retrograde IBC part of the trajectory is often well defined on the basis of combined mineralogical and microstructural reaction features, the crucial prograde stage of the tectonothermal evolution is often poorly constrained.

Fig. 12.16, based on Spear (1993) and Harley (1989), shows reaction microstructures (insets) that give evidence for two different reaction types and two different P–T paths for the individual rocks concerned. Path A, based on reaction microstructure (corona) of inset (a), suggests a granulite facies rock that followed an IBC path. Sapphirine (centre of inset (a)) is separated from quartz, and thus interpreted to be out of equilibrium with quartz. Instead, an irregular corona of Opx + Sil + Crd has formed. This indicates that the reaction Opx + Sil + Crd $\rightleftharpoons$ Spr + Qtz has moved from right to left (i.e. down temperature). Path B, based on the reaction microstructure (symplectite) of inset (b), is quite different from the previous case. In this example, the reaction microstructure relates to a continuous divariant reaction (Grt + Qtz $\rightleftharpoons$ Crd + Opx) on an ITD path, with the position of the reaction curve in P-T space varying as a function of composition. The sub-horizontal lines of Fig. 12.16, labelled 0.3 to 0.9, represent the Fe/(Fe + Mg) composition of garnet in the assemblage Grt + Crd + Opx + Qtz at given P–T. The lines show how the Fe : Mg ratio of garnet increases during decompression by the continuous reaction Grt + Qtz $\rightleftharpoons$ Crd + Opx. Simultaneous with the change in the Fe : Mg ratio, garnet is being steadily consumed, as newly formed cordierite and orthopyroxene grow to define an intricate symplectic intergrowth separating the corroded garnet core from the original coarse orthopyroxene and quartz (for further details, see Harley, 1989; Perchuk, 1989). The same reaction features were also used as evidence for isothermal decompression by Hisada & Miyano (1996), in a study of granulite facies rocks from the Botswanan Limpopo belt.

There are many other well documented examples in which P–T paths of granulites have been constructed utilising information

213

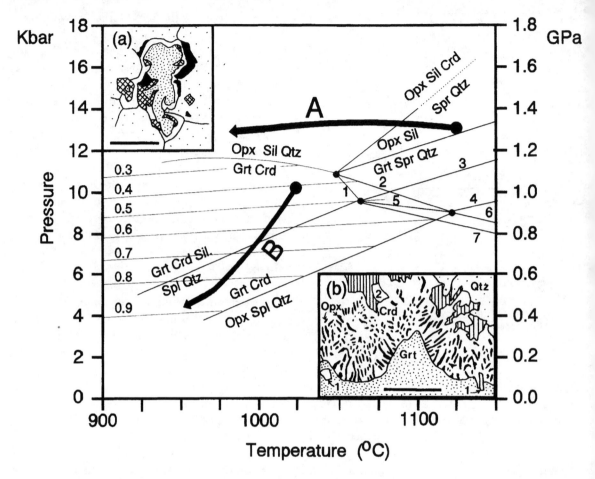

FIG. 12.16 A composite diagram based on Harley (1989) and Spear (1993) to show examples of specific reaction textures associated with IBC and ITD paths in granulites. The lines labelled 0.3 to 0.9 represent the Fe/(Fe + Mg) composition of garnet in the assemblage Grt + Opx + Crd + Qtz. The garnet compositions become enriched during decompression associated with the continuous reaction Grt + Qtz → Crd + Opx. The reaction lines numbered 1 to 7 are as follows (right-hand side of equation represents higher-P assemblage): (1) Grt + Crd + Sil ⇌ Spr + Qtz; (2) Grt + Crd ⇌ Opx + Spr + Qtz; (3) Spl + Qtz ⇌ Grt + Spr + Sil; (4) Spl + Opx + Qtz ⇌ Grt + Spr; (5) Spl + Crd ⇌ Grt + Spr + Qtz; (6) Spl + Crd ⇌ Spr + Opx + Qtz; (7) Crd + Spl + Sil ⇌ Spr + Qtz. *Inset* (a): a multi-corona reaction texture between Spr and Qtz in granulite from Enderby Land, Antarctica, representing an IBC path (path A). High-density stipple = Spr; low-density stipple = Qtz; black = Opx; white = Crd; cross-hatch = Sil. Scale = 1 mm. *Inset* (b): the reaction texture of an ITD path (path B) for granulite from the Sharyzhalgay Complex, Lake Baikal. 1 = Bt; 2 = Ilm; all other minerals labelled on inset. Scale = 1 mm.

from symplectite and corona reaction features in conjunction with petrogenetic grids. For example, Waters (1989), in a study of the Namaqualand granulites of southern Africa, uses the observation of Sil + Grt coronas around spinel as evidence for the Spl + Qtz ⇌ Grt + Sil reaction being crossed during the retrograde isobaric cooling stage of an anti-clockwise *P–T–t* path. In contrast, Clarke & Powell (1991), in a study of granulites from the Musgrave Complex, central Australia, record coronas and symplectites defining the reaction

P–T–t paths

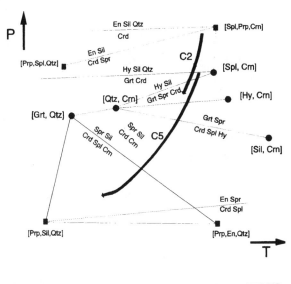

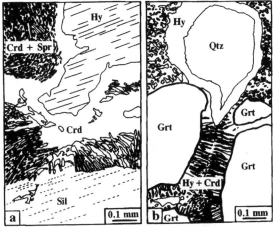

FIG. 12.17 An ITD retrograde reaction sequence, determined for metapelitic granulites from In Ouzzal, Algeria. *Inset* (a): the reaction Hy + Sil ⇌ Crd + Spr (sample C2). Scale = 0.1 mm. *Inset* (b): the reaction Grt + Qtz ⇌ Hy + Crd (sample C5). Scale = 0.1 mm (modified after Bertrand *et al.*, 1992).

Grt + Sil → Crd + Spl, which they use as evidence for retrogression during isothermal decompression. The detailed study by Bertrand *et al.* (1992) on Precambrian granulites from the In Ouzzal craton, southern Algeria, also used corona and symplectite relationships to infer an array of retrograde reactions, and from this deduced an isothermal decompression retrograde stage of a clockwise P–T–t trajectory (Fig. 12.17).

A study of silica-undersaturated granulites from the Strangways Range, central Australia (Goscombe, 1992) used coronitic and symplectitic reaction textures in the FMAS system to define the reaction sequence (1) Crd + Spl + Crn → Spr + Sil, (2) Crd + Spl → Opx + Spr + Sil, (3) Spr + Spl + Sil → Opx + Crn and (4) Spr + Sil → Crd + Opx + Crn, as evidence for an 'anticlockwise' P–T–t trajectory involving a prograde heating stage followed by a retrograde near-isobaric cooling. In contrast, a study of silica-deficient granulites from the Bamble region, southern Norway (Kihle & Bucher-Nurminen, 1992), defines a clockwise P–T–t path, with evidence for a retrograde period of isothermal decompression based on symplectites and coronas that give evidence for the discontinuous FMAS reaction Opx + Sil → Spr + Crd + Crn. In high-pressure garnet–sapphirine granulites from the Central Limpopo Mobile Belt, Zimbabwe, Droop (1989) uses symplectitic and coronitic reaction textures as evidence for the retrograde reactions Krn → Spr + Crd and Grt → Ged + Crd + Spr + Spl (Fig. 12.18). This in turn is used as evidence for near-isothermal decompression during a period of rapid uplift. A point to add in relation to the use of reaction textures and petrogenetic grids as an approach to defining a path of near-isothermal decompression or near-isobaric cooling in granulite facies rocks is that, in the absence of good thermobarometric data, it is often difficult to define with any accuracy the angle at which a given reaction curve is crossed. This point was emphasised by Vernon (1996), and while many authors show near-isothermal or near-isobaric paths, the truth of the matter is that although the direction of the arrow is usually well-defined, the precise positions of the paths may not be defined so accurately, and to some extent they are schematic.

Deciphering polydeformed and polymetamorphosed rocks

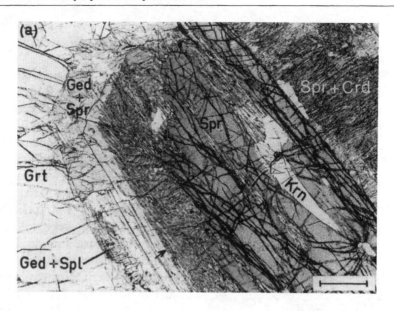

FIG. 12.18 (a) Symplectites and coronas associated with decompression-related reactions (see text for details) in high-pressure Grt–Spr granulite from the Central Limpopo Mobile Belt, Zimbabwe (modified after Fig. 2a of Droop, 1989).

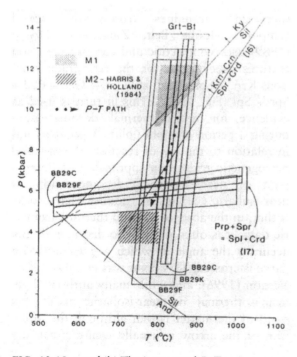

FIG. 12.18 contd (b) The interpreted $P-T$ trajectory on the basis of the petrographic evidence of (a) (Fig. 10 of Droop, 1989; courtesy of Blackwell Science).

12.3.5 Blueschist facies $P-T-t$ paths

Ernst (1988), Perchuk (1989) and Spear (1993) have graphically summarised the typical style of $P-T-t$ trajectories of rocks from subduction zone complexes (Fig. 12.19). Not suprisingly, the mineralogical and microstructural evidence suggests that the typical path of a blueschist facies rock involves an initial period of rapid pressure increase at low temperatures, as the rocks are subducted to deep levels. Following this, the rocks heat up to some extent prior to exhumation. In most cases, blueschist facies assemblages display a greenschist or low-amphibolite facies metamorphic overprint (e.g. Holland & Ray, 1985; Kryza et al., 1990). This is perhaps best explained in terms of rapid exhumation and isothermal decompression while still at temperatures in the range 300–500°C. Reactions to promote such a metamorphic overprint are greatly facilitated if fluids infiltrate the rocks during exhumation. Whatever the case, the most characteristic

P—T—t paths

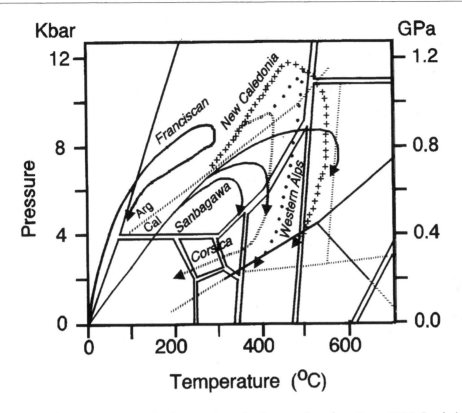

FIG. 12.19 Blueschist *P–T–t* trajectories from various classic areas (based on Ernst, 1988; Perchuk, 1989; Spear, 1993). The boxed areas shown on the diagram are the metamorphic facies designated in Fig. 2.1. Al_2SiO_5 triple-point and univariant curves are also shown. Arg $\rightleftharpoons$ Cal transformation (after Johannes & Puhan, 1971; Crawford & Hoersch, 1972).

FIG. 12.20 Lawsonite pseudomorphs in blueschist. Ile de Groix, Brittany, France. The coin is 24 mm in diameter.

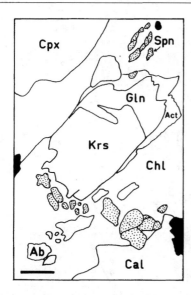

FIG. 12.21 Actinolite rimming glaucophane: an example of a glaucophane-bearing assemblage being overprinted by greenschist facies metamorphism from the Kaczawa Complex, Sudetes, Poland (modified after Kryza et al., 1990). Scale = 0.1 mm. Krs, Kaersutite (a brown amphibole with high Ti, typical of igneous amphiboles in alkali basic rocks); Gln, glaucophane; Act, actinolite; Spn, sphene; Chl, chlorite; Ab, albite; Cal, calcite.

$P-T-t$ trajectory of blueschist facies rocks has a 'clockwise' sense. In the western Alps, the path described above is particularly common (e.g. Chopin, 1984; Massone & Chopin, 1989), and Ernst (1988) refers to this $P-T-t$ trajectory as the 'western Alpine type'. Blueschists of the Sanbagawa belt, Japan (e.g. Takasu, 1989; Banno & Sakai, 1989; Otsuki & Banno, 1990), the Seward Peninsula, Alaska (e.g. Forbes et al., 1984), New Caledonia (e.g. Brothers, 1985; Brothers & Yokoyama, 1982), and many other blueschist terrains, also show this type of evolution. Ernst (1988) interprets the western Alpine type of trajectory with its near-isothermal decompression and greenschist or epidote–amphibolite facies overprint as being due to the nature of the convergence. In particular, he interprets the trajectory in terms of a switch from subduction of oceanic crust to a situation of continental convergence, with crustal thickening followed by rapid exhumation through a combination of erosion and tectonic processes.

Some of the key reaction microstructures that help to constrain the $P-T$ evolution of the 'western Alpine type' trajectory include Czo + Pg pseudomorphs after lawsonite (Fig. 12.20) as evidence for increasing temperature, decreasing pressure or a combination of both (e.g. Forbes et al., 1984). The development of reaction rims or epitaxial overgrowth of actinolite on glaucophane (or crossite) (Fig. 12.21) is used as evidence of destabilisation of glaucophane during uplift, and transition into the greenschist facies by the reaction Gln + Ep ± $CaCO_3$ → Act + Chl + Ab (Kryza et al., 1990). Holland & Ray (1985) describe blueschists from the Tauern Window, Austria, where petrographic studies have revealed that the assemblage Cros + Pg + Ep became unstable during decompression and reacted to give the greenschist facies assemblage Chl + Ab + Act + Mag + Tlc.

A second type of $P-T$ trajectory displayed by blueschist facies rocks is seen in the Franciscan complex of California, USA. In this case (the 'Franciscan type' of Ernst, 1988) the retrograde $P-T$ path seems almost to retrace the same path as the prograde trajectory, and gives an overall 'hairpin' trajectory (Fig. 12.19): that is, cooling accompanies decompression, rather than rapid near-isothermal decompression as in the 'western Alpine type' trajectory. The evidence for significant cooling accompanying decompression during uplift comes from the fact that many rocks still contain aragonite (high-pressure polymorph of $CaCO_3$). The fact that aragonite is still present suggests that the Arg ⇌ Cal transition is crossed at temperatures (perhaps < 100°C), at which the reaction kinetics are sufficiently slow that aragonite does not react out (Carlson & Rosenfeld, 1981). There is still much debate regarding the tectonic process(es) by which the Franciscan blueschists

P—T—t paths

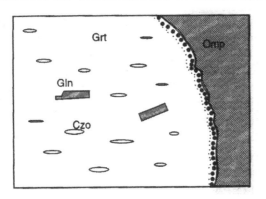

FIG. 12.22 Schematic illustrations of glaucophane (Gln) and clinozoisite (Czo) inclusions in eclogite facies garnets (Grt) surrounded by omphacitic pyroxenes (Omp); based on descriptions of Pognante et al. (1987), for a suite of eclogites from the Western Alps.

seemingly have an uplift trajectory that retraces the prograde burial path. Whatever the case, the preservation of aragonite seems to suggest that compared to most other blueschist terrains, the uplift path, although probably with a clockwise trajectory, is somewhat different. The presence of jadeiitic pyroxene in parts of the Franciscan complex, and in certain other blueschist facies terrains indicates that the reaction Ab → Jd + Qtz has taken place and that the highest-pressure part of the blueschist facies has been attained.

12.3.6 Eclogite facies P–T–t paths

Although the details of uplift trajectories for eclogites are moderately well known due to retrograde reaction assemblages and microstructures, the prograde path of such rocks is usually poorly constrained. Eclogites such as those of the Najac–Carmaux thrust unit, Massif Central, France (Burg et al., 1989), have garnets that preserve inclusions of blue–green calcic clinoamphibole, zoned from glaucophane cores to barroisite rims. In some instances, pure glaucophane inclusions occur in garnet cores (e.g. Barnicoat & Fry, 1989), but this is usually the only clear evidence to indicate that the prograde path of an eclogite has been through the glaucophane stability field. Pognante et al. (1987), studying part of the Western Alps, similarly reported glaucophane inclusions within eclogitic garnets (Fig. 12.22). It would be wrong to suggest that all eclogites have followed this trajectory, but some at least seem to have experienced rapid burial and a clockwise P–T–t path comparable to western Alpine type blueschists, but attaining higher P–T at peak metamorphism. Other blueschist facies rocks that have passed up temperature into the low-temperature part of the eclogite facies (see the review by Schielstedt, 1990), include the high-pressure rocks on the Greek islands of Sifnos and Syros (e.g. Ridley, 1984; Schielstedt, 1986; Schielstedt & Matthews, 1987) and some of the high-P rocks from New Caledonia (e.g. Brothers, 1985). As with many blueschist facies rocks, those rocks that have entered the low-temperature eclogite facies typically experience a greenschist facies overprint during exhumation.

Eclogites are known from various localities throughout the Alps. Many of these eclogites have experienced an initial high-P (c. 8–30 kbar), high-T (c. 400–450°C to 700–800°C) phase of eclogite facies metamorphism, subsequently overprinted by blueschist facies assemblages, comprising abundant glaucophane (± lawsonite), and in turn followed by a low-pressure greenschist facies retrograde event. Although the prograde part of the trajectory is not well constrained, Droop et al. (1990) report that some rocks of the 'Eclogitic Micaschist Complex' of the Sesia Zone preserve as inclusions in eclogite facies minerals, an earlier foliation defined by glaucophane and epidote needles, indicating that the prograde path passed through the glaucophane + epidote stability field. However, a 'clockwise' trajectory is not the only possibility for passing from early blueschist conditions to peak metamorphic eclogite facies conditions. For rocks from the Piemonte zone, western Alps,

Deciphering polydeformed and polymetamorphosed rocks

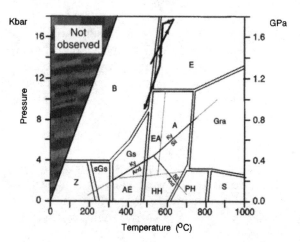

FIG. 12.23 A schematic illustration of an eclogite facies "hairpin' P–T–t trajectory (style of trajectory based on Barnicoat & Fry, 1989).

Barnicoat & Fry (1989) presented evidence for a tight 'hairpin' trajectory, that they considered to be 'anticlockwise' in character.

Some eclogite terrains show evidence of having experienced ultra high pressure metamorphism at moderate or high temperatures. As well as theromobarometric studies, one of the key lines of evidence for ultra high-P conditions ($\geq$ 20–25 kbar) comes from the preservation of coesite (a high-P polymorph of SiO_2), or quartz pseudomorphs after coesite, as inclusions in garnet or jadeiitic clinopyroxene. The preservation of coesite or quartz pseudomorphs after coesite has now been recognised in a number of ultra high P eclogite facies terrains, including the Western Alps (Chopin, 1984) and the Dabie Shan and Su-Lu terrains, China (e.g. Okay, 1995; Zhang et al., 1995). The recognition of preexisting coesite is often revealed by radial fractures emanating from an inclusion comprising minute radiating quartz crystals and perhaps a small core of remnant coesite (Fig. 7.2). The diagnostic radiating fractures in the host mineral around the inclusion are believed to result from the stresses caused by volume increase associated with the transformation coesite → quartz. Apart from the fact that ultra high-P conditions have been obtained, the precise nature of the P–T–t trajectory is not always well constrained, but the evidence, based on included mineral assemblages, geothermobarometry and retrograde assemblages, seems to suggest steep slopes for both prograde and retrograde paths (e.g. Zhang et al., 1995).

There is no doubt that eclogite facies metamorphism occurs at deep crustal levels, and rocks preserving such metamorphic assemblages are generally considered to have been rapidly exhumed. However, in most cases the available evidence is insufficient tightly to constrain the full P–T evolution. Because of this, it is usually difficult to say with any certainty whether an individual eclogite has experienced a 'clockwise' or 'anti-clockwise' trajectory. Indeed, it seems that a key feature of many eclogites may be deep burial, followed by rapid exhumation, to define what may best be termed a 'hairpin' trajectory (Fig. 12.23).

12.4 Final comments

The key to deciphering polydeformed and polymetamorphosed rocks lies in careful observation, both in the field and during petrographic work, whether by standard transmitted light microscopy or by more advanced techniques. Throughout this chapter it has been emphasised that interpretation of the P–T–t evolution of a metamorphic rock depends on the correct identification of equilibrium assemblages and reaction sequences, based on detailed observation. The detailed petrogenetic grids that now exist allow P–T conditions to be constrained reasonably well if certain key minerals and assemblages have been identified. When coupled with other approaches to geothermometry and geobarometry, as outlined in Section 12.3.1,

Final comments

and with additional geochronological information, it is often possible to constrain P–T–t paths with a fair degree of detail. When integrated with structural information, it is possible to gain a more complete understanding of tectonic processes and the tectonometamorphic evolution of a particular portion of the Earth's crust at some time in the geological past.

Despite the potential for major insight into crustal processes based on a firm understanding of metamorphic assemblages and their microstructural interrelationships, there are a few points that should be noted. First, the cause of a change in equilibrium assemblage may not necessarily be changing P–T conditions, but could be changing fluid composition or some other factor. As previously stated, it should be realised that many petrogenetic grids are constructed on the basis of the fluid being pure H_2O and that $P_f = P_{H_2O}$. During the metamorphism of carbonate and calc-silicate rocks, it is well established that the fluid typically varies between pure H_2O and pure CO_2, and that the exact composition may vary as reactions proceed. For this reason, it is crucial to appreciate the influence of different fluid compositions in controlling equilibrium assemblages at particular P–T conditions. For example, Cartwright & Buick (1995) describe the formation of wollastonite-bearing layers in granulite facies marbles due to the infiltration of water-rich fluids at peak metamorphic temperatures of around 700°C, rather than due to changing P–T conditions. The wollastonite-forming reaction is Cal + Qtz $\rightleftharpoons$ Wo + CO_2. The fact that adjacent marbles contain up to 11% modal quartz, in equilibrium with calcite, provides the evidence that some layers have been infiltrated by fluid and others have not, and that it is influx of aqueous fluid that has caused the reaction rather than changing P–T conditions.

A second point to bear in mind is that the prograde path of a P–T–t trajectory for a particular rock is generally the most difficult to constrain. In most cases the only thing to go on is whatever minerals may be preserved as included phases in the core of peak metamorphic minerals. In most cases, the minerals preserved may be phases such as quartz, which are stable over a wide range of P–T conditions, and even if a more useful phase such as glaucophane or chloritoid is preserved, the lack of a complete precursor assemblage makes it difficult to constrain the P–T path in anything other than general terms. Evidence for the exhumation or retrograde P–T–t trajectory may be better preserved, due to incomplete reactions, fluid inclusion studies, geothermobarometry and metamorphic cooling ages.

It is always important to give clear consideration to the errors on any part of the P–T–t trajectory, and to consider other possible interpretations that may exist. It is important to avoid the temptation of simply joining up a continuous line between various constrained points in a rock's P–T evolution in order to obtain a P–T trajectory without giving proper consideration to the processes involved and the time interval between given points that have been defined. Vernon (1996) emphasises this problem, and also discusses the problem of inferring the direction of a P–T path from the crossing of a single reaction curve in P–T space. Although the particular reaction may have been identified with a high degree of confidence, there is a 180° range of directions for the P–T path when crossing a constant slope (straight line) reaction curve, and when the reaction curve is convex outwards in the direction in which the reaction is proceeding, there is an even greater range of possible directions. Because of this range of possible interpretations when crossing individual reaction curves in P–T space, it is always advisable (where possible) to constrain the path to a higher degree of confidence by identifying several reaction curves that have been crossed and by using other approaches to geothermobarometry to quantify particular metamorphic events. As Spear (1993) states, *erroneous paths are worse than useless because they are misleading and will certainly create more confusion than clarity*, so be warned, and take care.

References

Allen, T. & Chamberlain, C.P. (1991) Metamorphic evidence for an inverted crustal section, with constraints on the Main Karakorum Thrust, Baltistan, northern Pakistan. *Journal of Metamorphic Geology*, 9, 403–418.

Anderson, M.W., Barker, A.J., Bennett, D.G. & Dallmeyer, R.D. (1992) A tectonic model for Scandian terrane accretion in the northern Scandinavian Caledonides. *Journal of the Geological Society*, 149, 727–741.

Armstrong, T.R., Tracy, R.J. & Hames, W.E. (1992) Contrasting styles of Taconian, Eastern Acadian and Western Acadian metamorphism, central and western New England. *Journal of Metamorphic Geology*, 10, 415–426.

Banno, S. & Sakai, C. (1989) Geology and metamorphic evolution of the Sanbagawa metamorphic belt, Japan, in *Evolution of Metamorphic Belts* (eds J.S. Daly, R.A. Cliff & B.W.D. Yardley). Geological Society Special Publication No. 43, 519–532.

Barker, A.J. (1989) Metamorphic evolution of the Caledonian nappes of north central Scandinavia, in *The Caledonide Geology of Scandinavia* (ed. R.A. Gayer). Graham & Trotman, London, 193–204.

Barker, A.J. (1995) Diachronous fluid release and veining associated with out-of-sequence thrusts in the north Scandinavian Caledonides. *Australian Journal of Earth Sciences*, 42, 311–320.

Barker, A.J. & Anderson, M.W. (1989) The Caledonian structural-metamorphic evolution of south Troms, Norway, in *Evolution of Metamorphic Belts* (eds J.S. Daly, R.A. Cliff & B.W.D. Yardley). Geological Society Special Publication No. 43, 385–390.

Barnicoat, A.C. & Fry, N. (1989) Eoalpine high-pressure metamorphism in the Piemonte zone of the Alps: south-west Switzerland and north-west Italy, in *Evolution of metamorphic belts* (eds J.S. Daly, R.A. Cliff & B.W.D. Yardley). Geological Society Special Publication No. 43, 539–544.

Barrow, G. (1893) On an intrusion of muscovite–biotite gneiss in the southeastern Highlands of Scotland, and its accompanying metamorphism. *Quarterly Journal of the Geological Society of London*, 49, 330–358.

Bennett, D.G. & Barker, A.J. (1992) High salinity fluids: the result of retrograde metamorphism in thrust zones. *Geochimica et Cosmochimica Acta*, 56, 81–95.

Bertrand, P., Ouzegane, K.H. & Kienast, J.R. (1992) P–T–X relationships in the Precambrian Al–Mg-rich granulites from In Ouzzal, Hoggar, Algeria. *Journal of Metamorphic Geology*, 10, 17–31.

Bohlen, S.R. (1987) Pressure–temperature–time paths and a tectonic model for the evolution of granulites. *Journal of Geology*, 95, 617–632.

Bohlen, S.R., Wall, V.J. & Boettcher, A.L. (1983) Experimental investigations and geological applications of equilibria in the system FeO–TiO_2–Al_2O_3–SiO_2–H_2O. *American Mineral-ogist*, 68, 1049–1058.

Brothers, R.N. (1985) Regional mid-Tertiary blueschist–eclogite metamorphism in northern New Caledonia. *Géologie France*, 1, 37–44.

Brothers, R.N. & Yokoyama, K. (1982) Comparison of the high-pressure schist belts of New Caledonia and Sanbagawa, Japan. *Contributions to Mineralogy and Petrology*, 79, 219–229.

Burg. J.P., Delor, C.P., Leyreloup, A.F. & Romney, F. (1989) Inverted metamorphic zonation and Variscan thrust tectonics in the Rouergue area (Massif Central, France): P–T–t record from mineral to regional scale, in *Evolution of metamorphic belts* (eds J.S. Daly, R.A. Cliff & B.W.D. Yardley). Geological Society Special Publication No. 43, 423–439.

Burton, K.W., Boyle, A.P., Kirk, W.L. & Mason, R. (1989) Pressure, temperature and structural evolution of the Sulitjelma fold-nappe, central Scandinavian Caledonides, in *Evolution of metamorphic belts* (eds J.S. Daly, R.A. Cliff & B.W.D. Yardley). Geological Society Special Publication No. 43, 391–411.

Carlson, W.D. & Rosenfeld, J.L. (1981) Optical determination of topotactic aragonite–calcite growth kinetics: Metamorphic implications. *Journal of Geology*, 89, 615–638.

Cartwright, I. & Buick, I.S. (1995) Formation of wollastonite-bearing marbles during late regional metamorphic channelled fluid flow in the Upper Calcsilicate Unit of the Reynolds Range Group, central Australia. *Journal of Metamorphic Geology*, 13, 397–417.

Chatterjee, N.D. (1972) The upper stability limit of the assemblage paragonite + quartz and its natural occurrence. *Contributions to Mineralogy and Petrology*, 34, 288–303.

Chatterjee, N.D. & Johannes, W. (1974) Thermal stability and standard thermodynamic properties of synthetic $2M_1$-muscovite, $KAl_2[AlSi_3O_{10}(OH)_2]$. *Contributions to Mineralogy and Petrology*, 48, 89–114.

References

Chopin, C. (1984) Coesite and pure pyrope in high grade pelitic blueschists of the Western Alps. *Journal of Petrology*, 22, 628–650.

Clarke, G.L. & Powell, R. (1991) Decompressional coronas and symplectites in granulites of the Musgrave Complex, central Australia. *Journal of Metamorphic Geology*, 9, 441–450.

Crawford, W.A. & Hoersch, A.L. (1972) Calcite–aragonite equilibrium from 50°–100°C. *American Mineralogist*, 57, 995–998.

De Yoreo, J.J., Lux, D.R. & Guidotti, C.V. (1989) The role of crustal anatexis and magma migration in the thermal evolution of regions of thickened continental crust, in *Evolution of metamorphic belts* (eds J.S. Daly, R.A. Cliff & B.W.D. Yardley). Geological Society Special Publication No. 43, 187–202.

Diella, V., Spalla, M.I. & Tunesi, A. (1992) Contrasting thermochemical evolutions in the Southalpine metamorphic basement of the Orobic Alps (Central Alps, Italy). *Journal of Metamorphic Geology*, 10, 203–219.

Dodson, M.H. (1973) Closure temperatures in cooling geochronological and petrological systems. *Contributions to Mineralogy and Petrology*, 40, 259-274.

Droop, G.T.R. (1989) Reaction history of garnet–sapphirine granulites and conditions of Archaean high-pressure granulite-facies metamorphism in the Central Limpopo Mobile Belt, Zimbabwe. *Journal of Metamorphic Geology*, 7, 383–403.

Droop, G.T.R., Lombardo, B. & Pognante, U. (1990) Formation and distribution of eclogite facies rocks in the Alps, in *Eclogite facies rocks* (ed. D.A. Carswell). Blackie, Glasgow, 225–259.

Ellis, D.J. (1987) Origin and evolution of granulites in normal and thickened crusts. *Geology*, 15, 167–170.

England, P.C. & Richardson, S.W. (1977) The influence of erosion upon the mineral facies of rocks from different metamorphic environments. *Journal of the Geological Society*, 134, 201–213.

England, P.C. & Thompson, A.B. (1984) Pressure–temperature–time paths of regional metamorphism, Part I: Heat transfer during the evolution of regions of thickened continental crust. *Journal of Petrology*, 25, 894–928.

Ernst, W.G. (1988) Tectonic history of subduction zones inferred fron retrograde blueschist P–T paths. *Geology*, 16, 1081–1084.

Essene, E.J. (1982) Geologic thermometry and barometry, in *Characterization of metamorphism through mineral equilibria* (ed. J.M. Ferry). Mineralogical Society of America, Reviews in Mineralogy No. 10, 153–206.

Essene, E.J. (1989) The current status of thermobarometry in metamorphic rocks, in *Evolution of metamorphic belts* (eds J.S. Daly, R.A. Cliff & B.W.D. Yardley). Geological Society Special Publication No. 43, 1–44.

Faure, G. (1986) *Principles of isotope geology*. John Wiley, New York, 464 pp.

Forbes, R.B., Evans, B.W. & Leyreloup, A.F. (1984) Regional progressive high-pressure metamorphism, Seward Peninsula, Alaska. *Journal of Metamorphic Geology*, 2, 43–54.

Frost, B.R. & Chacko, T. (1989) The granulite uncertainty principle: limitations on thermobarometry in granulites. *Journal of Geology*, 97, 435–450.

Ghent, E.D. (1976) Plagioclase – garnet – Al_2SiO_5 – quartz: a potential geobarometer– geothermometer. *American Mineralogist*, 61, 710–714.

Gillott, J.E. (1955) Metamorphism of the Manx Slates. *Geological Magazine*, 92, 141-154.

Goscombe, B. (1992) Silica-undersaturated sapphirine, spinel and kornerupine granulite facies rocks, NE Strangways Range, Central Australia. *Journal of Metamorphic Geology*, 10, 181–201.

Hames, W.E., Tracy, R.J., & Bodnar, R.J. (1989) Post-metamorphic unroofing history deduced from petrology, fluid inclusions, thermochronometry, and thermal modelling: an example from southwestern New England. *Geology*, 17, 727–730.

Hames, W.E., Tracy, R.J., Radcliffe, N.M. & Sutter, J.F. (1991) Petrologic, structural and geochronologic characteristics of the Acadian metamorphic overprint of the Taconian zone in part of southwestern New England. *American Journal of Science*, 291, 887–913.

Harley, S.L. (1989) The origins of granulites: a metamorphic perspective. *Geological Magazine*, 126, 215–247.

Haugerud, R.A. & Zen, E.-an (1989) Metamorphic path studies – a critique and prospectus, in *Advances in Geochemistry (D.S. Korzhinskiy)* (ed. L.L. Perchuk). Springer- Verlag, New York.

Hirschberg, A. & Winkler, H.C.F. (1968) Stability relations between cordierite, chlorite, and almandine during metamorphism. *Contributions to Mineralogy and Petrology*, 18, 17– 42.

Hisada, K. & Miyano, T. (1996) Petrology and microthermometry of aluminous rocks in the Botswanan Limpopo Central Zone: evidence for isothermal decompression and isobaric cooling. *Journal of Metamorphic Geology*, 14, 183–197.

Holdaway, M.J. (1971) Stability of andalusite and the aluminium silicate phase diagram. *American Journal of Science*, 271, 97–131.

Holland, T.J.B. & Ray, N.J. (1985) Glaucophane and pyroxene breakdown reactions in the Pennine units of the Eastern Alps. *Journal of Metamorphic Geology*, 3, 417–438.

Hoschek, G. (1969) The stability of staurolite and chloritoid and their significance in metamorphism of pelitic rocks. *Contributions to Mineralogy and Petrology*, 22, 208–232.

Hsu, L.C. (1968) Selected phase relationships in the system Al–Mn–Fe–Si–O–H: a model for garnet equilibria. *Journal of Petrology*, 9, 40–83.

Inger, S. & Harris, N.B.W. (1992) Tectonothermal evolution of the High Himalayan Crystalline Sequence, Langtang Valley, northern Nepal. *Journal of Metamorphic Geology*, 10, 439–452.

Johannes, W. & Puhan, D. (1971) The calcite–aragonite transition reinvestigated. *Contributions to Mineralogy and Petrology*, 31, 28–38.

Kihle, J. & Bucher-Nurminen, K. (1992) Orthopyroxene–sillimanite–sapphirine granulites from the Bamble granulite terrane, southern Norway. *Journal of Metamorphic Geology*, 10, 671–683.

Kisch, H.J. (1987) Correlation between indicators of very low-grade metamorphism, in *Low temperature metamorphism* (ed. M. Frey). Blackie, Glasgow, 227–304.

Kryza, R., Muszynski, A. & Vielzeuf, D. (1990) Glaucophane-bearing assemblage overprinted by greenschist-facies metamorphism in the Variscan Kaczawa complex, Sudetes, Poland. *Journal of Metamorphic Geology*, 8, 345–355.

Massonne, H.-J. & Chopin, C. (1989) P–T history of the Gran Paradiso (western Alps) metagranites based on phengite barometry, in *Evolution of metamorphic belts* (eds J.S. Daly, R.A. Cliff & B.W.D. Yardley). Geological Society Special Publication No. 43, 545–549.

Mezger, K., Rawnsley, C. Bohlen, S. & Hanson, G. (1991) U–Pb garnet, sphene, monazite and rutile ages: Implications for the duration of high grade metamorphism and cooling histories, Adirondack Mts., New York. *Journal of Geology*, 99, 415–428.

Naggar, M.H. & Atherton, M.P. (1970) The composition and metamorphic history of some aluminium silicate-bearing rocks from the aureoles of the Donegal granites. *Journal of Petrology*, 11, 549–589.

Okay, A.I. (1995) Paragonite eclogites from Dabie Shan, China: re-equilibration during exhumation? *Journal of Metamorphic Geology*, 13, 449–460.

Otsuki, M. & Banno, S. (1990) Prograde and retrograde metamorphism of hematite-bearing basic schists in the Sanbagawa belt in central Shikoku. *Journal of Metamorphic Geology*, 8, 425–439.

Oxburgh, E.R. & Turcotte, D.L. (1974) Thermal gradients and regional metamorphism in overthrust terrains with special reference to the eastern Alps. *Schweiz Mineralogische und Petrographishe Mitteilungen*, 54, 641–662.

Parrish, R.R., Carr, S.D. & Parkinson, D.L. (1988) Eocene extensional tectonics and geochronology of the southern Omineca belt, British Columbia and Washington. *Tectonics*, 7, 181–212.

Pecher, A. (1989) The metamorphism in the Central Himalaya. *Journal of Metamorphic Geology*, 7, 31–41.

Perchuk, L.L. (1989) P–T–fluid regimes of metamorphism and related magmatism with specific reference to the granulite facies Sharyzhalgay complex of Lake Baikal, in *Evolution of metamorphic belts* (eds J.S. Daly, R.A. Cliff & B.W.D. Yardley). Geological Society Special Publication No. 43, 275–291.

Perchuk, L.L. & Lavrent'eva, I.V. (1983) Experimental investigation of exchange equilibria in the system cordierite–garnet–biotite, in *Kinetics and equilibrium in mineral reactions* (ed. S.K. Saxena). Springer-Verlag, New York, 199–239.

Pognante, U., Talirico, F., Rastelli, N. & Ferranti, N. (1987) High pressure metamorphism in the nappes of the Vall dell'Orco traverse (Western Alps collisional belt). *Journal of Metamorphic Geology*, 5, 397–414.

Rao, B.B. & Johannes, W. (1979) Further data on the stability of staurolite + quartz and related assemblages. *Neues Jahrbuch für Mineralogie*, **1979**, 437–447.

Richardson, S.W., Gilbert, M.C. & Bell, P.M. (1969) Experimental determination of the kyanite–andalusite and andalusite–sillimanite equilibria; the aluminium silicate triple point. *American Journal of Science*, 267, 259–272.

Ridley, J. (1984) Evidence of a temperature-dependent 'blueschist' to 'eclogite' transformation in high-pressure metamorphism of metabasic rocks. *Journal of Petrology*, 25, 852–870.

Ridley, J. (1989) Vertical movement in orogenic belts and the timing of metamorphism relative to deformation, in *Evolution of metamorphic belts* (eds J.S. Daly, R.A. Cliff & B.W.D. Yardley). Geological Society Special Publication No. 43, 103–115.

Schielstedt, M. (1986) Eclogite–blueschist relationships as evidenced by mineral equilibria in high-pressure metabasic rocks of Sifnos (Cycladic Islands), Greece. *Journal of Petrology*, 27, 1437–1459.

Schielstedt, M. (1990) Occurrence and stability conditions of low-temperature eclogites, in *Eclogite facies rocks* (ed. D.A. Carswell). Blackie, Glasgow, 160–179.

Schielstedt, M. & Matthews, A. (1987) Transformation of blueschist facies rocks as a consequence of fluid infiltration, Sifnos (Cyclades), Greece. *Contributions to Mineralogy and Petrology*, 97, 237–250.

Schumacher, J.C., Schumacher, R. & Robinson, P. (1989) Acadian metamorphism in central Massachusetts and south-western New Hampshire: evidence for contrasting P–T trajectories, in

Evolution of metamorphic belts (eds J.S. Daly, R.A. Cliff & B.W.D. Yardley). Geological Society Special Publication No. 43, 453–460.

Selverstone, J. (1988) Evidence for east–west crustal extension in the Eastern Alps: implications for the unroofing history of the Tauern Window. *Tectonics*, 7, 87–105.

Selverstone, J. & Spear, F.S. (1985) Metamorphic P–T paths from pelitic schists and greenstones from the Southwest Tauern Window, Eastern Alps. *Journal of Metamorphic Geology*, 3, 439–465.

Selverstone, J., Spear, F.S., Franz, G. & Morteani, G. (1984) High-pressure metamorphism in the Southwest Tauern Window, Austria: P–T paths from hornblende–kyanite–staurolite schists. *Journal of Petrology*, 25, 501–531.

Spear, F.S. (1986) P-T PATH: a FORTRAN program to calculate pressure temperature paths from zoned metamorphic garnets. *Computers in Geoscience*, 12, 247–266.

Spear, F.S. (1993) *Metamorphic phase equilibria and pressure–temperature–time paths*. Mineralogical Society of America Monograph, Mineralogical Society of America, Washington, DC, 799 pp.

Spear, F.S. & Cheney, J.T. (1989) A petrogenetic grid for pelitic schists in the system SiO_2–Al_2O_3–FeO–MgO–K_2O–H_2O. *Contributions to Mineralogy and Petrology*, 101, 149–164.

Spear, F.S., Selverstone, J., Hickmott, D., Crowley, P. & Hodges, K.V. (1984) P–T paths from garnet zoning: a new technique for deciphering tectonic processes in crystalline terranes. *Geology*, 12, 87–90.

Takasu, A. (1989) P–T histories of peridotite and amphibolite tectonic blocks in the Sanbagawa metamorphic belt, Japan, in *Evolution of metamorphic belts* (eds J.S. Daly, R.A. Cliff & B.W.D. Yardley). Geological Society Special Publication No. 43, 533–538.

Thompson, A.B. & England, P.C. (1984) Pressure–temperature–time paths of regional metamorphism II. Their inference and interpretation using mineral assemblages in metamorphic rocks. *Journal of Petrology*, 25, 929–955.

Thompson, A.B. & Tracy, R.J. (1979) Model systems for anatexis of pelitic rocks. II. Facies series melting and reactions in the system CaO–$KAlO_2$–$NaAlO_2$–Al_2O_3–SiO_2–H_2O. *Contributions to Mineralogy and Petrology*, 70, 429–438.

Vernon, R.H. (1996) Problems with inferring P–T–t paths in low granulite facies rocks. *Journal of Metamorphic Geology*, 14, 143–153.

Waters, D.J. (1989) Metamorphic evidence for the heating and cooling path of Namaqualand granulites, in *Evolution of metamorphic belts* (eds J.S. Daly, R.A. Cliff & B.W.D. Yardley). Geological Society Special Publication No. 43, 357–363.

Winslow, D.M., Bodnar, R.J. & Tracy, R.J. (1994) Fluid inclusion evidence for an anticlockwise metamorphic P–T path in central Massachusetts. *Journal of Metamorphic Geology*, 12, 361–371.

Zeitler, P.K. (1989) The geochronology of metamorphic processes, in *Evolution of metamorphic belts* (eds J.S. Daly, R.A. Cliff & B.W.D. Yardley). Geological Society Special Publication No. 43, 131–147.

Zhang, R-Y., Hirajima, T., Banno, S., Bolin Cong & Liou, J.G. (1995) Petrology of ultrahigh-pressure rocks from the southern Su-Lu region, eastern China. *Journal of Metamorphic Geology*, 13, 659–675.

Appendix I: Abbreviations

Mineral abbreviations

Mineral abbreviations used in the text and diagrams are those of Kretz (1983), with the addition of those marked with * (after Barker, 1990, and this edition), and those with + (after Bucher & Frey, 1994).

Ab	albite
Act	actinolite
*Ads	andesine
Alm	almandine
An	anorthite
And	andalusite
Ank	ankerite
Anl	analcite
Ann	annite
Arg	aragonite
Ath	anthophyllite
*Bar	barroisite
+Brc	brucite
Bt	biotite
Cal	calcite
Chl	chlorite
Cld	chloritoid
+Cp	Fe/Mg-carpholite
Cpx	Ca-clinopyroxene
Crd	cordierite
*Cros	crossite
Crs	cristobalite
+Crn	corundum
Cs	coesite
Czo	clinozoisite
+Cum	cummingtonite
Di	diopside
Dol	dolomite
En	enstatite
Ep	epidote
Fo	forsterite
+Ged	gedrite
Gln	glaucophane
Gr	graphite
Grs	grossularite
Grt	garnet
+Hc	hercynite
Hem	haematite
Hbl	hornblende
Hul	heulandite
*Hy	hypersthene
Ill	illite
Ilm	ilmenite
Jd	jadeite
Kfs	K-feldspar
Kln	kaolinite
+Krn	kornerupine
+Krs	kaersutite
Ky	kyanite
Lmt	laumontite
Lws	lawsonite
Mag	magnetite
Mc	microcline
Mgs	magnesite
Mnt	montmorillonite
Mrg	margarite
Ms	muscovite
+Mul	mullite
Ol	olivine

Appendix I

*Olg	oligoclase		
Omp	omphacite		
Opx	orthopyroxene		
Or	orthoclase		
Osm	osumilite		
Pg	paragonite		
+Phe	phengite		
Phl	phlogopite		
Pl	plagioclase		
Pmp	pumpellyite		
+Prh	prehnite		
Prl	pyrophyllite		
Prp	pyrope		
Py	pyrite		
Pyx	pyroxene		
Qtz	quartz		
Rt	rutile		
Sa	sanidine		
+Scp	scapolite		
+Ser	sericite		
Sil	sillimanite		
*Smc	smectite		
Spl	spinel		
Spn	sphene		
+Spr	sapphirine		
Sps	spessartine		
Srp	serpentine		
St	staurolite		
Stp	stilpnomelane		
Tlc	talc		
Tr	tremolite		
+Trd	tridymite		
+Ves	vesuvianite (idocrase)		
Wo	wollastonite		
Wr	wairakiite		
Zeo	zeolites		
Zo	zoisite		
Zrn	zircon		

Additional abbreviations

CSD	crystal size distribution
CPO	crystallographic preferred orientation
ecc	extensional crenulation cleavage
EF	elastico-frictional
$f(O_2)$	oxygen fugacity
P	pressure
P_l	lithostatic pressure
P_f	fluid pressure
PPL	plane-polarised light
QP	quasi-plastic
SEM	scanning electron microscopy
T	temperature
t	time
TEM	transmission electron microscopy
T_h	homogenisation temperature (used with respect to fluid inclusions)
T_t	trapping temperature (used with respect to fluid inclusions)
μ_i	chemical potential of i
µm	micrometres (microns)
var.	variety
XCO_2	proportion of CO_2 in fluid in relation to H_2O (where $XH_2O + XCO_2 = 1.0$)
XH_2O	proportion of H_2o in fluid in relation to CO_2 (where $XH_2O + XCO_2 = 1.0$)
XRD	X-ray diffraction
XPL	cross-polarised light
σ_3	minimum compressive stress
σ_1	maximum compressive stress
Δf_H	enthalpy of formation
ΔT	change in temperature
>	greater than
≥	greater than or equal to
<	less than
≤	less than or equal to
≠	not equal to

Appendix II: Glossary

Absorption
The process by which additional units are chemically bonded into the structure of a growing crystal to form an integral part of it (modified after Spry, 1969).

Acicular
A term used to describe needle-like crystals (Figs 5.12(a) & (b); Plate 8(e)).

Activation energy
That energy required before a given reaction or diffusive process can proceed.

Activity
(Symbol = a). One of several ways to describe the behaviour of given component in solution (solid or fluid). The relationship between the activity of a given component (a_i), and the concentration of that component (X_i) is given by $\alpha_i = a_i/X_i$, where α_i is the activity coefficient. For gases $a_i = f_i/f_i^o$, where f_i is the fugacity (or thermodynamic pressure) of gas i, and f_i^o is the fugacity of i in a standard state. For an ideal solution, $f_i = a_i = X_i$. However, in real solutions, a_i (and hence f_i) may differ greatly from X_i (after Vernon, 1976).

Adsorption
The loose attachment (or bonding) of matrix or fluid phases to the surface of a growing crystal. (Fig. 6.3).

Amphibolite
A metamorphosed basic igneous rock with a mineral assemblage comprised largely of amphibole and plagioclase, usually with quartz and epidote (Fig. 4.5(a)).

Anatexis
The process of partial melting of high-grade metamorphic rocks in the presence of H_2O. (This process produces granitoid melts and typically operates in the middle to lower crust during orogeny.)

Anchimetamorphism
Sub-greenschist facies, very low grade metamorphism. (The 'limits' of anchimetamorphic conditions have been variably defined based on illite crystallinity.)

Anhedral
A term used to describe metamorphic crystals (especially porphyroblasts) with irregular form (Fig. 5.9(c)).

Annealing
A recovery process in deformed rocks while still at high temperature. It involves static recrystallisation, and the formation of new strain-free grains.

Antiperthite
A feldspar intergrowth comprising K-feldspar inclusions enclosed within plagioclase (the converse of perthite).

Glossary

Antitaxial vein
Vein with filling material grown from vein centre towards the vein walls (Ramsay & Huber, 1983) (Fig. 11.8(b)).

Atoll structure
A structure common to granulites consisting of a core of one mineral entirely surrounded by a rim of another mineral (e.g. garnet forming a core entirely surrounded by plagioclase) (Plate 2(a)).

Augen gneiss
A gneissose rock with abundant *augen* (German: eyes) represented by porphyroclasts (typically K-feldspar) enveloped by the foliation (Fig. 4.3).

Blastomylonite
A term used for mylonites with extensive mineral growth synchronous with shearing or else widespread static recrystallisation immediately after deformation.

Blueschist
A metamorphosed mafic rock indicative of high-P/low-T subduction-related metamorphism. It contains large quantities of sodic (blue) amphibole (glaucophane/crossite), and has a pronounced schistosity.

Bow-tie structure
A term used to describe aggregates of elongate prismatic, and acicular crystals that are arranged to give the appearance of a 'bow-tie'. It is commonly exhibited by amphiboles in *garbenschiefer* that have grown in the foliation plane under low differential stress (Fig. 5.13(b)).

Burgers vector
A vector defining the amount and direction of lattice displacement associated with an intracrystalline dislocation (Fig. 8.5).

Burial metamorphism
Low-grade metamorphism in response to ambient geothermal gradient generated in thick basinal sequences.

Calc-silicate rock
A rock with a chemistry dominated by calcium and silica, consisting of hydrous or anhydrous calc-silicate minerals such as tremolite, diopside and grossular. (Carbonate minerals are also commonly present.)

Cataclasis
A deformation mechanism in which crystal structure remains undistorted, but grains or groups of grains become cracked and the fragments may exhibit frictional sliding with respect to one another (Rutter, 1986) (Fig. 8.1).

Cataclasite
A cohesive largely unfoliated fault-rock containing angular clasts of variable size in a fine-grained matrix of similar composition.

Chemical potential
One of several expressions for the behaviour of a component in solution. The chemical potential of a component i in an ideal solution is given by $\mu_i = G^o_i + RT \ln X_i$ where G^o_i is a constant (the partial molar Gibbs free energy of pure i in the standard state), R is the gas constant, X_i is the molar concentration of i in solution and T is temperature (based on Vernon, 1976).

Chemical softening
Enhanced ductility in a deforming rock due to changes in the trace element content of a mineral (e.g. water weakening of quartz) (modified from White *et al.*, 1980).

Chemical zoning
Regular or abrupt changes in mineral chemistry from mineral core to rim.

Chlorite–mica stack
Interlayered chlorite and mica (typically phengite or muscovite) to form a lens or barrel-shaped phyllosilicate stack, often as a replacement of detrital biotite in sub-greenschist facies metasedimentary rocks (especially slates) (Figs 7.7(a)–(c)).

Cleavage
A sub-parallel set of closely spaced approximately planar surfaces produced during rock deformation. The rock preferentially splits along such surfaces due to the alignment of platy or elongate grains (usually phyllosilicate minerals) within the rock (Fig. 4.8).

Cleavage dome
An unusual feature involving arching or doming of matrix cleavage in areas immediately adjacent to faces of certain euhedral porphyroblasts. These domes (often comprising muscovite and graphite), are interpreted in terms of physical displacement of insoluble matrix phases ahead of a growing porphyroblast in a bulk hydrostatic stress field. They represent one of the few lines of evidence used as an indicator of porphyroblast growth by displacement rather than replacement.

Closure temperature
Symbol = T_c. The temperature at which (for a given mineral) the rate of increase of radiogenic daughter product relative to parent becomes constant for a given isotopic system (Table 12.1).

Component
The fundamental chemical constituents of a rock system (e.g. SiO_2, Al_2O_3 or K_2O) are described as components.

Contact aureole
The zone of rocks surrounding a plutonic intrusion which are thermally metamorphosed in response to the intrusion. The contact aureole can usually be subdivided into a series of concentric zones of varying metamorphic grade based on mineral assemblages.

Continuous reaction
See **Sliding reaction**.

Core-and-mantle microstructure
A characteristic feature of many mylonitic and protomylonitic rocks, also referred to as 'mortar structure'. It comprises a porphyroclast of a single crystal, mantled by a fine-grained recrystallised aggregate of the same mineral phase (Fig. 8.2(d)).

Corona
A monomineralic or polymineralic rim totally surrounding a core of another mineral phase. It represents an arrested reaction between the core phase and other components, often during retrogression of the core phase (Figs 12.16(a) & 12.17(b); Plates 2(a) & (b)).

Crack–seal vein
A vein built up by successive development of microcracks, followed by successive periods of cementation (Ramsay & Huber, 1983) (Fig. 11.4; Plates 8(a)–(d)).

Creep
Continuous, usually slow deformation of a rock or individual crystal resulting from relatively low stress acting over a long period of time.

Crenulation
Microfolding on a cm-scale, best developed in quartz–mica-dominated metasedimentary rocks with well developed primary lamination or earlier continuous tectonic foliation.

Cross-hatched twinning
'Cross-hatched twinning', 'tartan pattern' or 'gridiron twinning' is one of the most characteristic features of microcline (Fig. 5.19(b)). It results from the high-angle intersection of multiple albite and pericline twins.

Glossary

Crystalloblastic series
Based on original work by Harker (1939), the *crystalloblastic series* (Table 5.1) is a mineralogical sequence, with those minerals at the top having the greatest tendency towards euhedral form. The sequence reflects decreasing surface energy, and minerals higher in the sequence always tend to form euhedral faces against minerals lower in the sequence.

Decarbonation reaction
A reaction resulting in the liberation of CO_2 from the reactant(s).

Decussate structure
A term used to describe interlocking, randomly oriented, elongate, prismatic or sub-idioblastic crystals, generally of one species (modified after Spry, 1969) (Fig. 5.16).

Deformation bands
Distinct bands of deformation in crystals. Optically they are more sharply defined than undulose extinction, and represent higher strain. They can terminate either at grain boundaries or inside grains (Fig. 8.13(a)).

Deformation lamellae
Sets of very narrow planar or tapered lenticular structures in quartz and many other silicates. They terminate inside grain boundaries, and show a variety of orientations, although they are frequently perpendicular to bands of undulose extinction (after Vernon, 1976) (Fig. 8.10).

Deformation mechanism map
A plot of stress versus homologous temperature defining fields over which particular deformation mechanisms dominate during deformation of a particular material (e.g. quartz) of specified grain size (Fig. 8.7).

Deformation twins
Twins developed in crystals in response to deformation. They are especially common in calcite, plagioclase and pyroxene. Deformation twins terminating within the crystal have tapered ends (Figs 5.18(b) & (c); Plates 3(a) & (b)).

Dehydration reaction
A reaction liberating H_2O from the reactant(s).

Diagenesis
All physical, chemical and biological changes experienced by sediments during lithification prior to metamorphism. The boundary between diagenesis and low-grade metamorphism is transitional and poorly defined.

Differential stress
Usually defined as the difference between maximum (σ_1) and minimum (σ_3) compressive stresses in a non-hydrostatic (lithostatic) stress regime.

Diffusion
The process by which atoms, molecules or ions move from one position to another within a fluid or solid, under the influence of a chemical potential gradient. In solids this means the periodic jumping of atoms from one site in the structure to another (modified after Jensen, 1965; Vernon, 1976).

Dihedral angle
The interfacial angle formed at the triple-junction of a particular mineral phase, A, in contact with two adjacent crystals of a different mineral phase, B.

Discontinuous reaction
A reaction involving the complete breakdown of one mineral or minerals and formation of another mineral or minerals in response to changing P, T and or fluid/chemical conditions.

Disequilibrium
An incompatible association of mineral phases and/or a combination of textures and structures incompatible with prevailing conditions.

Glossary

Dislocations
Line defects in crystals, produced during growth or deformation, and which are thermodynamically unstable (Fig. 11.16).

Dislocation tangle
During intracrystalline deformation, the intersection of different slip systems leads to entanglement of migrating dislocations. Such dislocation tangles make further deformation of the crystal increasingly difficult, and greatly contribute to overall strain (work) hardening.

Dissolution
The process by which minerals are chemically corroded and the constituent elements pass into solution.

Dynamic recrystallisation
Deformation-induced reworking of the grain sizes, shapes or orientations with little or no chemical change (Poirier & Guillope, 1979).

Eclogite
A dense, high-grade metamorphic rock of mafic composition with an essential assemblage of clinopyroxene (omphacite) and garnet.

Edge dislocation
A type of dislocation whereby the crystal has an additional half lattice plane (Fig. 8.5(a)).

Endothermic reaction
A reaction that consumes energy, such that heat energy must be added to the system for the reaction to proceed (e.g. prograde dehydration reactions in pelitic schists).

Enthalpy
The energy that is associated with heat (e.g. the heat evolved when substances react is the enthalpy of reaction) (Powell, 1978).

Entropy
A property that reflects the degree of disorder in a system: the more disordered the system is, the higher is its entropy (Powell, 1978).

Epitaxial growth
The oriented growth of one mineral phase in optical continuity with another due to structural similarities between the two phases (e.g. growth of secondary amphibole rims on earlier amphiboles of different composition) (Fig. 12.21; Plate 4(a)).

Equilibrium
That state of a rock system in which the phases present are in the most stable, low-energy arrangement, and in which all phases are compatible with the given P, T and fluid conditions.

Euhedral
A term used to describe metamorphic crystals (especially porphyroblasts) with well developed crystal faces and form (synonymous with idioblastic) (Fig. 5.9(a)).

Exothermic reaction
A reaction that gives off heat as it proceeds. It is the converse of endothermic reactions, which consume heat.

Exsolution
The process whereby an initially homogeneous solid solution separates into two (or possibly more) distinct crystalline phases (typically during cooling) without the addition or removal of material (modified after Bates & Jackson, 1980).

Fabric
The geometric and spatial relationships between the crystal components making up a rock. The fabric can relate to the preferred orientations of grain shapes, to grain sizes, and to crystallographic orientations of the components. Preferred orientations of platy- and needle-shaped crystals can give rise to planar (S-fabrics), linear (L-fabrics) or to mixed linear-planar (L–S fabrics) (Ramsay & Huber, 1983) (Fig. 4.4).

Glossary

Faserkiesel
See **Fibrolite**.

Fault breccia
A cataclastic rock associated with fault zones. It consists of coarse angular fragments of variable size in a fine-grained, often silicified, matrix.

Fault gouge
A soft and incohesive fine-grained fault-rock. It is usually rich in clay minerals resulting from the chemical breakdown of adjacent wall rocks.

Fibrolite
A fine fibrous or hair-like variety of sillimanite common in amphibolite facies schists and gneisses. It often occurs in matted aggregates or knots, termed *faserkiesel*.

Fluid inclusion
A term for microscopic and sub-microscopic (and rarely macroscopic) inclusions of fluid trapped in minerals during primary crystallisation or fracture healing. They are typically < 50 μm in size (Section 11.6; Figs 11.11–11.13; Plate 8(f)).

Foliation
A set of closely spaced planar surfaces produced in a rock as a result of deformation (e.g. schistosity or cleavage) (modified after Park, 1983) (Figs 4.4–4.6).

Fugacity
An expression for the behaviour of a gas in a solid or fluid medium. The fugacity or thermodynamic pressure of a gas i (f_i) is related to activity by $a_i = f_i/f^o_i$ where f^o_i is the fugacity of i in a standard state. For ideal solutions, $f_i = a_i = X_i$. However, in real solutions, a_i (and hence f_i) may differ greatly from X_i (X_i = mole fraction of i in solution).

Garbenschiefer
Regionally metamorphosed impure calcareous sediments and meta-tuffs commonly give rise to g*arbenschiefer*. These are rocks with abundant amphibole prisms lying within the foliation plane in 'bow-tie' and radiating aggregates. Spry (1969) defines the porphyroblasts as being 'the size of caraway seeds', but in practice the same term is also used for rocks with much larger amphiboles (Fig.5.13(b)).

Geometric softening
Enhanced ductility in deforming rocks caused by lattice reorientation of deforming grains. It is most pronounced in materials with limited slip systems such as granitoid mylonites, and generally involves reorientation of grains so that their slip directions approach parallelism with the shear direction.

Geotherm
A curve expressing the thermal gradient throughout the lithosphere. The nature of the geotherm varies from place to place, and with time.

Geothermal gradient
The rate of change of temperature with depth in the lithosphere. Many factors influence the geothermal gradient, such that it varies greatly from one place to another and at different depths in the lithosphere.

Gibbs free energy
The Gibbs free energy for a closed system is given by $G = E + PV - TS = H - TS$, where E is the internal energy, V is the volume, T is the absolute temperature, S is the entropy and H is the enthalpy of reaction. The driving force of metamorphic reactions is the change in Gibbs free energy (ΔG). For equilibrium, the Gibbs free energy of a system would be at its minimum so for reactions to proceed the reaction should give rise to a lowering of G for the system (i.e. ΔG is negative) (based on Vernon, 1976).

Gneiss
A coarsely banded high-grade metamorphic rock consisting of alternating, mineralogically distinct (usually felsic and mafic) layers.

Grain-boundary migration
To help minimise the energy of the rock system in response to changing P–T conditions, the atoms forming the contacts between individual grains are rearranged to a more stable configuration. To a large extent this is achieved by the process of grain-boundary migration. This involves movement at a high angle to the plane of the grain boundary. During prograde metamorphism, such a process produces regular interfaces and a polygonal aggregate of grains. (The aggregate shown in Fig. 5.14(b) probably developed from an aggregate such as Fig. 5.14(a) by this process.)

Grain-boundary sliding
Movement within the plane of the grain boundary. It can be envisaged in terms of the physical movement of individual grains past each other under an applied shear stress.

Granoblastic structure
An aggregate consisting of equidimensional crystals of approximately equal sizes. In many cases the crystals are rounded to anhedral, but granoblastic-polygonal aggregates are equally common. Granoblastic structure is especially characteristic of granulites, eclogites and many hornfelses (Figs 5.14(b) & 5.15).

Granofels
Non-foliated medium- to coarse-grained granoblastic metamorphic rock.

Granulite
The term *granulite* has been used in a number of ways, and some confusion has arisen because of different usage. In strict terms, *granofels* is the best term to use when describing an even-grained granular rock with no implication regarding metamorphic facies, and the term *granulite* should strictly be reserved for a granulite facies rock with an essential assemblage of pyroxene (typically hypersthene) and anorthitic plagioclase (Plate 2(a)). It is commonly granoblastic.

Greenschist
A low-grade mafic rock with schistose texture and a mineral assemblage consisting largely of actinolite, chlorite, epidote, albite, quartz and accessory sphene.

Growth twin
Primary (or growth) twins represent twins present in a given crystal that formed at the time of crystal growth (Fig. 5.18(a)).

Helicitic structure
Strictly, this refers to an organised, semi-symmetrical, S-shaped arrangement of inclusions within poikiloblastic crystals (e.g. garnet (Plate 5(c)) or staurolite).

Heterogeneous nucleation
Non-random nucleation on some pre-existing substrate, such as new crystals preferentially nucleating at pre-existing grain boundaries.

Homologous temperature
The expression of temperature (in Kelvin) for a particular mineral (e.g. a mineral or rock) as a proportion of the melting temperature for that particular material (i.e. T/T_m). For example, an homologous temperature of 0.6 means 60% of the melting temperature (T_m) for the material (mineral or rock) in question.

Hornfels
A hard, fine- to medium-grained granoblastic rock produced by high-grade contact metamorphism.

Hour-glass structure
A structure common in chloritoid porphyroblasts and some other minerals, consisting of a

Glossary

dense mass of fine-grained (usually opaque) inclusions arranged in the form of an 'hourglass' (Plate 1(b)). It forms due to the influence of the host crystal structure.

Hydrostatic pressure
The pressure exerted by a vertical column of water, and acting equally in all directions. Hydrostatic pressure increases at approximately 0.1 kbar (10 MPa) per kilometre depth.

Idioblastic
See **Euhedral**.

Idiotopic
A structural term used to describe rocks comprised almost entirely of idioblastic crystals.

Illite crystallinity
The degree of ordering of the structure of illite. Illite becomes increasingly 'crystalline' from anchimetamorphic to greenschist facies conditions. This change shows a regular relationship with temperature and has been calibrated for use as a geothermometer.

Inclusion
A solid or fluid phase totally enclosed within a mineral (e.g. fluid inclusions in vein minerals, or inclusions of matrix phases in porphyroblasts) (Fig. 6.1; Plate 1(a)).

Inclusion trail
A regular shaped arrangement of inclusions (fluid or solid) in a crystal (commonly a porphyroblast) to define a distinct fabric (Fig. 11.11; Plates 5(a)–(f)).

Index mineral
A mineral characterising a particular stage or zone of a progressive metamorphic sequence (e.g. sillimanite is an index mineral for upper amphibolite facies metamorphism).

Interfacial energy
See **Surface energy**.

Intergranular slip
A deformation process involving slip between individual crystals of the assemblage.

Interstitial
A point defect represented by an extra atom or molecule in the crystal lattice (Fig. 8.4).

Intracrystalline plasticity
A deformation mechanism in which grains become internally distorted through dislocation motion and deformation twinning (Rutter, 1986).

Inversion twins
Twins formed due to a mineral changing its crystal structure in response to changing P–T conditions. (Compare with deformation twins, which form in response to superimposed stress) (Figs 5.19(a) & (b)).

Ionic reaction
A reaction involving ionic exchange between several different reaction sites within the system in order to facilitate a particular transformation (Fig. 1.5).

Isobaric cooling (IBC)
A temperature decrease at near-constant pressure. It is a feature common to certain granulite facies retrograde P–T paths.

Isobaric reaction
A reaction occurring at constant pressure, irrespective of temperature.

Isochemical system
A system maintaining constant bulk composition during metamorphism; a closed system with no introduction or loss of chemical components.

Glossary

Isograd
A surface joining points of equal metamorphic grade in a unit of metamorphic rocks. It is usually marked by the first appearance of a particular index mineral in a progressive metamorphic sequence.

Isothermal decompression (ITD)
A pressure decrease at near-constant temperature. It is a feature common to the early stages of many granulite facies uplift trajectories.

Isothermal reaction
A reaction occurring at constant temperature, irrespective of pressure.

Kelyphitic rim
A symplectic intergrowth mantling a core of an earlier, typically high-temperature mineral phase. The term *kelyphitic rim* is often used to refer to intergrowths of pyroxene and spinel developed around olivines in gabbros subjected to slow cooling at deep crustal levels.

Latticed-preferred orientation (LPO)
An anisotropic spatial arrangement of the crystal lattices in a population of mineral grains (after Shelley, 1993).

Leucosome
The light-coloured (leucocratic) component of migmatites. It is usually represented by melt-derived granitoid material.

Lineation
A recognisable linear component in rocks, usually formed in response to deformation. Lineations are often contained within a foliation, and in metamorphic rocks are usually defined by alignment of elongate minerals such as amphiboles and micas (Fig. 4.4).

Lithostatic pressure
The vertical pressure at a point in the Earth's crust, equal to the pressure caused by the weight of the column of overlying rock (after Bates & Jackson, 1980).

L-tectonite
A deformed rock containing a recognisable linear structure (Fig. 4.4).

L–S tectonite
A deformed rock containing recognisable linear and planar structural components (Fig. 4.4).

Marble
A metamorphic rock comprised largely of calcite (Plate 3(a)).

Melanosome
The dark-coloured (typically biotite-rich) component of migmatites, often seen as selvages around layers and lenses of leucosomes (Fig. 4.12).

Mesoperthite
The term given to exsolved feldspars with subequal volumes of intergrown K-feldspar and plagioclase.

Mesosome
A portion of migmatites that is intermediate in colour between the leucosome (light) and melanosome (dark) components. The mesosome is often considered to represent the residual unmelted fraction of high-grade metamorphic rock, but this need not necessarily be the case.

Metamorphic facies
A metamorphic facies is a subdivision of metamorphic conditions in P–T space on the basis of diagnostic mineral assemblages which have been shown by experimental and field observations to characterise a specific range of P–T conditions for a particular compositional group of rocks (in association with H_2O fluid). This concept was introduced by Eskola (1915, 1939), and is especially useful for the study of metapelites and metabasites (Fig. 2.1).

Glossary

Metamorphic facies series
An observed sequence of progressively changing metamorphic facies representing increasing metamorphic grade across a given terrane (e.g. the change from greenschist to epidote amphibolite and finally amphibolite facies is a metamorphic facies series).

Metamorphic grade
A general term used to describe the relative rank of metamorphism of rocks in a given area, without specifying a particular metamorphic facies (e.g. rocks with slaty appearance would be described as low-grade, while those with migmatitic appearance would be described as high-grade).

Metamorphism
The mineralogical, chemical and structural adjustment of solid rocks to physical and chemical conditions which have generally been imposed at depth below the surface zones of weathering and cementation, and which differ from the conditions under which the rocks in question originated (Bates & Jackson, 1980).

Metasomatism
Metamorphism involving modification of the bulk rock chemistry by influx or removal of chemical components via a fluid phase (e.g. widespread potassium metasomatism associated with granite intrusions is a common occurrence).

Mica-fish
'Fish'- or 'lozenge'-shaped mica porphyroclasts/porphyroblasts aligned within a finer-grained schistose matrix. They are characteristic of phyllonites and highly deformed schists, and can be used to determine shear sense (Figs 10.9–10.12).

Microstructure
The geometric arrangement and interrelationships between grains and internal features of grains.

Migmatite
A coarse-grained heterogeneous rock type characteristically with irregular and discontinuous interleaving of leucocratic granitoid material (leucosome) and residual high-grade metamorphic material (restite). Migmatites are often intensely folded and heavily veined. General opinion considers most migmatites to have developed by *in situ* anatexis (Figs 4.11 & 4.12).

Millipede microstructure
A term introduced by Bell & Rubenach (1980) for particular syntectonic usually elongate porphyroblasts, that for a given porphyroblast margin where the external foliation (S_e) passes into the internal foliation (S_i), shows S_e deflected in opposite directions. The pattern is repeated approximately symmetrically across the porphyroblast to give an appearance likened to the arrangement of legs on a millipede. Originally taken to be indicative of bulk coaxial shortening, Johnson & Moore (1996) have suggested that they can form during various types of deformation (Plate 5e).

Mimetic growth
Crystal growth that is influenced by and reproduces a pre-existing rock or crystal structure. (e.g. (1) micas preferentially growing along pre-existing foliation; or (2) pseudomorphing, where the nucleating and growing mineral has its form influenced by the structure of the old phase it is replacing rather than taking on its normal habit).

Mortar structure
A term used to describe the structure especially common in highly deformed, dynamically recrystallised quartz-rich rocks, in which large porphyroclasts of quartz have abundant small sub-grains and new grains developed around their margins (Fig. 8.6(d)) (see also **Core-and-mantle microstructure**).

Mylonite
A cohesive, foliated and usually lineated rock produced by tectonic grain-size reduction via crystal–plastic processes in narrow zones of intense deformation. It contains abundant porphyroclasts (10–50% of the rock), which characteristically are of similar composition to the matrix minerals. The term *mylonite* is not restricted to a specific compositional range of rocks. It is thus possible to have granitoid mylonites, carbonate mylonites, amphibolitic mylonites, and so on, depending on the observed mineralogy. Wise *et al.* (1984) use the term mylonite for all rocks in the spectrum protomylonite to ultramylonite and suggest the term orthomylonite for rocks in the middle of the range. The definition preferred here still reserves the term for those rocks in the middle of the range containing 10–50% porphyroclasts (Figs 8.2 & 8.6(c)–(e) and 8.13(b)).

Myrmekite
A symplectic intergrowth of vermicular quartz and plagioclase resulting from the retrograde replacement of K-feldspar (Fig 6.13).

Nabarro–Herring creep
The dominant grain-scale rock deformation process at high temperatures and low shear stress. Nabarro–Herring creep is a type of diffusion creep involving a combination of grain-boundary sliding and diffusional transport of matter through the crystal lattice and along grain boundaries.

Neoblasts
Crystals that are more newly formed compared to others in the rock.

New grains
New grains develop during recrystallisation of highly deformed rocks and are typically located at grain boundaries and the margins of porphyroclasts. They are stable, strain-free areas that have significant optical discontinuity with the parent grain and, unlike 'sub-grains', have sharply defined boundaries (Fig. 8.6(c)).

Omphacite
A sodium-rich variety of the clinopyroxene augite: an essential and characteristic mineral of eclogites.

Overgrowth
The nucleation and growth of the same or a different mineral on the outer surface of another crystal (Figs 12.7(a, b) & 12.21; Plates 4(a) & 6(d, e)).

Paragenesis
In metamorphic petrology, a term used synonymously for the evolution of the mineral assemblage characterising a given rock.

Pelite
A rock of argillaceous composition with a mineral assemblage dominated by phyllosilicate minerals. Original sedimentary rocks are mudstones and siltstones, which when metamorphosed become slates, phyllites, schists, and so on (Fig. 4.6(a)).

Peristerites
Sodic plagioclases consisting of microscopic and sub-microscopic intergrowths of albite and oligoclase. Such feldspars characterise the greenschist facies–amphibolite facies transition.

Perthite
A feldspar intergrowth with plagioclase inclusions enclosed within K-feldspar. It is common in high-grade metamorphic rocks and plutonic igneous rocks (Figs 6.7 & 6.8).

Petrogenetic grid
A diagram the co-ordinates of which are intensive parameters characterising the rock-forming environment (e.g. pressure and temperature) on which may be plotted equilibrium curves

Glossary

delimiting the stability fields of specific minerals and mineral assemblages (Bates & Jackson, 1980) (e.g. Fig. 12.8).

Phase
A real chemical entity composed of one or more components. The various minerals and fluids present in a given system are described as phases.

Phyllite
A well cleaved pelitic rock characterised by a distinctive sheen on the planar surface. It is generally of intermediate grain size and of metamorphic grade between slate and schist.

Phyllonite
An intensely sheared phyllosilicate-rich rock with synchronously developed 'S–C fabrics', the intersection of which gives rise to so-called oyster shell or button-schist appearance in outcrop. Phyllonites are characteristically associated with thrust zones, shear zones and tectonic slides, and are often developed over broad areas (Fig. 10.7).

Pinnitisation
The retrogression, especially of cordierite, to an ultra fine grained green or yellow felty mixture of muscovite and chlorite (modified from Deer et al., 1966).

Pleochroism
The ability of an anisotropic crystal differentially to absorb various wavelengths of transmitted light in various crystallographic directions, and thus show different colours in different directions (i.e. as the microscope stage is rotated in plane-polarised light the mineral shows a colour change, and is said to be pleochroic) (modified after Bates & Jackson, 1980) (Plate 2(d)).

Poikiloblastic
A term used to describe porphyroblasts with abundant mineral inclusions (Plate 1(a)).

Polygonal structure
A term for rocks containing crystals (usually equigranular) with polygonal shapes (commonly five- or six-sided) and dominantly straight boundaries meeting at triple-points (modified from Spry, 1969) (Figs 5.14(b) & 5.15).

Polymorph
One of two or more crystallographic forms of the same chemical substance. For example, there are three common polymorphs of Al_2SiO_5, namely, andalusite, kyanite and sillimanite (Fig. 7.9).

Porphyroblast
A metamorphic mineral that has grown to a much larger size than minerals of the surrounding matrix (Fig. 5.4; Plate 1).

Porphyroblastesis
The growth of porphyroblasts.

Porphyroblastic
A term used to describe a metamorphic rock with large crystals (porphyroblasts) grown within a finer-grained matrix (Fig. 5.4).

Porphyroclast
A large relict crystal, or crystal fragment, in a fine-grained matrix of a deformed rock (Figs 8.3, 8.13(a, b), 9.3(a, b) & 10.15–10.19).

Porphyroclastic
A term used to describe rocks with abundant porphyroclasts (e.g. mylonites) (Figs 8.3, 8.13(a, b) & 9.3(a, b)).

Post-tectonic growth
The growth of minerals or parts of minerals (e.g. porphyroblast rims) after deformation of the rock has ceased. It is often deduced from porphyroblast–foliation relationships (Plate 6).

Glossary

Pressure shadow
See **Strain shadow**.

Pressure solution
A deformation process whereby material under stress goes into solution at a localised point in a material. This material is transported by flow or diffusion and is usually deposited at some other locality in the rock system; a process termed *solution-transfer* (Ramsay & Huber, 1983).

Pre-tectonic growth
Mineral growth before deformation.

Primary twins
Twins developed during crystal growth (Plates 2(c)–(f)).

Product
A phase produced as a consequence of a given reaction.

Prograde metamorphism
Metamorphic changes resulting from increasing temperature and/or pressure conditions.

Protomylonite
A mylonitic rock with porphyroclasts comprising more than 50% of the rock (Fig. 8.13(a)).

Psammite
A metamorphosed impure sandstone with a mineral assemblage comprised largely of quartz with lesser amounts of feldspar and mica.

Pseudomorph
A mineral or aggregate of minerals having the form of another mineral phase being replaced. A pseudomorph is described as being 'after' the mineral the outward form of which it has (e.g. chlorite after garnet). Pseudomorphing is a gradual process such that at the arrested stage seen in thin section the pseudomorph may be 'partial' or 'complete' (Plates 4(c)–(e)).

Pseudotachylite
A glassy rock produced by frictional melting in a fault or thrust zone (Mason, 1978) (Fig. 8.12).

Pure shear
An irrotational strain where the area dilation is zero (Ramsay & Huber, 1983).

Reactant
A phase being consumed in a given reaction.

Reaction rim
A monomineralic or polymineralic rim (commonly of hydrous phases) surrounding the core of another phase in the process of retrogression (Plate 4(b)).

Reaction softening
Enhanced ductility in a deforming rock due to a metamorphic reaction (based on White et al., 1980).

Recovery
Any of the processes through which the number of grain dislocations (i.e. strain energy) produced during rock deformation can be reduced (e.g. sub-grain and new grain development) (after Bates & Jackson, 1980) (Figs 8.6(b) & 8.10).

Recrystallisation
Solid state textural modification in rocks, involving intra- and inter-crystalline rearrangement to produce a more stable lower-energy system. This typically involves grain-boundary migration and the production of more regular crystal faces (e.g. polygonization of quartz). During the prograde metamorphism of quartz-rich rocks, abundant smaller quartz grains generally amalgamate to produce fewer but larger crystals (Figs 8.6(c) & (d), 8.11 & 5.14(b)).

Regional (orogenic) metamorphism
Metamorphism affecting large areas of the Earth's crust and commonly associated with

Glossary

collisional orogeny. Regional metamorphic rocks commonly exhibit complex interrelationships between mineral growth and deformation.

Resorption
A process involving partial or complete chemical modification of earlier formed crystals (e.g. porphyroblasts) that are no longer in equilibrium. This typically involves diffusion at the crystal margins penetrating to varying depths within the crystals. Distinct jumps in rim chemistry of zoned porphyroblasts (e.g. Mn-rich rims in garnet) are evidence of resorption (Fig. 5.23(b)).

Retrograde metamorphism
Metamorphic changes in response to decreasing pressure and/or temperature conditions (Plates 5(b)–(h)).

Retrogression
Modification of the primary mineral assemblage due to waning P–T conditions and/or changing fluid chemistry. This process typically involves partial or complete replacement of high-grade largely anhydrous phases by lower-grade hydrous phases (Plates 4(b)–(h)).

Rheology
A branch of physics that gives a phenomenological account of mechanical behaviour of matter, which involves its material properties (after Poirier, 1985).

Ribbon quartz
With intense ductile deformation associated with high strains (especially at high T), individual quartz crystals often become exceptionally elongate within the mylonitic fabric. This is known as ribbon quartz (Fig. 8.6(e)).

Saussuritisation
The retrogressive replacement of anorthitic plagioclase by a fine-grained aggregate of epidote group minerals and sericite (± calcite) (Section 7.1.3; Plates 4(f) & (g)).

S–C fabrics
Mylonites and phyllonites commonly exhibit two fabrics (S–C fabrics) that simultaneously developed during the intense shearing that formed such rocks. First described by Berthé *et al.* (1979), the C-surfaces are parallel to the shear zone margin, while S-surfaces are oblique to this (typically by about 30°). In phyllonites the intersection of these two fabrics gives rise to a texture that has been described as fish-scale, button schist or oyster shell because of its distinctive appearance (Figs 10.6–10.8).

Schist
Metamorphic rock commonly of pelitic composition, with a well developed schistosity (e.g. Plate 5(c)) (modified after Mason, 1978).

Schistosity
A planar structure defined by the alignment of inequant minerals such as micas and amphiboles and where individual minerals are discernable in hand specimen. Rocks showing this structure are termed schists. Such a structure is common in regional (orogenic) metamorphic and blueschist facies metapelites and metabasites (Plate 5(c); Fig. 4.5(a)).

Schlieren
Streaks or elongate segregations of non-leucosome (usually biotite-rich) material within the leucosome component of migmatites (Fig. 4.11). The *schlieren* represent entrained restite that has not been entirely separated from the melt.

Screw dislocation
A type of dislocation in which part of the crystal is displaced by a lattice unit, giving a twisted lattice at the line of dislocation, but elsewhere the lattice planes line up (Fig. 8.5(b)).

Glossary

Secondary twins
Twins that have formed subsequent to crystal growth. This includes *inversion twins* formed due to change in crystal habit as a result of instability of the initial structure with changing P–T, and *deformation twins* formed in response to deformation of the crystal lattice (Figs 5.18 & 5.19; Plates 3(a) & (b)).

Sericitisation
The alteration of a mineral or minerals to an aggregate of fine-grained white mica, known as sericite (Fig. 7.6; Plate 4(g)).

Serpentinite
A retrogressed ultramafic rock with a mineral assemblage comprised largely of serpentine minerals (Fig. 7.3).

Serpentinisation
A process involving the conversion of high-temperature minerals (especially olivine) to an aggregate of serpentine. This is common in retrogressed ultramafic rocks, and takes place at temperatures below 500°C and often <300°C in the presence of aqueous fluids.

Sieve-structure
Synonymous with **Poikiloblastic** (Plate 1(a)).

Simple shear
Rotational, constant volume, plane strain deformation (essentially two-dimensional).

Skarn
A rock formed during contact metamorphism/metasomatism by reaction between carbonate rocks and fluids rich in elements such as iron and silica. Skarns have an assemblage dominated by calc-silicate minerals, often in association with magnetite.

Skeletal crystals
Individual crystals that have nucleated and grown between the grain boundaries of other crystals (particularly quartz) to form a skeletal or mesh-like network of thin interconnected strands (Fig. 5.11).

Sliding reaction
A continuous, or sliding, reaction is one not involving the production of new minerals but involving a gradual change in chemistry of a particular phase or phases in response to changing metamorphic conditions (e.g. biotites of pelitic rocks typically show an increasing Mg:Fe ratio with increasing temperature).

Snowball structure
A term used to describe spiralled or S-shaped inclusion fabrics in syntectonic porphyroblasts (e.g. garnet). It was originally considered to form due to physical rotation of the porphyroblast during growth, but it is now largely accepted as resulting from differential rotation of the fabric with respect to the porphyroblast during growth, and/or crenulation overgrowth (Fig. 9.4; Plate 5(d)).

Solid solution
A process involving the substitution of one or more elements between two (or more) end-member phases to produce a complete range of mineral compositions between the end-members. For example, olivine forms a solid-solution series between pure forsterite (Mg_2SiO_4) and pure fayalite (Fe_2SiO_4).

Spherulitic
A term describing a sub-spherical mass of acicular crystals radiating from a common point (Fig. 5.12(c)).

Stacking fault
The surface defining the zone of mismatch between partial dislocations and the adjacent ordered crystal lattice.

Static recrystallisation
The modification of the grain structure (size, shape and orientation of grains) that occurs

Glossary

during high-temperature annealing following deformation (after Poirier, 1985).

Steatisation
A process involving the retrogression of ultrabasic rocks to an assemblage comprised largely of talc.

S-tectonite
A deformed rock containing recognisable foliation planes (Fig. 4.4).

Strain
The change in size, shape and volume of a body resulting from the action of an applied stress field (i.e. the deformation resulting from stress) (modified after Park, 1983).

Strain hardening
A variety of processes such as pile-up of dislocations (= dislocation tangles) lead to situations in which, with increasing strain, greater differential stress is required to maintain a given strain rate. This is strain hardening.

Strain partitioning
The heterogeneous nature of rocks and their deformation means that, on all scales from macro- to micro-, strain is partitioned between areas of higher strain and lower strain. Estimates for bulk strain obtained (on whatever scale) represent the average of these partitioned strains (Fig. 9.7).

Strain shadow
A region of low strain protected from deformation by a rigid or competent object in a rock of lower competence. The rock matrix and object are frequently detached along their contact, the space being filled with crystals showing fibrous growth forms (if not recrystallised at higher temperatures) (Ramsay & Huber, 1983) (Figs 9.3(c) & 10.20–10.26).

Strain softening
Softening, in mechanical terms, can be expressed as a reduction in differential stress to maintain constant strain rate or an increase in strain rate at constant stress. 'Strain softening' can occur by a number of processes, including change in deformation mechanism, geometric softening, continual recrystallisation, reaction softening and chemical softening (after White et al., 1980).

Stylolite
Irregular, serrated or jagged pressure solution surfaces, especially common in massive carbonate and quartzite units from diagenetic to greenschist facies conditions. They generally develop perpendicular to principal compressive stress, and are commonly defined by thin (typically 0.5–3.0 mm) seams of dark insoluble material (Fig. 8.8).

Sub-grains
Sub-grains develop in deformed crystals, and are areas misoriented by a few degrees relative to the parent grain. They are picked out by standard optical microscopy since they pass into extinction at a slightly different position to the parent grain, although lacking sharply defined boundaries. Sub-grains are strain-free, and form during recovery (Fig. 8.6(b)).

Subhedral
A term used to describe metamorphic crystals (especially porphyroblasts) with moderately good form, and some well-formed crystal faces (Fig. 5.9(b)).

Surface energy
Atoms at the crystal edge do not have all of their bonds satisfied, and consequently the surface of the crystal is less stable and has excess energy, termed *surface energy* (also referred to as *interfacial energy* or *grain-boundary energy*).

Glossary

Symplectite
Complex and intimate intergrowth of two or sometimes three simultaneously co-nucleating phases, one of which is usually vermicular or rodded in form. Such structures are especially common in retrogressed granulites (Figs 6.10–6.11, 12.16 & 12.17).

Syntaxial vein
Vein with filling material grown from walls towards the vein centre (Fig. 11.8(a)) (Ramsay & Huber, 1983).

Syntectonic growth
Mineral growth synchronous with deformation (Figs 9.5, 9.6, 9.8 & 9.9; Plate 5).

Texture
A specific type of microstructure in which the component parts show a preferred orientation.

Tiltwall
A sub-grain boundary comprising an array of edge dislocations with the same Burgers vector (Fig. 11.16).

Topotactic transformation
Replacement of a primary phase by a secondary phase with minimal rearrangement of the original lattice. The replacement of biotite by chlorite (Plate 4(h)) is a good example of such a transformation at an arrested stage (Fig. 7.7; Plate 4(h)).

Translational gliding
A process of crystal deformation involving translation of a layer in the crystal lattice relative to its neighbour by a unit amount which leaves the crystal portions on each side similarly oriented both before and after displacement (Fig. 5.20(a)).

Twin
A polycrystalline unit composed of two or more homogenous portions of the same crystal species, mutually oriented according to certain simple laws (Cahn, 1954) (Plates 2(c)–(f)).

Ultramylonite
A mylonite with a fine-grained matrix and less than 10% porphyroclasts (Fig. 8.13(c); Plates 7(a) & (b)).

Undulose extinction
A term describing a property of deformed crystals (especially quartz) in which poorly defined zones of extinction are seen to sweep across individual crystals when the microscope stage is rotated. This feature is caused by deformation/distortion of the crystal lattice (Fig. 8.6(a)).

Uralitisation
A process of alteration involving the pseudomorphic replacement of primary igneous pyroxenes by secondary metamorphic/hydrothermal amphiboles (often tremolite, actinolite or hornblende).

Vacancy
A point defect represented by a vacant site in the crystal lattice (Fig. 8.4).

Vein
A fracture or microfracture filled with crystalline material (Figs 11.1–11.10; Plates 8(a)–(e)).

Appendix III: Key mineral assemblages

The following key mineral assemblages are associated with the major rock compositions for each of the metamorphic facies. The list is based on information in Yardley (1989), Spear (1993), Bucher & Frey (1994), Miyashiro (1994) and Raymond (1995), and on the author's own observations.

Zeolite facies

Metapelite: Qtz–Kln–Ill/Smc–Chl–Ab–(±Anl)–(±Cal)–opaques (Py, Mag, Hem, Gr)

Metabasite: **Zeo–Qtz**–Ab–Chl–(±Prh)–Spn
Qtz–Ab–Chl–clays–Cal

Meta-calcsilicate: Cal–Dol–Qtz

Meta-ultramafic: **Srp–Mag–Tlc**–Cal
Srp–Mag–Dol
Srp–Mag–Chl–Dol
Srp–Mag–Mgs–Dol

Sub-greenschist facies

Metapelite: Qtz–Kln–Ill/Smc–Chl/Smc–Ab–opaques (Py, Mag, Hem)
Qtz–Ab–Chl–Phe–Stp–Hem–(±Prh–Pmp)

Metabasite: **Prh–Pmp**–Qtz–Ab–Chl–Spn–(±Stp)
Qtz–Ab–Chl–clays–Cal
Prh–Pmp–Qtz–Ep–Chl

Meta-calcsilicate: Cal–Dol–Qtz

Appendix III

Meta-ultramafic: **Srp–Mag–Tlc**–Cal
Srp–Mag–Dol
Srp–Mag–Chl–Dol
Srp–Mag–Mgs–Dol
Srp–Brc

Greenschist facies

Metapelite: **Ms(Phe)–Chl**–Qtz–Ab–opaques–(±Mc)–(±Cal)–(±Gr)–(±Pg)–(±Stp)
Ms(Phe)–Chl–Ctd–Qtz–Ab–Ilm
Ms(Phe)–Chl–Bt–Qtz–Ab–Ilm–(±Ep)–(±Cal)–(±Mc)
Prl–Ms–Qtz–Chl–Ctd [high-Al pelite]

Metabasite: **Ep/Czo–Qtz–Act–Ab–Chl**–Spn–opaques–(±Ms)–(±Cal)
Ab–Act–Ep–Pmp–Cal
Chl–Ep–Qtz–Ab–Cal–Prh–Act
Ep–Qtz–Act–Ab–Chl–Spn–Cal

Meta-calcsilicate: Dol–Cal–Qtz–(±Tlc)
Ank–Cal–Qtz–Ab–Ms–Chl–(±Bt)–(±Py, ±Ilm, ±Mag)

Meta-ultramafic: Srp–Mgs–Mag
Srp–Chl–Tlc–Tr–Mag
Srp–Brc

Epidote–amphibolite facies (Barrovian 'Garnet Zone')

Metapelite: Qtz–Olg–Ms–**Bt–Grt(Alm)**–Ilm–(±Chl)–(±Ctd)–(±Ep)

Metabasite: Qtz–Ab/Olg–**Ep–Hbl(blue–green)**(±Act)–(±Chl)–Spn/Rt

Meta-calcsilicate: Cal–Tr–Qtz–Py
Dol–Qtz
Zo–Cal–Bt
Zo–Hbl

Meta-ultramafic: Srp–Fo

Amphibolite facies

Metapelite: Qtz–Pl–Ms–**Bt–Grt–St**–Ilm (lower amph. facies – 'St zone')
Qtz–Ms–Chl–**Ctd–St**–(±Grt) (high-Al pelite, lower amph. facies)
Qtz–Pl–Ms–**Bt–Grt–Ky**–(±St)–Ilm/Rt (mid. amph. facies – 'Ky zone')
Qtz–Ms–**St–Ky**–Bt–(±Grt) (high-Al pelite, mid. amph. facies)
Qtz–Pl–Bt–Grt–**Sil**–(±Ky)–Rt–(±Kfs) (upper amph. facies – 'Sil zone')

Eclogite facies

	Qtz–Ms–St–Sil–Bt–(±Grt)	(high-Al pelite, upper amph. facies)
	Qtz–Pl–Bt–Grt–**Sil–Kfs**–Rt [No Ms!]	(uppermost amph. facies)

Metabasite: **Hbl(green–brown)–Pl(Ads)**–(±Grt)–Qtz–Ilm/Rt–(±Ep)
Hbl(green–brown)–Bt–**Pl(Ads)**–Qtz–Spn/Ilm

Meta-calcsilicate: **Cal–Tr–Qtz**–Py   (low amphibolite facies)
Cal–**Di**–**Tr**–Qtz–Py   (middle/upper amphibolite facies)
Cal–Di–Grs–Qtz
Dol–Cal–Tr
Hbl–An
Di–Pl–Mc–Qtz

Meta-ultramafic: Fo–Ath–Tlc–Chl–Mag
Ath–Tr–Tlc–Chl–Mag
Fo–En–Chl–Tr
Fo–Di–Tr
En–Ath–Tr

Granulite facies

Metapelite: Qtz–Pl(An)–Bt–Grt–**Sil–Kfs**–Rt [No Ms!]
Sil–Crd–Grt–Kfs–Pl(An)–Qtz
Spl–Qtz assemblages
Spr–Qtz assemblages   [high-*P*–*T* granulite]
Opx–Sil–Qtz assemblages   [high-*P*–*T* granulite]
Osm–Grt assemblages   [high-*P*–*T* granulite]

Metabasite: Grt–An–Cpx–(±Qtz)–(±Opx)   Garnet granulite [high-*P*]
Cpx–Opx(Hy)–An–(±Ol)–(±Qtz)   Pyroxene granulite [low- /med-*P*]
Cpx–Opx(Hy)–An–Qtz–(± brown Hbl)

Meta-calcsilicate: Fo–Cal–Qtz
Fo–Cal–Dol–(± Spl, Phl, An)
Di–An–Cal–(±Qtz)
Wo–Di–Cal
Wo–Scp–Cal

Meta-ultramafic: Fo–En
Fo–En–Cpx–Spl

Eclogite facies

Metapelite: Tlc–Phe–Ky–Qtz–(±Grt)
Tlc–Phe
Tlc–Phe–Ky–Cld

Appendix III

Metabasite:	**Omp–Grt**–Rt–(±Ky)–(±Zo)–(± Qtz, or Cs at very high P) [Pl absent!]
	Bar–Jd/Omp–Grt–Rt
Meta-calcsilicate:	Di–Qtz–Cal
	Dol–Phl
Meta-ultramafic:	Fo–Opx–Cpx–Grt

Blueschist facies

Metapelite:	Phe–Chl–Ctd–Qtz–**Gln**–Spn–Ep(±Lws)–(±Stp)	
	Phe–Qtz–Lws–Gln/Cros–Hem–(±Sps)	
	Cp–Qtz	(Mg-rich assemblages)
Metabasite:	**Ep–Gln**(±Act)–Chl–Ab–Qtz–Spn–(±Cal)	
	Lws–Gln–Chl–Ab–Qtz–Spn–(±Cal)	
	Jd–Qtz–Gln–Arg–Mag–Phe–Spn	(high-P blueschist)
Meta-calcsilicate:	Arg/Cal–Qtz–Chl–Hem	
Meta-ultramafic:	Srp–Mag–Mgs–Dol	

Albite–epidote hornfels facies (& low-P orogenic metamorphism: And–Chl; Chl– Bt zone)

Metapelite:	Qtz–Ms–Prl–Ctd–Chl–Ab–opaques (Py, Mag, Hem)	
	Qtz–Ms–Ctd–Chl–Bt–Ab–opaques (Py, Mag, Hem)	
	Qtz–Ms–Chl–Bt	
	Qtz–Ms–Chl–And–Ctd	[high-Al pelite]
Metabasite:	Ep/Czo–Qtz–Ab–Chl–Act–Spn–(±Cal)–(±Bt)	
Meta-calcsilicate:	Tlc–Qtz	
	Tr–Cal–Qtz	
	Dol–Qtz	
Meta-ultramafic:	Srp–Brc	

Hornblende hornfels facies (& low-P orogenic metamorphism – And ± St, Crd zone)

Metapelite:	Qtz–Ms–**And**–Bt–Olg–Ilm–(±Grt)
	Qtz–Ms–**And**–**Crd**–Bt–Olg–Ilm
	Qtz–Ms–**And**–**St**–Bt–Olg–Ilm
	Qtz–Ms–**Crd**–**Chl**–Bt–Olg–Ilm

Sanidinite facies

| | Qtz–Ms–**And–St**–Chl–(±Grt) | [high-Al pelite] |
| | Qtz–Ath–Crd | [Mg-rich assemblages] |

Metabasite: Hbl–Qtz–Olg–Ep(Zo)–(±Grt)–(±Cum)–(±Bt)–Spn/Rt

Meta-calcsilicate: Qtz–Cal–Di
Tr–Di–Qtz
Grs–Qtz–Cal or Ves–Qtz–Cal [Al_2O_3–H_2O–calc-silicates]

Meta-ultramafic: Tlc–Fo

Pyroxene hornfels facies (& low-*P* orogenic metamorphism – sillimanite zone)

Metapelite: Qtz–Kfs–**Sil–Grt–Crd**–Ilm–(±Pl)
Qtz–Kfs–**Grt–Crd–Bt**–Ilm–(±Pl)
Qtz–Kfs–**Sil–Crd–Bt**–Ilm–(±Pl)
Qtz–Ath–Crd (Mg-rich assemblages)

Metabasite: Qtz–Pl–Kfs–**Grt–Bt–Opx**–Spl
Qtz–Pl–Hbl–Cpx(±Opx)–(±Bt)–opaques
Cpx–Opx–Pl–(±Ol)–(± brown Hbl)
Opx–Cpx–Pl–Qtz

Meta-calcsilicate: Wo–Qtz
Wo–Cal
Wo–An–Di
Fo–Cal
En–Di–Qtz

Meta-ultramafic: Ath–Fo

Sanidinite facies

Metapelite: Qtz(±Crs ±Trd)–(Sil, Mul, Crn)–Hc–Crd
Crn–Mag–An

Metabasite: Cpx–Opx–Pl–Sa–Qtz(±Trd)

Meta-calcsilicate: Fo–Cal
En–Di–Qtz

Meta-ultramafic: Fo–Cpx–Opx–An

References for Appendices I–III and the Plates

Barker, A.J. (1990) *Introduction to metamorphic textures and microstructures* (1st edn). Blackie, Glasgow, 162 pp.

Bates, R.L. & Jackson, J.A. (1980) *Glossary of geology*. American Geological Institute, Falls Church, Virginia, 751 pp.

Bell, T.H. & Rubenach, M.J. (1980) Crenulation cleavage development – evidence for progressive, bulk inhomogeneous shortening from 'millipede' microstructures in the Robertson River Metamorphics. *Tectonophysics*, **68**, T9–T15.

Berthé, D., Choukroune, P. & Jegouzo, P. (1979) Orthogneiss, mylonite and non-coaxial deformation of granites: the example of the south Armorican shearzone. *Journal of Structural Geology*, **1**, 31–42.

Bucher, K. & Frey, M. (1994) *Petrogenesis of metamorphic rocks*. Springer-Verlag, Berlin, 318 pp.

Cahn, R.W. (1954) Twinned crystals. *Advances in Physics*, **3**, 363.

Deer, W.A., Howie, R.A. & Zussman, J. (1966) *An introduction to the rock forming minerals*. Longman, London, 528 pp.

Eskola, P. (1915) On the relations between the chemical and mineralogical composition in the metamorphic rocks of the region. *Bulletin de la Commission Geologique de Finlande*, 44.

Eskola, P. (1939) Die metamorphen Gesteine, in *Die Entstehung der Gesteine* (eds T.F.W. Barth, C.W. Correns & P. Eskola). Julius Springer, Berlin (reprinted, 1960, 1970), 263–407.

Harker, A. (1939) *Metamorphism – a study of the transformations of rock masses*. Methuen, London, 362 pp.

Jensen, M.L. (1965) The rational and geological aspects of solid diffusion. *Canadian Mineralogist*, **8**, 271–290.

Johnson, S.E. & Moore, R.R. (1996) De-bugging the 'millipede' porphyroblast microstructure: a serial thin-section study and 3-D computer animation. *Journal of Metamorphic Geology*, **14**, 3–14.

Kretz, R. (1983) Symbols for rock-forming minerals. *American Mineralogist*, **68**, 277–279.

Mason, R. (1978) *Petrology of the metamorphic rocks*. George Allen & Unwin, London, 254 pp.

Miyashiro, A. (1994) *Metamorphic petrology*. UCL Press, London, 404 pp.

Park, R.G. (1983) *Foundations of structural geology*. Blackie, Glasgow, 135 pp.

Poirier, J.-P. (1985) *Creep of crystals*. Cambridge University Press, Cambridge, 206 pp.

Poirier, J.-P. & Guillope, M. (1979) Deformation-induced recrystallization of minerals. *Bulletin Mineralogique*, **102**, 67–74.

Powell, R. (1978) *Equilibrium thermodynamics in petrology*. Harper & Row, London, 284 pp.

Ramsay, J.G. & Huber, M.I. (1983) *The techniques of modern structural geology; Volume 1: Strain analysis*. Academic Press, London.

Raymond, L.A. (1995) *Petrology: the study of igneous, sedimentary, and metamorphic rocks*. Brown, Dubuque, Iowa, 742 pp.

Rutter, E.H. (1986) On the nomenclature of mode of failure transitions in rocks. *Tectonophysics*, **122**, 381–387.

Schoneveld, Chr. (1977) A study of some typical inclusion patterns in strongly paracrystalline garnets. *Tectonophysics*, **39**, 453-471.

Shelley, D. (1993) *Igneous and metamorphic rocks under the microscope*. Chapman & Hall, London, 445 pp.

Spear, F.S. (1993) *Metamorphic phase equilibria and pressure–temperature–time paths*. Mineralogical Society of America Monograph, Mineralogical Society of America, Washington, DC, 799 pp.

Spry, A. (1969) *Metamorphic textures*. Pergamon Press, Oxford, 350 pp.

Vernon, R.H. (1976) *Metamorphic processes*. George Allen & Unwin, London, 247 pp.

White, S.H., Burrows, S.E., Carreras, J., Shaw, N.D. & Humphreys, F.J. (1980) On mylonites in ductile shear zones. *Journal of Structural Geology*, **2**, 175–187.

Wise, D.U., Dunn, D.E., Engelder, J.T., Geiser, P.A., Hatcher, R.D., Kish, S.A., Odom, A.L. & Schamel, S. (1984) Fault-related rocks: suggestions for terminology. *Geology*, **12**, 391-394.

Yardley, B.W.D. (1989) *Introduction to metamorphic petrology*. Longman, Harlow, 248 pp.

INDEX

Numbers in **bold** refer to pages where figures (diagrams/photographs) illustrate the item indexed. Numbers in *italics* refer to Tables and Appendices.

Absorption 86–88, *229*
Acicular crystals *see* Crystal form; acicular
Actinolite *see* Amphibole; actinolite
Activation energy *see* Energy; activation
Activity 13, 194, *229*
Adsorption 86–88, **88**, *229*
Al$_2$SiO$_5$ polymorphs *see* Andalusite, Kyanite, Sillimanite
Albite *see* Plagioclase
Albitisation 108
Alkali feldspar *see* K-feldspar, Plagioclase; albite
Allanite 86
Almandine *see* Garnet; almandine
Alteration *see* Retrogression
Amphibole
 actinolite 34–35, **72**, 105–106, 138, 218
 anthophyllite 36–37
 barroisite 36, 219
 chemical zonation 80
 crossite 36, 219
 edenite 11
 glaucophane 36, 110, 197, 218–219
 hornblende 35, 94, 105–106, 137–138, 158, 202
 inclusion fabrics 158
 pargasite 11
 tremolite 11, 32, 36, 105
 twinning 76
Amphibolite *229*
 see also Metabasite
Analcite *see* Zeolites; analcite
Anatexis *229*
 see also Partial melting
Anchimetamorphism 185, *229*

Andalusite
 chiastolite 31, 88, **89**, 210
 reactions 31, 93, 110, 209
 stability 8, 9, 31, 181, 209
Annealing 64, 141, *229*
Anorthite *see* Plagioclase; anorthite
Antiperthite 92, *229*
Apatite *202*
Aragonite 12, 36, 68, **217**, 218–219
Argillic alteration 103
Atoll structure 98, *230*
Augen 41
Augen gneiss 41, **44**, *230*
Axinite **74**

Banding 41
Barroisite *see* Amphibole; barroisite
Barrovian zones 21, 199, 202, 203–204, 210
Biotite 28, 68, 104, 154, 187, 204
Blastomylonite 135, *230*
Blueschist *see* Facies; blueschist
Bonding (Bonds) 143
Boudinage (Boudins) 143, 151, 164, 179, **180**
Bow-tie structure *see* Crystal form; bow-tie
Brecciation 101, **118**
Brittle-ductile transition 16–17, 125, 132
Brucite 105
Bulk rock chemistry 16, 21–22, 25, 27–37, 57, 198
Bulk rock permeability *see* Permeability
Bulk rock porosity *see* Porosity
Burgers vector 120, *230*
Button schist 135, 166

Calc-silicate rocks 32–33, 42, *102*, *230*
Calcite
 Cal-Dol thermometry 90–91
 crystallographic preferred orientation 176–177
 deformation 79, 127, 139–141
 reactions 32–33, 218
 size & form **68**, 79
 twinning 77, 79, 139–141
 veins 180, 188
Captive minerals 190
Carbonate rocks
 cathodoluminescence & SEM 82, 182
 crystallographic preferred orientation 176–177
 deformation 139–141, 171
 fluids 32, 42, 192
 reactions 32–33, 192
 stylolites 125
Carbon 87
Carbon dioxide (CO$_2$)
 effect on reactions 10, 32–33, 36–37
 fluid inclusions 190–191
Cataclasis (Cataclastic flow) 117, 132, 138, 144, 145, *230*
Cataclasite 132, *230*
Cation-exchange *see* Reactions; cation-exchange
Cathodoluminescence 141, 181
CH$_4$ *see* Methane
Chemical potential
 definition 11, *230*
 gradients 50, 51–52, 94, 99, 125, 143
Chemical system
 ASH 9, 11
 CASH 34

255

Index

Chemical system *cont'd.*
  CMASH 36–37
  CFMASH 36–37
  CMHSC 32
  component 8, 11, 25, 93, 143, 231
  degrees of freedom 9
  divariant field 9–10, 11, 25
  energy 10–11
  FASH 30
  FMAS 214–215
  invariant point 10
  KFMASH 27, 28, 29–31
  NCFMASH 34
  phase 8–9, 10–11, 25, 201, *260*
  univariant curve 9–10, 12–13
  variance 9
Chemical zoning *See* Zoning; chemical
Chiastolite 31, 88, **89**, 210, **211**
Chlorite 28, 34, 36, 48, **68**,
Chlorite-mica stack 109, 110, 151, 231
Chloritisation 101, 106
Chloritoid 29–30, *68*, 87, 88, 197, 206
Clay minerals 28, 33, *103*, 106
Cleavage 44–50, *231*
  classification of 44–45
  continuous 45, **46**
  crenulation 45, **47**, 49–50, 125, 157, 168–169
  crystallographic 89, 104, 155
  definition of 44
  disjunctive 45
  domains 45, **47**
  domes 87, 231
  formation 45–51
  P-domains **47**, 49–50
  pressure solution 47
  process of formation 45–51
  Q-domains **47**, 49
  refraction 48, **49**
  rough 45
  seams 50
  slaty 45, **46**, **47**, 210–211
  spaced 45, 50
  stylolitic 45
  time to form 48
Climb *see* Dislocation; climb
Clinochlore 36
Clinopyroxene *see* Pyroxene; clinopyroxene
Clinozoisite 93, 218
Closed system *see* Isochemical System

Closure temperature *202*, *231*
CO_2 *see* Carbon dioxide
Coarsening *see* Grain coarsening
Coesite 7, 8, **104**, 220
Component *see* Chemical System; component
Compositional layering 41, **42**, 51, 61
Conduction 8, 203
Contact aureole 231
  see also Metamorphic aureole
Cooling 65, 95, 103, 199, 202, 218
Cordierite 31, 57, 76–77, 78, 94, 213–215
Core-and-mantle structure **122**, 130, 136, 139, 141, 231
Corona 94, 98–99, 103, 206, 213–216, 231
Corundum 215
Crack-seal veins *231*
  see also Veins; crack seal
Creep *231*
  coble 125
  dislocation 121–122, 124–125, 130–131, 136, 139, 141
  diffusion 51, 125, 127, 145
  mechanisms 123–125
  Nabarro-Herring **124**, 125, *238*
  power-law 123–124
  steady-state 123
Crenulation *231*
  hinge region 144
  limb region 143, 144
  overgrowth 157–159, 160
Crossite *see* Amphibole; crossite
Crush breccia 132
Crustal thickening 207–209, 213, 218
Crustal thinning 5, 203
Crystal form 65–75
  acicular 68, 70, *229*
  anhedral 65, **67**, *230*
  bow-tie 68, **72**, *230*
  dendritic 68
  decussate 74, **74**, *232*
  euhedral 65, **66**, 67–68, 183, 205, *233*
  fascicular bundles 68, **72**
  fibrous *see* fibrolite
  idioblastic **66**, 87
    see also euhedral
  needles *see* acicular
  rodded 68, **72**, 89
  skeletal 68, **69**, *243*
  spherulitic 68

subhedral 65, **66**, 74, 205, *244*
vermicular 95, **97**
Crystal growth 60–62
Crystal lattice 77, 87–88, 91, 119, 122–123, 151
Crystal-plastic processes 51, 117
  see also Deformation; crystal-plastic
Crystal shape 65–75, 125, 131
Crystal size 61–65
Crystal size distributions 64
Crystallographic cleavage *see* Cleavage; crystallographic
Crystallographic inversion *see* Twinning; inversion
Crystallographic misorientation *see* Lattice misorientation
Crystallographic preferred orientation (CPO) 43, 51, 131–132, 176–177
  calcite 131, 140–141, 177
  in mafic/ultramafic rocks 138, 177
  olivine 176
  quartz 120, 131, 142, 177
Cut-effects 89, 149–151, 160, 163

Daughter minerals *see* Fluid inclusions; daughter minerals
Decarbonation *see* Reaction; decarbonation
Decay constant 202
Decompression 94, 98, 104, 203, 208–209, 218
Decrepitation *see* Fluid inclusions; decrepitation
Deerite 70
Defects 117–119
  energy contribution 17
  point defects 118, **119**
    see also Vacancies, Interstitials
  line defects *see* Dislocation
  surface 18, 67, 86
Deformation
  bands 123, 136–137, 141, *232*
  brittle 16–19, 117, 132, 136–138, 164–165, 167
  crystal-plastic 123, 132, 135, 138–139, 168
  ductile 16–19, 51, 117, 120, 135, 141
  effect on reactions 17, 18, 143
  intracrystalline 117–132, 138, 139, 151, 193, *236*
  lamellae 128, 136–137, *232*
  mechanisms 117, 136, 139–141, 145

Index

Deformation *cont'd*
  mechanism map **124**, *232*
  partitioning 143, 154, 158
  pure shear *see* Shear; pure
  simple shear *see* Shear; simple
  twins 77–79, 131, 137, 139–141, *232*
Dehydration *see* Reaction; dehydration
Dendritic crystals *see* Crystal form; dendritic
Detrital phases 28, 32, 80, **109**, 110, 151
Devolatilisation *see* Reaction; devolatilisation, decarbonation, dehydration
Diachroneity *see* Metamorphic; diachroneity
Diagenesis 7, 110, 125, *232*
Diffusion
  coefficients 15
  definition 15, *232*
  Fick's laws 15, 118
  grain boundary 15, 16, 94, 125, 127, 145
  halo 60, 63, **64**
  intracrystalline (volume) 82, 93, 118, 125, 143, 197
  pathways 94
  process 15, 42, 60–61, 202
  rates of 15, 64, 82, 91, 94, 98
  to reaction sites 61, 63–64, 98, 187
Diffusive mass transfer
  associated with deformation 16, 132, 140
  facilitating reactions 51, 98, 112, 127, 143
Dihedral angle 71, 142, *232*
Diopside *see* Pyroxene; diopside
Disequilibrium 82, 98, 103–104, 107, *232*
Dislocation
  climb 121, 124, 127, 130, 136, 193
  creep *see* Creep; dislocation
  definition 17, 119, *233*
  density 59, 122, 128, 130, 136–137, 142–144
  edge 119, **120**, *233*
  location 119, 139, 193
  migration of 120–121, 128–130, 136
  screw 60, 119, **120**, *242*
  tangles 120–121, 144, *233*
  walls 128, 130

Dissolution 46, 48, **50**, 125, 143–144, *233*
Dolomite 32–33, 68, 77, **90–91**, 201
Dolomitization 101
Ductile shear *see* Deformation: ductile
Dunite 36
Dynamic recrystallisation *see* Recrystallisation; dynamic

Eclogite 5, 98, 130, 218–220, *233*
Energy
  activation 11, 58–60, 87, 143, 202, 229
  chemical 143
  free 17, 69, 85
  Gibbs free 10–11, 58–60, 62, 89, 112, *234*
  grain boundary *see* surface
  interfacial *see* surface
  lattice 67, 144
  mechanical 142, 143
  stored elastic 59, 119, 128
  strain 58–59, 120, 142, 143
  surface
    contribution to free energy 17, 58–60, 62
    definition 17, *244*
    enhancement 142–144
    of inclusions 86
    minimisation 67–71, 112, 129, 131
  of system 10–11
  thermal 10, 11
Enstatite *see* Pyroxene; enstatite
Enthalpy 10, 12, 89, 142, *233*
Entropy 10, 12, 200, *233*
Epidote 34–35, 68, 107, 110, 154, 181
Epitaxial overgrowth 112, 218, *233*
Equigranular 61
Equilibration 39–40, 51, 69, 80
Equilibrium
  assemblages 8–10, 39–40, 199, 200–201, 213
  definition of 39, *233*
  conditions of 60, 143
Erosion 203, 208, 213, 218
Euhedral crystals *see* Crystal form; euhedral
Exhumation 203, 213, 216, 219–220
Exsolution 89–92, *233*

Extensional crenulation cleavages 166–167
Extensional thinning (of crust) 5, 203

Fabric
  definition of *233*
  grain-shape 43, 51, 131, 140–141, 176–177
  inclusion trails 150–161
  mylonitic 141–142
  see also Cleavage, Schistosity
Facies (metamorphic) *237*
  albite-epidote hornfels *250*
  amphibolite 29–30, 34, 103, 137–138, 193, 248–249
  blueschist 6, 32, 33, 36, 216–219, *250*
  concept of 22
  definition of 22
  eclogite 137, 219–220, 249–250
  epidote-amphibolite 29, 34, 218, *248*
  granulite 35, 37, 94–95, 194, 212–215, *249*
  greenschist
    deformation 125, 132, 137–138, 140–141, 172, 183
    metabasites 34–35, 137–138, 210–211, *248*
    metapelites 28–29, 106–107, 110, *248*
  hornblende hornfels 250–251
  prehnite-actinolite 34
  prehnite-pumpellyite 22, 34
  in P-T space **23**
  pyroxene hornfels 35, *251*
  sanidinite *251*
  sub-greenschist 23, 28, 34, 183–185, 210, 247–248
  zeolite 22, 34, *247*
Fascicular bundles *see* Crystal form; fascicular bundles
Faserkiesel *see* fibrolite
Fault zone 132–135, 163, 179, 199
Fault breccia 132, *234*
Fault gouge 132, *234*
Feldspar *see* K-feldspar, Orthoclase, Plagioclase, Sanidine
Fe-Ti phases 13, **14**, 88
Fibrolite 31, 112, 144, *234*
Fission tracks 202
Fluid
  aqueous (H_2O) 12–13, 105–106, 187, 190, 191–194, 221
  carbonic (CO_2) 13, 190–194

257

Index

Fluid cont'd.
  see also effect on reactions
  chemistry 8, 16, 25, 103,
    105–106, 221
  circulation 101
  effect on reactions 13, 16, 22,
    32–34, 101, 221
  flow/flux/migration 101, 143,
    179
  grain boundary see Grain
    boundary; fluid
  immiscibility 191–192
  infiltration 103, 106, 108, 199,
    215, 221
  interstitial/intergranular 15, 16,
    48, 50, 103, 126
    see also Grain boundary; fluid
  pressure 101, 145, 179,
    182–183, 200, 221
  salinity 32, 193–194
  temperature 103
  wetting characteristics 191–
    193
Fluid inclusions 187–194, 201,
  209, *234*
  chemistry 190–194
  implosion 188
  leakage 128, 192–193
  necking-down **191**, 192
  occurrence 128, 184, 187–190,
    192
  overpressuring 188, 193
  primary 188, 190
  pseudosecondary 188
  secondary 188, **189**, 190
  shape 190
  size 188
Fluid-rock interaction 181
Fluorite 188
Folding 163–165
Foliation 43–44, 132, 150–151,
  155, 179, *234*
  see also Cleavage, Fabric,
    Schistosity
Forsterite see Olivine
Fractures 151, 164, 179–182, 183,
  187
  see also Micro-cracks,
    Microfractures
Free energy see Energy; free, Gibbs
  free
Frictional heating 144, 206
Frictional melting 144
Fugacity *234*

Garbenschiefer 68, *234*

Garnet
  in amphibolites 34
  almandine **66**, 80, 206, 208, 211
  deformation of 138
  in eclogites and granulites 35, 94,
    97–98, 213–215, 219
  grossular 33
  growth times 65
  inclusion fabrics 150–161
  in pelites 28
  porphyroblasts **62**, **66–67**,
    150–157, 204–206, 211, 212
  reaction involvement 94, 197
  size 64–65
  spessartine 28, **65**, 211
  thermometry 201, 206–207
  zonation 80, **81**, **82**, 204
Geobarometry 28, 200–201, 204,
  206
Geochronology see Radiometric
  dating
Geotherm 8, 203, *234*
Geothermal activity 34
Geothermal gradient 5, 8, *234*
Geothermometry 91, 200–201,
  204, 206
Gibbs free energy see Energy;
  Gibbs free
Glass 7, 129, 132
Glaucophane see Amphibole;
  glaucophane
Glide 120
Gneiss 41, **52**, 96–99, 187, *235*
  flecky 98–99
  origin of banding 51–52
  orthogneiss 51
  paragneiss 51, 53
Gneissosity 41, 51
Grain boundary 125–128
  bulging 130, 140
  diffusion see Diffusion; grain
    boundary
  dilatancy 125, 143
  fluid 15–16, 50, 126, 129, 143,
  192
  migration 71, 127, 129–131,
    136, 140–141, *235*
  nature of 119, 126
  reaction 18, 94, 98, 110
  serrated 130, 136, 139–140, 187
  sliding 47, 127, 132, 140–141,
    145, *234*
  sutured 130, 136
  triple points 71, **73**, 74–75, 126
Grain coarsening 17, 51, 61–62,
  127, 140, 188

Grain-scale dilatancy see Grain
  boundary; dilatancy
Grain-shape see Crystal form
Grain-shape fabric 43, 51, 131,
  140–141, 176–177
Grain size 61–65, 79, 125,
  139–140, 154
Grain size reduction 133, 136,
  140, 142, 144–145
Grain-size sensitive flow 139, 145
Granitoid see Metagranitoid
Granoblastic 51, 74, *235*
Granoblastic-polygonal 69, 73
Granofels 75, *235*
Granulite 71, 94, 98, 188,
  212–215, *235*
Graphite 65, 87–88, 154
Greenschist *235*
  see also Facies; greenschist
Grossular see Garnet; grossular
Growth of crystals 60–61, 154,
  198, 205
  absolute times 65
  controls 60–61, 75
  rate 60–61, 62–65, 68, 85–87, 142

H_2O see Fluid; aqueous (H_2O)
Halite 190, 193
Harzburgite 36
Heat 10, 101, 142, 210, 215
  conduction 203
  flow 5, 31, 61
  frictional 144, 206
  transfer 144, 199, 203, 206
Heating rate 208
Helicitic structure *235*
Heulandite see Zeolites; heulandite
Homologous temperature 124, *235*
Hornblende see Amphibole;
  hornblende
Hornfels 64, 71, 74–75, 77, *235*
Hour-glass structure 88, *235*
Hydration reactions see Reaction;
  hydration
Hydraulic fracturing 145, 179,
  182–183
Hydrostatic pressure 87
Hydrothermal activity see
  Metamorphism; hydrothermal
Hypersthene see Pyroxene;
  hypersthene

Idioblastic see Crystal form;
  idioblastic
Idiotopic *236*
Idocrase 33, 68

Index

Illite 28
Illite crystallinity 28, 201, *236*
Ilmenite 68, 154, 206
Immiscibility 89, 191–192
Impingement structures 69, 75, 82
Inclusion *236*
 bands 183, 185
 fluid 184, 187–194, 208, 209, *234*
 solid 85–98, 104, 197, 200, 204, 221
 trails (fabrics) 150–151, 154–161, *236*
Index minerals 21, *236*
Intergrowths 91, 93–98
Interstitial 118, **119**
Inversion
 aragonite-calcite **217**, 218
 cordierite 77
 orthoclase-microcline 77
Ion exchange/mobility 103, 144
Ionic reaction *see* Reaction; ionic
Iron oxides *see* Fe-Ti phases
Isobaric cooling 209, 212–213, 214, *236*
Isochemical system 8, 16, *236*
Isograd 21, 29, 31, *237*
Isothermal decompression 212–215, *237*

Jadeite *see* Pyroxene; jadeite
Jogs 165

Kaolinite **9**
Kelyphitic rim 94, *237*
K-feldspar 31, 91–93, 97, 106, 144, 187
 see also Microcline, Orthoclase, Sanidine
Kinematic indicators *see* Shear; sense indicators
Kink bands/kinking 49, 123, 153–154, **153**
Kornerupine 215
Kyanite
 ionic reactions 14
 polymorphic reactions 12, 110, *111*
 shape & form 68
 stability **9**, 31, 204, 206, 210

Laser Raman 194
Lattice misorientation 123, 125, 126, 130
Laumontite *see* Zeolites; laumontite

Lawsonite 36, *68*, **217**, 218
Layering *see* Compositional layering
Leucosome 53, 82, 99, 187, *237*
Lherzolite 36
 see also Peridotite
Lineation 43, *237*
Lithostatic pressure *see* Pressure; lithostatic
L-S tectonite 43, 131, *237*
L-tectonite **45**, 237

Magnesite 36
Magnetite 68
Marble 32, 130, 139–141, 179, 221, *237*
Melanosome 53, **54**, 187, *237*
Melt (melting)
 composition 7, 31
 environment of 7, 31, 132–133, 187, 209–210
 frictional 133, 144
 segregations 31, 187
 temperature 7, 31
 see also Anatexis, Partial melting
Mesoperthite 92, *237*
Mesosome 53, 99, 187, *237*
Metabasites 33–36, *102*, *110*, 130, 137–139
Metagabbro 138
Metagranitoid 37, 97, *102*, 117
Metamorphic
 aureole 6, 31, 53, 64, 142, 210–211
 diachroneity 198, 206
 facies *see* Facies
 facies series 199, *238*
 grade 23, *238*
 peak 198, 199–201, 203–204
Metamorphism *238*
 Barrovian *see* Barrovian zones
 Buchan-type 209
 burial 6, 34, *230*
 contact/thermal 6, 65, 88, 105, 160, 210–211
 definition of 3
 duration 5
 dynamic 7
 environments of 3–7
 hydrothermal 6, 101, 103, 107
 limits of 7–8
 ocean floor **4**, 5, 34, 105, 107
 orogenic 4–5, 65, 151, 198, 203–211, *241*
 prograde 9, 12, 188, 189, 197, 199, *241*

regional *see* Metamorphism; orogenic
 relationship with deformation 18, 142–145
 retrograde *see* Retrogression
 shear zone 6–7, 105
 shock 7
 subduction zone **4**, 5–6, 101, 216–219
 timing 198–201
Metapelite *see* Pelite
Metasomatism 83, 103, 107, 181, *238*
Metastability 12, 91, 112
Metavolcanic rocks 68, 93, 194
Methane (CH_4) 8, 190–191, 194
Mica 86, 202
Mica-fish **167**, 168, 176, *238*
Microcline 77, **78**, 152
Microfold *see* Crenulation
Microfractures 50, 57, 104, 138, 151, 182
Microlithon *see* Cleavage; domains
Microstructure
 definition of 39, *238*
 original 61
Mid-ocean ridge 45
Migmatite 5, 31, 41, 187, *238*
 diatexite 53
 metatexite 53, **54**
 process of formation 53–55, 181
 schlieren 53, **54**
 stromatic 53, 187
 see also Leucosome, Melanosome, Mesosome, Restite
Millipede microstructure 159–160, *238*
Mimetic growth 51, *238*
Monazite 86
Mortar-structure **122**, *238*
Muscovite 65, 68, 93, 97, 110, 204
Mylonite
 carbonate 139–141
 definition of 133, 141–142, *239*
 fabric 131, 136, 141–142, 165–166, 168
 granitic 118, **134**, 135–137, 145, 169, 176
 mafic 132, **137**–139, 145
 occurrence 132, 133, 135, 142
 process of formation 6, 44, 132, 136
 quartzitic **122**–123, 136–137
Myrmekite 95–97, **97**, 144, *239*

Index

N₂ *see* Nitrogen
Natrolite *see* Zeolites; natrolite
Neoblasts 239
New grains **121**, 130, 136–137, 145, *239*
Nitrogen 8, 190, 194
Nucleation 57–60
  controls on 58–60
  crystallographic control 112
  embryos 58–59
  at grain boundaries 57, 59, **59**, 68, 110,
  heterogeneous 57, 59, *235*
  homogeneous 57, 59
  rate of 60, 61, 62–63, 65, 103, 142
  sites of 59, 71, 79, 103, 143, 181
  theory 58–60
Nucleii
  abundance of 62–63, 68, 93
  distribution of 57
  site saturation 63, 103

Old grains *see* Detrital phases
Oligoclase *see* Plagioclase; oligoclase
Olivine 36, 94, 98, 137, 176–177
  alteration 105
  forsterite 33, 37, *68*, 105
Omphacite *see* Pyroxene; omphacite
Ophiolites 36
Orthoclase 77
Orthopyroxene *see* Pyroxene; orthopyroxene
Ostwald ripening 62, 64–65
Overgrowth
  of calcite 141
  of fabric 154–161
  prograde 110–112
  of quartz 182
Oxidation *see* Reaction; oxidation
Oyster-shell texture 135, 166

P-domains **47**, 49
Paragenesis 239
Paragonite 218
Paramorphic replacement 93
Partial melting 7, 31, 41, 51–54, 210
Pelite 27–31, *102*, 106, 194, 206, *239*
Peridotite 177
Peristerite gap 29, 34, *239*
Permeability 142–143, 144, 145, 179, 183, 192

Perthite 91, **91–92**, 144, *239*
Petrogenetic grids 23–25, 98, 207–209, 214–216, *239*
Phase *see* Chemical system; phase
Phase rule 8–10, 40
Phengite 28, 32, 48
Phlogopite 33
Phyllite 50, *240*
Phyllonite 44, 134, **166**, 167, *240*
Pinnitisation *240*
Plagioclase
  albite 12, 29, 34–36, 92, 138, 218
  alteration *102*, 104
  anorthite 33, 35, 93, 107, 138
  deformation of 117, 136–138
  gefüllte plagioklas 93, **94**
  as inclusions 91–92, 204
  oligoclase 29, 35
  in symplectite 94, 95, **96, 97**
  twinning 76
Pleochroic haloes 86, 87
Pleochroism *240*
Poikiloblastic 85, 93, 154, *240*
Polygonal aggregates 71, 131, 180, 185, *240*
Polymetamorphism 161, 197–198
Polymorphic transformation 12, 60, 99, 110, **111–112**, *240*
Porosity 34, 142, 145
Porphyroblast **62**, *240*
  abundance 63
  atoll 98
  definition of 61
  foliation-relationships 149–161, 198, 201
  growth 60–61, 85–87, 154, 157, 198, 210
  growth rates 63–65, 154, 158
  inclusion trails 149–151, 154–160, 169, 199, 205
  non-rotation 157–159, 161
  nucleation 57, 61, 63, 143, 157
  post-tectonic 160, 198, *240*
  pre-tectonic 151–154, *241*
  retrogression 103, 210
  rotation 150, 154–159
  serial sections 150–151
  shape 65–75
  shear-sense indicators 169
  size 63, 151
  straight inclusion trails 154–155, 205
  S-trails 149–150, 154–159, 160, 169

  syn-tectonic 154–160, 161, 198, *245*
Porphyroblastesis 69, 154–155, 158–159, 198, 211, *240*
Porphyroclastic microstructure 135, 141–142, **170**, *240*
Porphyroclasts 133, **134**, 135–138, *240*
  core-and-mantle 130
  δ-type 169–171, **171**
  shear-sense 169–171
  σ-type 169–171, **170**
Potassic alteration *see* Sericitisation
Prehnite 34, *68*, 181
Pressure
  confining *see* Pressure; hydrostatic, lithostatic
  fluid 22, 145, 179, 183, 221
  hydrostatic 34, 182, *236*
  influence on reactions 24, 36, 187
  lithostatic 8, 34, 182, *237*
Pressure shadow *see* Strain; shadow
Pressure solution 50, **124**, 125, 143, *241*
Products of reactions *241*
Propyllitic alteration 103
Protomylonite **134**, 135, 141, 152, *241*
Psammite 27, 75, 142, 198, *241*
Pseudomorph 103–104, 112, 160, 210, 218, *241*
Pseudotachylite 132, **133**, 144, *241*
P-T-t path 199–221
  anticlockwise 209, 213–215, 220
  blueschist facies 216–218
  clockwise 203–209, 214–217, 219
  eclogite facies 219–220
  granulite facies 212–215
  hairpin trajectory 218, 220
  orogenic metamorphism 203–209
  with thermal overprint 209–211
Pumpellyite 34
Pure shear *see* Shear; pure
Pyrite 68, 153, **172–174**, 175
Pyrophyllite **9**
Pyroxene 82, 137–138
  aegerine 33
  chemical zonation 80
  clino- 35, 94
  diopside 10, 33, 36
  enstatite 10, 37, 105
  exsolution 89, 91
  ferrosilite 10

Index

hedenbergite 10, 33
hypersthene 35
jadeite 12, 33, 181, 219
omphacite 35, *239*
ortho- 94, 97–98, 120, 213–215
twinning 76–77
Pyroxenite 36

Q-domains 47, 49
Quartz
  crystallographic preferred orientation 131–132, 176–177
  deformation of 117–131, 136–137
  grain boundaries 126–131
  grain shape/size 176
  inclusions 85–89, 220
  in mylonites 122–123, 133–137
  in quartzo-feldspathic rocks 132, 136
  pressure shadows 172–174
  recovery 128–129, 136
  recrystallisation 129–131, 136, 180
  stability **104**
  veins 126, 180–187
  vermicular 95, **97**
Quartzite
  coarsening of quartz 62
  definition 27, 33
  deformation 125, 130, 136
  granoblastic 71, **73**
  permeability 179
Quartzo-feldspathic rocks 33, 97, 106, 130, 136, 169

Radioactive decay 65, 86
Radiometric dating 65, 201–203, 208
Reactants 63, *241*
Reaction
  cation-exchange 12, 14, 16
  continuous *see* Reaction; divariant
  decarbonation 12, 13, *232*
  dehydration 10, 12–13, 60, 144, 145, *232*
  devolatilisation 179
    see also Reaction; decarbonation, dehydration
  discontinuous 11, 206, *232*
  divariant 11–12, 213
  endothermic 10, 142, *233*
  enhanced ductility 145
  exothermic 10, *233*
  grain boundary 94, 98

grain size 17, 62
halo *see* Diffusion; halo
hydration 101, 103
interface 94
ionic 12, 14, 97, *112*, *236*
isothermal 12–13, *237*
kinetics of 94, 98, 142
net-transfer 12
overstepping 60
oxidation-reduction 12, 13, **14**
prograde 110–112, 197, 199, 201, 204
rates 14–15, 94, 143
retrograde *see* Retrogression
rim 103–104, 110, 217, *241*
sites of 18, 94, 98, 110, 144
sliding *243 see* Reactions; divariant
solid-solid 12, 60
types 11–14
univariant 11, 200, 206
volume changes 16, 105, 145
Recovery 123, 128–129, 132, 135–137, 193, *241*
Recrystallisation
  affect on fluid inclusions 128, 188–189, 193
  of calcite 139–141
  definition 129, *241*
  dynamic 129–131, 136, 139–141, 145, 176, *233*
  granoblastic microstructure 69
  of plagioclase 138
  of pressure shadows 175
  of previous features 41, 177
  process 123, 129–131, 135
  of quartz **130**, 136–137
  static 131, *243*
Reduction *see* Reaction; oxidation-reduction
Replacement 87, 92–94, 104–106
Resorption 82, *242*
Restite 53
Retrograde metamorphism *see* Retrogression
Retrogression 101–110, *102, 242*
  of Al_2SiO_5 *102*, 106, **108**
  of amphiboles *102*
  of biotite *102*, 106
  of blueschist 219
  disequilibrium 9, 94
  environments of 101–103, 199
  of feldspar 92, 95, *102*, 106–110
  fluids 194, 199
  of garnet 160, 194, 199, 203, 214–215, 242

  of olivine 102, 105
  of pyroxene 102, 105–106, **106**
  reactions 102–110, 214–215
  of staurolite 102, 103, **108**
Rheology 18, 132, 140, 145, *242*
Ribbon quartz **123**, 136–137, 141, 242
Rocksalt 126
Rolling structures 169–170
Rutile 35, 68, 206

S–C fabrics 134–135, 138, 141, 165–168, 169, 242
Sanidine 77
Sapphirine 213–215
Saussuritisation 107, *242*
Scapolite 68, 110
Schist 242
  calcareous 33, 68
  pelitic 8, 27–31, 51, 181, 198
  recognition of 141–142
  semi-pelitic 51, **73**, 75, 142, 198
Schistosity 45–51, **46**, 142, 151, 242
Schlieren 242 *see* Migmatite; schlieren
Sector twinning *see* Sector trilling
Sector trilling 76, 77, **78**
SEM back-scattered electron images **48**, **81**, **82**, **96**, **109–110**, 201
Sericite 37, 106–107, 210
Sericitisation 103, 105, 106–107, **108**, *243*
Serpentine
  minerals 36, 105
  serpentinisation 16, 105, **105**, 243
  serpentinite 36, *243*
Shape fabric *see* Grain-shape fabric
Shear
  band 168
  heating 142, 144
  pure 42, 49, 175–176, 177, *241*
  sense indicators 163–177
  simple 43, 139–140, 156–157, 164, 176–177, *243*
  strain 143–144, 158, 175–176
  stress 79
Shear zones
  deformation processes 6, 125, 138–139, 141–142, 176
  fluids 142, 179
  metamorphic processes 6–7, 101, 106, 138–139, 199

261

Index

Shear zones *cont'd*.
  shear-sense 163, 166–168, 176
  strain/strain rates 44, 48, 155
Sieve structure 85, *243*
Silicification 101
Sillimanite
  knots 112 *see also* 'Faserkiesel'
  nucleation sites 59, 110
  polymorphic reactions 8, 12, 14, 93, 209
  stability **9**, 31–32, 213–215
  shape/form 31–32, 68
Simple shear *see* Shear; simple
Skarn 42, 82, *243*
Skeletal crystals *see* Crystal form; skeletal
Slate 44, 51, 175, 181, 185, 210
Slip systems
  in calcite 139–141
  intergranular 236
  in mica 168
  in quartz 120, 123
Snowball structure *243*
Softening
  chemical 230
  geometric 120, *234*
  reaction 145, *234*
  strain 120, 145, *244*
Solid-solution 89, *243*
Solubility 125, 142–143, 144, 179
Solute 89–91, 183, 188
Solution transfer 139, 141
  see also Dissolution, Pressure Solution
Solvus 89–**90**, 91
Spessartine *see* Garnet; spessartine
Sphene 68
Spherulitic 68, *243*
Spinel 35, 37, 91, 93, 97, 214–215
Stacking fault 120, 139, *243*
Static recrystallisation *see* Recrystallisation; static
Staurolite 29–31, 68, 87–88, 197, 206
Steatisation *244*
S-tectonite **45**, *244*
Stilpnomelane 32, 68, **70**, 72
Stishovite 7
Strain *244*
  bulk 18, 43, 140, 175
  definition of 42, *244*
  effect on microstructure 48, 117, 136–137
  elastic 183
  ellipsoid 43–44, 48, 163–164
  gradient 43, 143, 157

hardening 121, 123, 145, *244*
partitioning 17, 43, 143, 157, 168, *244*
rate 48, 106, 120–123, 132, 135–138, 155
shadow 144, 152–153, 160, 171–176, 185, *244*
softening *see* Softening; strain
Stress
  control on grain size 128
  control on growth 86, **96–98**, definition 42
  deviatoric 183
  differential 79, 131, 135, 143, 232
  induced twins 79, 140
  shear 125, 127, 143
Stylolite 125, **126**, *244*
Subduction zone *see* Metamorphism; subduction zone
Sub-grain **121**, 128–130, 136–137, 139–140, 144, *244*
Substitution 14, 35
Sulphide minerals 71
Symplectite 93–98, **95–97**, 213–215, **216**, *245*
System *see* Chemical system

Talc 32, 36–37
Temperature
  effect on deformation 79, 117, 127, 131, 136, 139–140
  effect on reactions 8, 9–10, 12, 15, 89, 95
Texture
  definition of 39, *245*
  original 6, 34, 41
Thermal overprint 209–211, 216, 218
Thermodynamics
  first law 10
  second law 11
Thrust zones *see* Shear Zones
Thrusting 199, 203
Tiltwall 193, *245*
Topotactic transformation 60, 104, *245*
Tourmaline 68, 80
Tourmalinisation 103
Transformation kinetics 218
Translation gliding 77, **79**, *245*
Tremolite *see* Amphibole; tremolite
Triple-junctions 71, **73**, 74–75
Tschermak substitution 35

Twinned nucleii 76
Twinning 75–79, *245*
  albite 76
  annealing 76
  cross-hatched 77, **78**, 231
  cruciform 76
  definition of 75–76, *245*
  deformation 77–79, 139–141
  growth 76, *235*
  inversion 76, 77, *236*
  multiple 76
  polysynthetic 76, 77
  primary 76–77, *241*
  secondary 76–79, *243*
  sector *see* Sector trilling
  simple 76, 77
  transformation *see* Twinning; inversion

Ultracataclasite 132
Ultramafic rocks 36–37, *102*, 137
Ultramylonite **135**, 135, 138, 141–142, 176, *245*
Undulose extinction **121**, 122–123, 136–137, 151, **152**, *245*
Universal stage (U-stage) 131–132, 142
Unmixing 89
Unroofing 203, 208, 209
Uplift
  fluid inclusion studies 189
  mineral intergrowths 91, 94, 98, 215
  rates 201, 208
  retrogression 101, 103, 160–161, 218–219
Uralitisation 106, *245*

Vacancy 17, 118, **119**, 121, *245*
Valency 13
Vein 179–187, 188, *245*
  antitaxial 185, 186, *230*
  composite 186
  crack-seal 183–184, 185, 187
  fibrous 180, 183–184, 185–187
  formation of 164–165
  mineralised 181
  occurrence of 164–165
  ptygmatic 55
  as shear sense indicators 163–165
  stretched fibre (ataxial) 186–187
  syntaxial 185, 186, *245*

Index

Vein *cont'd*
time of development 164, 179–180
Vermicular intergrowths *see* Crystal Form; vermicular, Intergrowths
Vesuvianite *see* Idocrase
Vitrinite reflectance 201
Voids 126–127, 145
Volatiles *see* Fluid
Volcanic rocks *see* Metavolcanic rocks
Volume changes 16, 48–49, 105, 144, 145, 220
Volume diffusion *see* Diffusion; intracrystalline (volume)

Wairakite *see* Zeolites; wairakite
Water, effect on reactions *see* Fluid; aqueous
Wollastonite 13, 33, 68, 221
Work hardening 121
Wulff's theorem 67

X-ray element maps **81**

Zeolites
in amygdales 68, 71
analcite 33
heulandite 33
laumontite 33–34
natrolite **71**
radiating aggregates 68, **71**
wairakite 33–34
Zeolitisation 108
Zircon 86, **87**
Zoisite 68, 138
Zoning 79–83
chemical 79–83, 98, 107, 201, 204, 230
diffusion 80, 82
growth 80, 82, 199
oscillatory 82
reverse 82
sector *see* Sector trilling
textural 83, 150, 161

PLATE CAPTIONS

PLATE 1 (a) Abundant quartz inclusions in staurolite porphyroblasts, to give the characteristic poikiloblastic structure. Staurolite schist, Ghana. Scale = 0.5 mm (PPL). (b) Hour-glass structure in chloritoid porphyroblast. Chloritoid (ottrelite) phyllite, Belgium. Scale = 0.5 mm (PPL). (c) A cruciform inclusion arrangement shown by andalusite (var. chiastolite) in a slate from the Skiddaw Granite aureole, England. Scale = 0.5 mm (PPL). (d) A star-like arrangement of inclusions in garnet due to their concentration at interfacial boundaries. Garnet–mica schist, Troms, Norway. Scale = 1 mm (XPL).

PLATE 2 (a) A corona structure in granulite: a garnet core surrounded by plagioclase corona and in turn by orthopyroxene and magnetite. Saxony, Germany. Scale = 1 mm (PPL). (b) A corona structure in hornfels; garnet from an earlier regional metamorphic event is surrounded by a corona of cordierite formed during a later stage of contact metamorphism, while the matrix outside the corona comprises Bt + Qtz + Pl + Sil. Hornfels from the inner aureole of the Ross of Mull Granite, Scotland. Scale = 0.5 mm (PPL). (c) A simple twin developed in an albite porphyroblast. Greenschist, Start Point, Devon, England. Scale = 0.1 mm (XPL). (d) A cruciform twin in chloritoid: note also how the differently oriented crystals show the pale green to blue green pleochroic scheme of chloritoid. Chloritoid phyllite, Ghana. Scale = 0.5 mm (PPL). (e) Polysynthetic (repeated lamellar) twinning in chlorite. Pelitic schist, Troms, Norway. Scale = 0.1 mm (XPL). (f) Polysynthetic twinning in plagioclase from a metabasite of the hornblende hornfels facies. Scale = 0.1 mm (XPL).

PLATE 3 (a) Deformation twins in calcite. Marble, Ottawa, Canada. Scale = 0.5 mm (XPL). (b) Deformation twins in plagioclase. Migmatite, Ghana. Scale = 0.5 mm (XPL). (c) Chemical zonation in diopside crystals. Diopside skarn, Sutherland, Scotland. Scale = 1 mm (XPL). (d) Chemical zonation in tourmaline; the colour variation defining the zonation relates to variations in the Fe : Mg ratio. Tourmaline-bearing quartz vein, Ghana. Scale = 0.5 mm (PPL).

PLATE 4 Replacement features. (a) Colourless to pale green cummingtonite being overgrown by blue–green hornblende. Biotite–amphibole schist, Sierra Leone. Scale = 0.1 mm (PPL). (b) A reaction rim (or corona) of chlorite enclosing a heavily corroded core of garnet undergoing retrogression. Garnet–mica schist near thrust zone, Troms, Norway. Scale = 1 mm (XPL). (c) A chlorite pseudomorph after garnet porphyroblast. Retrogressed garnet–mica schist near to thrust zone, Troms, Norway. Scale = 1 mm (XPL). (d) Serpentine pseudomorphing olivine: note how some crystals are completely pseudomorphed, whereas others are almost entirely fresh. Forsterite marble, Sri Lanka. Scale = 1 mm (XPL). (e) Garnet retrogressing to an aggregate of biotite and chlorite. Garnet–mica schist, Troms, Norway. Scale = 0.5 mm (PPL). (f) Core replacement in plagioclase. Granite gneiss, Ghana. Scale = 0.5 mm (XPL). (g) Zone replacement in plagioclase. Partially altered granite, Ireland. Scale = 0.5 mm (XPL). (h) Biotite being replaced along cleavage by chlorite. Coaker Porphyry, Dunnage Zone, Newfoundland. Scale = 0.1 mm (PPL).

PLATE 5 Syntectonic porphyroblastesis. (a) A garnet porphyroblast with a straight inclusion fabric sharply discordant with the external fabric: note also the large chloritoid inclusion oblique to S_i. Scale = 1 mm (XPL). (b) A garnet porphyroblast with a straight inclusion fabric discordant with the external schistosity: the prominent quartz-rich bands that form part of the inclusion fabric reflect original variations in compositional layering, overgrown at the time of porphyroblastesis. Scale = 1 mm (XPL). (c) Garnet with a gentle-S inclusion fabric (S_i) sharply discordant with the external fabric S_e. Scale = 1 mm (XPL). (d) Garnet with a strongly spiralled core region, succeeded by an outer zone of late growth. Scale = 1 mm (PPL). (e) 'Millipede' microstructure developed in plagioclase porphyroblast from the Robertson River Metamorphics, Australia (photograph courtesy of S.E. Johnson, from Johnson & Moore (1996); the sample is from the original 'millipede' microstructure locality of Bell & Rubenach (1980)). The matrix of the rock is dominated by quartz and muscovite, and the inclusion trails are defined by quartz. The dashed white lines have been superimposed on the photograph to emphasise the 'millipede' microstructure as defined by the included S_1 fabric in the porphyroblast. This S_1 fabric is deflected in the matrix by the prominent S_2 fabric, which trends top to bottom across the photograph. Scale = 1 mm (XPL). (f) A large syntectonic garnet with an S-shaped form: note how the quartz-rich pressure shadow areas have changed position with time, to define two quartz-rich arcs, now being enclosed by garnet in a manner similar to that described by Schoneveld (1977). Scale = 1 mm (XPL). (With the exception of Plate 5(e), all of the examples shown in this plate are from schists of the Caledonian nappes of Troms, Norway.)

PLATE 6 Post-tectonic porphyroblastesis. (a) A post-tectonic garnet porphyroblast overgrowing schistosity: note that the external fabric does not deflect around the porphyroblast, the internal fabric is continuous with the external fabric and there is no pressure shadow. Garnet–mica schist, Troms, Norway. Scale = 1 mm (PPL). (b) Randomly oriented post-tectonic chloritoid porphyroblasts in contact-metamorphosed pelite. Chloritoid phyllite, Ghana. Scale = 1 mm (PPL). (c) Post-tectonic garnet overgrowing a crenulation to give an S-shaped (helicitic) internal fabric: as with (a), note the continuity of the fabric through the porphyroblast and the lack of a pressure shadow. Garnet–mica schist, Troms, Norway. Scale = 1 mm (XPL). (d) Late-stage rim development at the mica-rich edge of a garnet porphyroblast: note the lack of rim development at the quartz-rich margin (upper centre), and the change in style of the inclusion fabric from the core region to the rim. Garnet–mica schist, Troms, Norway. Scale = 1 mm (XPL). (e) Complete rim development of garnet on an earlier core, with a sharply defined boundary between the two zones. Garnet–mica schist, Troms, Norway. Scale = 0.5 mm (PPL). (f) Post-tectonic garnet overgrowing crenulated micaceous matrix: note the continuity of the matrix fabric through the garnet, and the fact that the external fabric does not deflect around the porphyroblast (compare and contrast with Plate 5(c)). Garnet–mica schist, Ox Mountains, Ireland. Scale = 1 mm (XPL).

PLATE 7 (a, b) An example to show how the *sensitive tint plate* can be used to demonstrate the degree of crystallographic alignment in mylonitic rocks. The example used here is the ultramylonite shown in Fig. 8.13(c). It represents a highly sheared rock, and has excellent crystallographic alignment of quartz grains. In both (a) and (b), the scale is 0.1 mm (XPL) with tint. The field of view is the same in each photograph, but (b) represents a rotation of the microscope stage (and slide) through 90° with respect to (a), with the tint plate in the same position. (c) Coarsely crystalline dolomite vein cutting through quartz: note the deformation twins in many of the crystals. Scale = 0.5 mm (PPL). (d) The same dolomite vein as (c), this time viewed under cathodoluminescence: note the detail of the growth zonation visible in the dolomite crystals (red), reflecting differing degrees of Fe and Mn substitution (the bands with more Fe substitution are darker). Scale = 0.5 mm.

PLATE 8 (a, b) Quartz vein cutting slate, and developed by the crack–seal mechanism: note the distinctive inclusion bands parallel to the vein walls, which represent successive stages of vein opening. Cornwall, England. Scale = 1 mm; (a) PPL, (b) XPL. (c) Detail of the inclusion bands in the vein depicted in (a) and (b). Scale = 0.1 mm (XPL). (d) A recrystallised crack–seal vein: note the lines of mica parallel to the vein wall, which represent recrystallised inclusion bands. Amphibolite facies schist, Norway. Scale = 0.5 mm (XPL). (e) Acicular crystals of tourmaline developed at the margins of a quartz vein. Tourmalinised slate, Cornwall, England. Scale = 0.1 mm (XPL). (f) A highly saline fluid inclusion in quartz vein: various daughter minerals are present, the largest of these being a cube of halite. Scale = 15 μm.

FRONT COVER CAPTION Domainal foliation development in mica schist (Ox Mountains, Ireland). S2 schistosity (diagonal) defined by muscovite, cuts preexisting crenulated S1 foliation seen in intervening Qtz-Ms-Ab domains. Late chlorite development (dark) is also aligned parallel to S2. Field of view (length) = 4 mm (XPL).